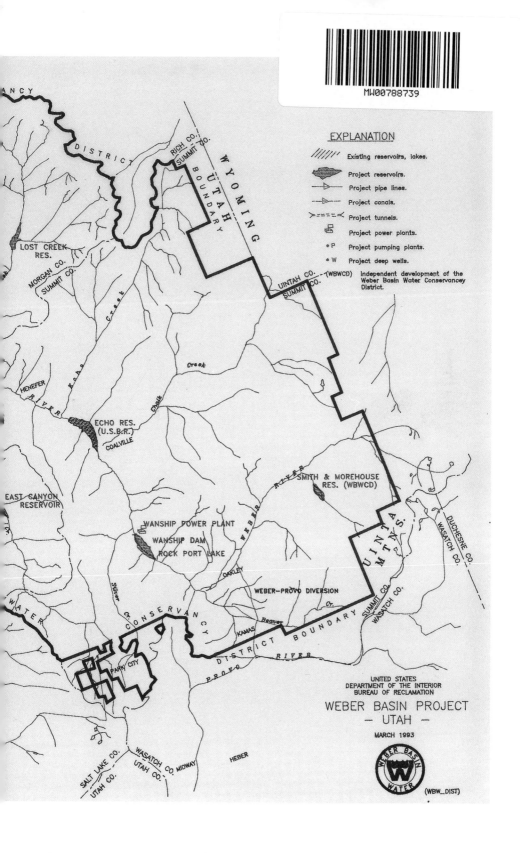

EXPLANATION

//////	Existing reservoirs, lakes.
	Project reservoirs.
—▷—	Project pipe lines.
---▷--	Project canals.
⟩==⊏==	Project tunnels.
🏭	Project power plants.
• P	Project pumping plants.
• W	Project deep wells.
(WBWCD)	Independent development of the Weber Basin Water Conservancey District.

DISTRICT

RICH CO.
SUMMIT CO.

W Y O M I N G

U T A H BOUNDARY

LOST CREEK RES.

MORGAN CO.
SUMMIT CO.

UINTAH CO. (WBWCD)
SUMMIT CO.

Creek

HENEFER

RIVER

KAMAS

Chalk

ECHO RES. (U.S.B.R.)
COALVILLE

SMITH & MOREHOUSE RES. (WBWCD)

WEBER RIVER

EAST CANYON RESERVOIR

U I N T A MTNS.

DUCHESNE CO.
WASATCH CO.

WANSHIP POWER PLANT
WANSHIP DAM
ROCK PORT LAKE

OAKLEY

SUMMIT CO.
WASATCH CO.

WEBER-PROVO DIVERSION

WATER

CONSERVANCY

Beaver Cr.

SUMMIT CO.
WASATCH CO.

PARK CITY

DISTRICT BOUNDARY

PROVO RIVER

UNITED STATES
DEPARTMENT OF THE INTERIOR
BUREAU OF RECLAMATION

WEBER BASIN PROJECT
— UTAH —

MARCH 1993

SALT LAKE CO.
UTAH CO.

WASATCH CO. MIDWAY
UTAH CO.

HEBER

WEBER BASIN WATER

(WBW_DIST)

The
Weber
River
Basin:

Grass Roots Democracy
and Water Development

The
Weber
River
Basin

Grass Roots Democracy and Water Development

Richard W. Sadler and Richard C. Roberts

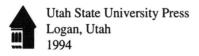
Utah State University Press
Logan, Utah
1994

Utah State University Press
Logan, Utah 84322-7800

Publication of this book was made possible through the support of
the Weber Basin Water Conservancy District.

Back cover photograph by Mel Davis. All cover photographs
courtesy U.S. Bureau of Reclamation
Jacket design by Mary Donahue

Library of Congress Cataloging-in-Publication Data
Sadler, Richard W., 1940–
      The Weber River Basin: grass roots democracy and water development /
Richard W. Sadler and Richard C. Roberts.
            p. cm.
      Includes index.
      ISBN 0-87421-164-6 (acid free)
      1. Water resources development—Economic aspects—Utah—Weber River
Watershed—History. 2. Water resources development—Law and legislation—
Utah—Weber River Watershed—History. 3. Weber River Watershed (Utah)—
History. 4. Weber Basin Water Conservancy District (Utah)—History. 5. Weber
Basin Project (U.S.)—History. I. Roberts, Richard C., 1932–. II. Title.
HD1694.U8S23 1994   94-7700
333.9'1'00979228—dc20   CIP

To Delore Nichols, Ezra Fjeldsted, Joe Johnson,
and other Weber River water pioneers.

 # Contents

Preface.. ix

Introduction.. 1

Chapters

1   Water and a New Kingdom...................................... 15

2   Early Water Development in Weber County 27

3   Davis County Water Developments 49

4   Water Development in Box Elder, Morgan, and Summit Counties....... 69

5   Transitions in Water Development from 1880 to 1940 103

6   The 1940s: An Era of Foresight 139

7   Construction of the Weber Basin Project, 1951–1969 169

8   The Weber Basin Water Conservancy District and Recent Water Use ... 197

Notes 229

Appendices

A   Weber Basin Water Conservancy District Directors and Managers 251

B   Water Companies in the Weber Basin Area...................... 253

C   Allocation of Weber Basin Water, 1992........................ 267

Index 271

 **Preface**

Water has always been a valuable quantity in the Great Basin. Irrigation and water use on the Weber River has been recorded in historical documents for more than a century and a half. The story of water use and development on the Weber River is one of struggle and community efforts, efforts to control flooding and to extend the agricultural season. Seldom does water use and development move to the front page, but it underpins any growth and development that this particular region might have. Water pioneers have been rather quiet in their day-to-day efforts to dig and build, dam and develop for themselves and for those who come after them.

We became interested in water development as we examined the development of northern Utah. Since beginning research on this manuscript, more than a decade ago, a number of individuals have been helpful. We originally began to take a snapshot of water development over the past half century, but soon decided that a more full treatment of the river system was needed. Charles Peterson, Craig Fuller, and Jay Haymond have offered suggestions and encouragement throughout the project. Three of our students—Jim White, Jerry Mower, and Steve Readdick—helped us with oral histories and background research. Water users and developers throughout the Weber Basin have assisted us with their expertise, reassurance, and time. These include Keith Jensen, Delore Nichols, Ezra Fjeldsted, Boyd Storey, Wayne Winegar, and Ivan Flint. Weber River water users associations and ditch companies have been helpful to us, including individuals from these groups: Floyd Baham, Donna Thurgood, Ed Southwick, Grant Salter, Sherrie Mobley, and Mark Anderson. Francis Warnick gave us many of his valuable insights gained in his years with the Bureau of Reclamation. Peggy Lee and John Alley have lived with the manuscript and have made many valuable suggestions for improving it. Since we began to write, JoAnn Reynolds, Marilee Sackolwitz, Dorothy Draney, Sheila Rasmussen, Cherise Gentry, Marsha Steele, and Jennifer Judd have assisted us in typing the manuscript. We are appreciative of assistance throughout the project from Weber State University and the Weber Basin Water Conservancy District. We alone are responsible for the final manuscript. We believe that it gives continuing evidence of the importance of individuals taking history into their own hands and shaping it.

Richard W. Sadler
Richard C. Roberts

 **Introduction**

In the late nineteenth century William E. Smythe wrote in his book, *The Conquest of Arid America*, that " . . . irrigation is the life blood of institutions in the Western half of the continent and its history is, in a marked degree the history of the people themselves."[1] The history of the Weber Basin confirms Smythe's idea; it is a history of the people involved in the development of water projects in the Weber River Basin and a portion of the Wasatch Front of Utah, which have been, in recent times, known as the Weber Basin Project. This story of grassroots democratic action started with Euro-American settlement during the pioneer days of the 1840s and continued through various changes and improvements, including the inception and completion of the Weber Basin Project.

In many ways it is a story of heroics—of people and communities overcoming the obstacles of geography and politics. It is the story of Luman Shurtliff who hand-dug and maintained water ditches in Harrisville; of Isaac Goodale who toiled and struggled in constructing the Ogden Bench Canal; of Perrigrine Sessions who did the first farming and irrigating of Bountiful; of Joe Johnson, Ezra Fjeldsted, and DeLore Nichols who in modern days worked to get the Weber Basin approved by the local population and the federal government. In all time periods and at all levels of political involvement, there is evidence of grassroots participation and politics. The people in communities such as Oakley, Morgan, North Ogden, and Centerville participated in planning and implementing their own water projects. At first they did so as individuals and families and later as communities, but the projects always benifited from broad citizen participation.

This is a history then where not only the individual is signal, but it is one that also demonstrates how individuals worked to develop institutions to further their goals. Nor is this the story of what Karl Wittfogel saw in Asia where the power over water created an Oriental despotism; but rather in the Weber Basin it was the individuals who put together, through democratic action, the institutions that governed water at different stages of development. First it was individual effort, then cooperative effort through the Mormon church organization, and finally county and municipal efforts channeled through charters that created ditch and irrigation companies. The county force did not really come into play until after World War II when greater demands for water led county governments to cooperate in the larger federal programs. Finally, it was the all-encompassing organization of the Weber Basin Project that fostered the modern-day institution that would govern water in the Weber Basin area.

This is a history of leadership in a federal fashion. In pioneer times it was a play of power between the central organization of the Mormon church, with Brigham Young at the head, and the local leaders, such as bishops and neighborhood church councils. Young's guidance regarding water policy during the early years came when he visited settlements and when he issued policy statements and epistles from church headquarters. Most of the actual decisions and practices, however, were carried out on a trial-and-error basis by the communities and their leadership who determined the projects and put forth the backbreaking effort to insure a continued supply of water. For example, it was Brigham Young who prophesied in 1856 that the "Weber be brought to the Hot Springs in North Salt Lake, there to meet the waters from the south and empty into the Jordan,"[2] but it was the communities along the route who determined how it would be accomplished and who would actually do the work. By completing the project with the development of the Weber Basin Project, they virtually fulfilled what Brigham Young foresaw.

As time passed, though the territorial government made secular laws for the territory, it was left to the local mutual companies and irrigation districts to plan and carry out the water projects. The territorial laws outlined a "prior appropriation" water law system, which in western United States history has been categorized as a system that falls under the "Colorado Doctrine" of prior appropriation. But even in this the peculiarities of Utah communal society made Utah's water development unique. According to Robert Dunbar, Utah's water policy system was different because it was

> molded by the Mormons' sense of ecclesiastical brotherhood. They divided rights in streams into two categories, early and late rights, or as they called them, primary and secondary. Within these categories water was divided coequally or proportionately among the water users. Division was not by measurement in miners' inches or cubic feet per second, but by fractions of the stream flow, delivered to each one. None were cut off, unless the stream went dry. All the "brothers" within each category were treated alike; their rights were correlative.[3]

Even under territorial law, the grassroots communal interest had enough impact to make Utah's practices distinct.

This grassroots democratic effect carried on and even influenced the modern-day federal projects such as the Weber Basin Project. Many of Utah's farmers, who had remained on relatively small farms as compared to other areas of the country, became involved in the decisions and projects to develop local water resources. They had organized into "water-user" or "conservancy" districts and had much to say and decide in the many meetings and hearings that were held to establish the projects. Their votes and participation were important in approving the projects and giving them direction. The realization of the Weber Basin Project depended on the relationship between the local communities and water organizations and the national Bureau of Reclamation. In this case it was a relationship instigated by the local leaders with the national leaders.

Although the communities and counties were compartmentalized by legislated boundaries, they were also drawn together by the way northern Utah's county boundaries had been drawn along the watershed borders. The boundaries of Summit, Morgan, and Weber counties were drawn along the crests of the Weber and Ogden River drainages. Davis and Box Elder County boundaries fell along the watershed for streams that drain into the Great Salt Lake. Whether they knew it or not, the creators of Utah's laws—and county boundaries—developed the borders somewhat like John Wesley Powell had advocated in the 1880s. These watersheds and tributaries had perpetuated their local control and practices until the time that larger projects were needed. The earlier division along watershed lines proved to be a major factor in getting local support from the communities and water districts that had common interest in the Weber Basin.

This study traces Weber Basin water history from the arrival in the 1840s of the Mormon settlers to the present time. It is a record of people struggling early on to bring water to the land in order to survive; of later settlers making improvements in water rights regulation and designing new techniques to deliver the water; and finally, of the efforts of a region uniting into the Weber Basin. Though this story has never been told in its entirety, the main themes have been presented in bits and pieces or as parts of larger comprehensive histories—but never with such detail or concentration on the subject of water use and development in the Weber Basin.[4] This history outlines the development of various laws that were passed to protect the rights of the people—as water development moved from primitive and simple projects to larger and more complex programs—and in response to certain demands. It is important to see the scope of this legal trail in order to see how the entire history fits into the water developments of the American West and the policy of the United States.

Utah's history can be broken into several distinct phases, but for the purposes of this introduction and to put water developments in historical perspective, the periods will be divided as follows: first, from Mormon settlement in 1847 to the completion of the transcontinental railroad in 1869; second, from 1869 to statehood in 1896; third, from 1896 through the end of World War II in 1945; and finally, from 1946 to the present when there has been a great demand on water due to the pressures of increasing population and new industry. The development of

Utah water policy parallels in many ways the historic and economic changes in these periods.

The first phase, 1847 to 1869, was a time of economic struggle for survival and establishment. Brigham Young had counseled that the Mormon experiment would be based on agriculture rather than commerce and manufacturing. It was a period when pioneers worked toward self-sufficiency and centered all their efforts inwardly rather than outwardly. Not surprisingly, the water laws of the time were legislated with that same emphasis. At first the water was controlled by the Mormon church under a theocratic system of government, but after the creation of the territory of Utah in 1850, the first territorial act dealing with water was passed into law February 4, 1852. This law created the county courts, which were composed of a probate judge and three select men and actually functioned more like a county commission than a court. Under this law the county court would control "all timber, water privileges or any water course or creek to grant mill sites, and exercise such powers as in their judgment shall best preserve the timber and subserve the interest of the settlement in their distribution of water for irrigation or other purposes. All grants or rights held under legislative authority shall not be interfered with."[5]

Although the county courts were authorized to control water rights and to settle water difficulties, these functions were, for the most part, settled under the direction of the Mormon church. In fact, county court officials and the local leaders of the church were often the same individuals. There was a unity of church and state in Utah, and the county courts were often used to put a secular stamp of approval on what was really a decision of ecclesiastical authority.[6]

In his advice to farmers at this time, Brigham Young encouraged them "to construct cooperative canals, open individual ditches and apply water to the land that native fruits and cereals might be produced in sufficient quantities to supply the demand for home consumption."[7] These cooperative efforts developed into water groups, which built the canals and spread the work of construction and maintenance among the users; these groups based the cost of labor on the amount used of the stream's allotment. This cooperative effort was codified in the Utah Irrigation District Act of 1865. This act provided for the organization of water districts within the county boundaries through the action of a mass meeting to create a district and then to submit the proposal to the vote of the residents within the district.[8]

Intertwined with territorial laws were the federal laws, including the Homestead Act of 1862 that allowed the claim of 160 acres of federal land for a minor fee, occupation, and development. The purpose of the Homestead Act was to make public land in the West available to the populace at a low cost in hopes that easterners would move onto the unclaimed areas. The law failed in its goal, but it did become the basis of the land policy under which much of the West was developed. In 1866 Congress passed another law, the General Mining Law, which recognized the local laws and customs of the various states with regard to waters that were not on the public domain and were not navigable streams controlled by interstate commerce. The 1866 law said "whenever, by priority of possession, rights to the use of water for mining, agricultural, manufacturing or other purposes, have vested and accrued, and the same are recognized and acknowledged by the local customs, laws and the decisions of courts, the possessions and owners of such vested rights shall be maintained and protected in the same."[9] This law gave sanction to the "prior appropriation" practices that had been created in the western United States. Thus in the first phase of Utah's history, basic water laws were established in a cooperative mode—first ecclesiastically and then secularly.

The second phase of Utah's growth was from 1869 to 1896. The highlights of this period include the completion of the transcontinental railroad through Utah—which ended its isolation from outside influence—and the death, in 1877, of Brigham Young, who had been a great advocate of economic cooperation. His attempt to hold off the move to a laissez-faire economy and free enterprise had been lost to the Utah liberal movement, which helped guide the Utah economy into the American mainstream. With Young, the Mormon warhorse, gone this transformation came much more quickly. By the 1890s, the speculative aspect of the "new movement" was evident as land and water rights were marketed in a venture fashion, which contrasted markedly from the previous economic phase.[10]

Utah water law changes paralleled the shift in economic philosophy and action to the free enterprise mode. With the passage of a territorial law in 1880, water rights were no longer publicly owned, but were transferred to private owners who were given the right to use specific shares. George Thomas wrote that "after 1880, water appropriated for a beneficial use became the property of the appropriator. With the abandonment of public ownership, water shares could be traded like any other piece of property."[11]

There is much debate as to why Utah water rights were turned over to private ownership in 1880. Some Utah historians see this law as a step backward in water law policy because it took water distribution out of the hands of public interests and placed it in the hands of private speculators. Some feel the law was designed to keep the control of water out of the hands of the federal government, while others note that the trend of the time was away from the cooperative economy towards the free enterprise philosophy. Thomas Alexander states in this regard that "the 1880 law did not benefit only the Mormon Church, but anyone appropriating water for any use, including mining and manufacturing companies. The 1880 law was evidence of the demise of the cooperative commonwealth characteristics of early Utah and the subsequent movement to a more capitalistic economy." Despite any of its adverse affects on Utah's progress in water development, it remained the main statute governing water until Utah became a state in 1896.[12]

With the 1880 water law, which did away with the policy of water rights being appurtenant to the land, Utah water became a part of the speculative business activity that was going on not only in Utah, but throughout the entire western United States. During this period commercial water projects seeking quick and high profits developed all over the West, especially in California. In the 1890s there was considerable investment of eastern money in these projects, and almost every one failed. The large amounts of money invested in the projects were not returned quickly enough or with enough profit to make this method of development feasible. The farmers were suspicious and reluctant to participate fully in these programs, and they too helped bring bankruptcy and failure to virtually all of these free enterprise ventures of the 1890s.[13]

In Utah the main investments at this time into commercial projects were the Bear River Irrigation Company and the Bear River Canal and Ogden Water Works Company of Ogden. Both of these projects failed to make money for the water speculators, and the projects changed hands to ownership that was more in the public interest.

Federal policy during the 1869 to 1896 period greatly influenced both Utah and the West. Much of western water policy derived out of several laws passed by Congress and the work of John Wesley Powell. Laws enacted to manage western water, or the lack of it, were the Timber Culture Act of 1873, the Desert Land Act of 1877, and the Timber and Stone Act of 1878. These laws were passed to encourage farmers to take up the land in the West and to develop larger farms which, because of the arid regions, required more land to be as productive as the eastern areas. The Desert Land Act had a water provision that allowed the acquisition of 640 acres of land if a settler would irrigate it within a three-year period. Another federal law of this time was the Carey Act of 1894, which turned one million acres of federal public domain over to each state to be used to finance irrigation. This program failed to raise enough money to support the possible projects so it was abandoned for a more direct approach to get government sponsorship for state projects—but this action will be discussed in the following time period.[14]

John Wesley Powell was a major influence on water law in this period. Powell, who had become acquainted with the West through two river expeditions, in 1869 and 1871–72, down the Green and Colorado rivers, published his conclusions in the *Report on the Arid Lands of the United States, with a More Detailed Account of the Land of Utah.* From these experiences, Powell dedicated his efforts to better understand and to develop a reasonable policy, though radically different, for the western United States. Powell's recommendations for mapping and classification of lands, his proposal for prohibiting private entry on the lands until studies were concluded, and his ideas for developing regions by watershed configurations got him into trouble with those who wanted to exploit the West at a rapid pace. Under much pressure because of his policies as director of the Geological Survey, Powell resigned in 1894.

Powell wanted each watershed district to develop its own system of technology and its own jurisdiction of laws and courts in relationship to the resource base. He also wanted to keep the water rights appurtenant to the land and prohibit the exploitation of water from one valley to another. Powell also planned to limit future settlement to eighty acres, which he hoped would put a strict curb on the land grabbers and corporate developments.[15]

There are many students of western water history who conclude that, although Powell's watershed unit program was not adopted, it had merit and, had it been sanctioned, might have avoided the present conditions of the American West water system. This system has proved to be extremely costly as it transfers water long distances from one watershed to other watersheds and places the power and profits in the hands of a few corporate organizations.[16]

The third phase of Utah history, from 1896 to 1946, brought to completion the move from stewardship economics to commercialism and speculation. It was a time when "competitive capitalism triumphed." During this time farming and mining activities greatly increased. In the first half of this era the growth in agricultural activity centered mainly on truck, field, orchard, and livestock farming. In 1890 there were 19,916 farms in Utah; by 1930 there were 30,000 farms operating, but the number of farms dropped off from that peak level. In 1910 there were 132,000 people on the Utah farms; by 1940 there were 94,000, and the Utah farm population continued to decline. Mining activities also spread as innovative processes made extraction of ore more efficient; but like agriculture, mining profits generally declined—with the exception of some major companies such as the Utah Copper Company and later the Kennecott Corporation. With these noticeable declines, Utah's industry expanded into a variety of configurations: woolen mills, clothing and shoe manufacturing, candy making, sugar production, evaporated milk production, soda-water bottling, and cigar making were among the major enterprises.

The big change in the latter part of this period, however, was the influx of government agencies into Utah. This was a notable change in attitude from the Brigham Young antifederal days. The agencies that produced a significant impact were the war industries of World War II, including Hill Air Force Base, Clearfield Naval Supply Depot, Utah General Depot, Tooele Ordinance Depot, Dugway Proving Grounds, Remington Arms, Kearns and Wendover air bases, Geneva Steel Corporation, and a whole variety of subsidiary industries. All of these activities meant a greater demand for increased water supplies, which led to the proposal of a greater number of water projects.

This demand conjured up new philosophies and methods to sponsor projects. This then became the era of the federal Bureau of Reclamation, which sponsored many of the western water projects. In Utah new water laws were legislated to take advantage of the federal policy. The federal government took the lead in establishing water for the West. Out of the United States Geological Survey came a group of scientists and engineers led by Frederick Haynes Newell who "approved water development from a fresh point of view." They incorporated a broad view of river planning into their program. To them "all possible uses of water should be considered so that rivers could produce the greatest possible benefit for man." The multiple-purpose river-basin development in later years grew directly from these ideas of the Geological Survey hydrographers. While Newell and his forces increased the federal interest in water systems, western leaders sought ways to obtain capital to finance reservoir and storage systems for their growing needs. After the irrigation boom and bust of the commercial enterprise period, westerners looked directly to the federal government for aid. The Carey Act of 1894 had been passed in response to this demand, but it would finance only a few projects. So westerners turned to the National Irrigation Congress to support their demands for a federal program.[17]

The National Irrigation Congress, and later the National Irrigation Association, played a major role in establishing a federal water policy. William Smythe outlined the growth and importance of the National Irrigation Congress. The movement started from western problems, which were plaguing the western states. Smythe became involved because of a drought that had hit Nebraska in 1890 when he was an editorial writer for the *Omaha Bee.* At a state irrigation convention at Lincoln, Nebraska, Smythe was made "chairman of a committee to arrange for a National Irrigation Congress . . . ." The first convention was held in Salt Lake City in 1891, "within sight of the historic ditch on City Creek where English-speaking men began the conquest of the desert."

The first meeting, hosted by Governor Arthur L. Thomas of Utah, and subsequent meetings centered on a resolution approved without a dissenting vote that the federal government should cede the arid lands to the states. Smythe noted that there was "no such sentiment for public ownership" in 1891 that would later prevail. At that time "speculation in water was considered as legitimate as speculation in land or mines." In succeeding years, Congress in its conventions in various parts of the West and through its journal, *The Irrigation Age,* changed its position on irrigation uses and pushed for a federal program. By 1897 there were other forces supporting this move. In that year Captain Hiram M. Chittenden of the United States Corps of Engineers published his *Report on the Reservoirs in the Arid Region,* in which he "recommended that the government should acquire full title and jurisdiction to any reservoir site which it might improve, and full right to the water necessary to fill the reservoir; also that it should build, own and operate the works, holding the stored water absolutely free for public use under local regulation." The Chittenden report had the effect of setting the direction for the irrigation movement.

In 1897 a new organization known as the National Irrigation Association was organized in Wichita, Kansas, by a young energetic lawyer from California, George H. Maxwell, who devoted his full time to reclaiming the American West.

Maxwell had been a member of the National Irrigation Congress, but saw the need of a new organization to "strengthen and supplement," not supplant the pioneer organization.

By 1900 the National Irrigation Congress, assembled in Chicago at its Ninth Congress, adopted "ringing resolutions in favor of a comprehensive national system" for the storage of floods and the reclamation of public lands. "Then in 1900, nine years after the first Irrigation Congress at Salt Lake and three years subsequent to the Chittenden report and the formation of the National Irrigation Association," the national political parties—Republican, Democratic, and the Silver Republican—placed in their platforms plans that called for the reclamation of the arid lands of the United States by the national government.

The political action was led in Congress by Representative Francis G. Newlands from Nevada. Newlands had been involved in the irrigation movement for several years. He played a leading role in the first National Irrigation Congress in 1891 and carried the banner for that movement through the decade of the 1890s. In 1901 Newlands proposed legislation that called for federal financing of irrigation projects through a Reclamation Fund derived from proceeds from the sale of public lands. Some westerners and easterners objected to the legislation, but because of skillful political maneuvering and compromising on the bill, western politicians, lobbyists from the various irrigation groups, and finally the strong support of President Theodore Roosevelt pushed the Newlands Reclamation Act of June 17, 1902, through Congress; the bill was then signed by President Roosevelt.[18]

This new law was important to the development of the West. It designated that all money derived from the sale of public lands in sixteen western states be put into a "reclamation fund" for the planning and construction of dams and other irrigation projects. To cover the cost of their projects and maintain a revolving fund to finance other projects, the users of project waters were to pay the fees, without interest.

The Reclamation Service, a new branch of the Geological Survey, was created to carry out the operation of the bill. By 1907, the Reclamation Service was made an independent bureau under the secretary of the interior, and in 1923 it became the Bureau of Reclamation; in 1944 it established seven regional offices, one of which was the Upper Colorado Region in Salt Lake City.

The Reclamation Service entered quickly into the business of water projects. The first project was the Truckee-Carson Ditch in Nevada, and by 1910 twenty-four other projects were under construction. In 1911 the Warren Act, sponsored by Senator Francis E. Warren of Wyoming, authorized the sale of surplus water to nonfederal lands or undivided holdings and promoted the construction of drainage for marsh lands. This law expanded greatly the projects of the bureau and set the policy for some of the major projects that followed. This led to the Colorado River Basin development, which centered on the Hoover Dam project;

the Columbia River Basin Project with the Grand Coulee Dam as its key; the Central Valley Project of California, which opened up the richest agricultural section of the United States; and the Missouri River Basin development with the Army Corps of Engineers. These were main developments on a federal level that transpired during the corresponding period of Utah history from 1896 to 1946. Lack of space and interest for this study prohibits a discussion of the pro and con arguments that are a part of Bureau of Reclamation history.[19] Utah's involvement in the federal projects began with the Strawberry Valley Project, which was approved in 1905 and completed in 1915. Other federally sponsored projects in Utah up to 1946 included Echo Dam in 1930 and Pineview Dam in 1937.

The Utah water law of 1880 that allowed private control of water scuttled the ingenious public control systems Utahns had instituted previously; but between the period of 1896 to 1903, the state legislators struggled with the problem of water use and rights. "Gradually the State returned to a somewhat modified secular version of the early Mormon system—public control of water resources, and a power to grant or refuse rights to water vested in a central authority." Laws passed in 1896, as well as ones enacted in 1897 and in 1901, expanded the state engineer's authority and provided for the first surveys of streams and canals; this promised to give order to a system that had been neglected for some time by the state government.[20]

The federal Reclamation Act of 1902 required conformity of the state laws to the federal policy. Officials of the Geological Survey "made it clear that reformation of water-right laws was a prerequisite to federal reclamation."[21] Utah met that requirement by the passage of the Utah Water Statute of 1903. This landmark law returned the control of water to the public and placed the supervision of water rights in the office of the state water engineer; it also required a hydrographic survey of the state, which would establish the previous claim and future claims on water as determined by the state engineer. After the entire legal process, including appeals, had been completed, each water user was issued a certificate that was recorded in the county recorder's office, and future water ownership transfers were made with deeds exactly like land transfers.[22] It was 1937 before the adjudication of water claims on the Weber River was finished under this program.

The fourth and final period of Utah water history, from 1946 to the present, showed a decrease in the number of Utahns involved in agriculture and mining. With a population that reached 1,454,330 by 1980, only 4 percent (58,173) would be engaged in farming and 3 percent (43,630) in mining. Industry related to agriculture, such as canning and sugar refining, dropped off markedly also, and new industries that focused on such resources as uranium and oil became significant for a time. The oil industry climbed to 31,350,000 barrels in 1978, but dropped off as the world supply improved.

Meanwhile, government employment continued to rise. In 1946, 21,550 Utah employees held jobs in federal government agencies. By 1974, 108,000 Utahns served in the federal ranks—giving the state the highest ratio—11 percent—of the number of federal employees to state population in the United States. Many of the military bases continued to function in Utah while several subsidiaries of the military-industrial complex came and departed. New businesses in this period were the Internal Revenue Service, the Hercules Powder Corporation, the Sperry Rand Corporation, Litton Corporation, and the Thiokol Chemical Corporation (later Morton-Thiokol).

Several water projects of the Colorado River and Central Utah Project got underway, including the Flaming Gorge and Glen Canyon projects. Not the least of these was the Weber Basin Project. As a result of these developments, recreation and tourism also became an important part of Utah's industry.

Population itself became a factor as communities along the Wasatch Front from Brigham City in the north to Provo in the south encroached upon the farm land and virtually drove agriculture out of these areas. In 1950 the state's population was 700,000, by 1970 it had surpassed the one million mark at 1,059,273, with over 80 percent of the population living in cities. During this same period, however, farm population dropped to 12,700 by 1970. To maintain the large population centers, water use shifted from primarily irrigation use to culinary use in these heavily populated areas. These population pressures again created a greater demand for water and eventually led to several projects that were meant to provide culinary water mainly along the Wasatch Front, including the Pineview Reservoir and the Weber Basin and Central Utah projects.

Water laws in Utah under the state water engineer remained much the same and reflected the philosophy of public control, though a few minor amendments were made to the Utah Water Code. One important innovation to state water law as it related to federal reclamation projects was the Water Conservancy Act passed in 1941. The main state-level activity during this time, which related directly to the federal reclamation project, was the organization of water-user associations and water conservancy districts; these were necessary to qualify for federal reclamation project funds. Examples of this and pertinent to this study are the South Ogden Water Conservancy District, the Weber-Box Elder Water Conservation District, Ogden River Water Users' Association, and the Weber Water Users' Association. These agencies negotiated the contracts and assumed the responsibility of repaying the project debts.

Though federal water policy experienced a major expansion of projects in the 1950s through the 1970s, by the late 1970s environmentalists, state rightists, and economists questioned the feasibility of federal projects and wondered whether these programs were still needed. President Jimmy Carter's administration threatened to call off several major projects in the western United States. After a period of protest and adjustment, some of the projects on the "hit list" were restored, but not before a change of attitude and philosophy had entrenched itself in the federal policy. The federal government committed itself to completing projects already begun and promised to continue maintenance on existing projects; but by 1987 the Bureau of Reclamation announced that there would be a slow down of further funding for water projects by the federal government. This meant that the Central Utah Project of the Colorado River would be completed, but other projects such as the Bear River would have to seek greater funding from local sources than had previously been the case.[23]

From this brief historical outline it becomes evident that Utah's water history is indeed unique and significant; in some ways it differs with and in other ways it parallels the national story of water history. It is, however, the people and events of this history that this book endeavors to present. Irrigation and water history after all is the "life-blood" of the institutions and "in a marked degree, the history of the people themselves."

REVIEW OF LITERATURE

The American West has had a plethora of works written about the problems of water. These works range from the individual treatment of each of the seventeen western states, questions of water law development in its riparian and prior appropriation format, the move from local development to federal development, the projects under the Bureau of Reclamation, the rise and fall of the Bureau of Reclamation, and a consideration of the future of water development—to name but a few of the many possible categories of water subjects.

Major works in these areas and as they pertain to this study are found in several key volumes. They include Leonard J. Arrington, *Great Basin Kingdom: An Economic History of the Latter-day Saints*, which covers early irrigation in Utah. Leonard J. Arrington, Feramorz Y. Fox, and Dean L. May, *Building the City of God: Community and Cooperation among the Mormons*, which shows some of the early irrigation techniques. G. Lowery Nelson, *The Mormon Village: A Pattern and Technique of Land Settlement* is another description of Mormon settlement irrigation techniques. Newbern I. Butt in a master's thesis, *The Soil as One Factor in Early Mormon Colonization*, shows the irrigation techniques in early Salt Lake City and credits the Mormons with demonstrating the feasibility of irrigation that could support a large population. Charles S. Peterson, "Imprint Agricultural Systems on the Utah Landscape," in *The Mormon Role in the Settlement of the West*, edited by Richard H. Jackson, is one of several studies of Peterson's on the agricultural growth in Utah. These are a few of the many works on early Utah history that show the importance of water and irrigation to Utah.

Several works are important to this study in regard to western and Utah water law. They include Elwood Mead, *Irrigation Institutions: A Discussion of the Economic and Legal Questions Created by the Growth of Irrigated Agriculture in the West*, which traces western development from riparian to prior appropriations and shows variations of the Colorado, California, and Wyoming practices. It also has a chapter on Utah water law changes, which finally culminate in the 1903 law when water resources are brought under the control of the state engineer. Charles H. Brough, *Irrigation in Utah*, traces some of the same history of the Utah water district laws and suggests a state-level system that would supervise and control public water, but would allow private development. George Thomas's work, *The Development of Institutions under Irrigation with Special Reference to Early Utah Conditions*, traces the Utah laws in detail with emphasis on the irrigation districts' developments to the 1903 law. William E. Smythe's, *The Conquest of Arid America*, praises the early "Mormon Commonwealth" for its creation of a successful water system through common effort and demonstrates the importance of federal participation in opening up the West as a productive area of small farms in a democratic society. Smythe saw the newly passed Reclamation Act of 1902 as the beginning of that movement. Robert G. Dunbar, in his *Forging New Rights in Western Water,* looks at water law systems in the western states and treats the Utah laws in a chapter on "the Mormon Experience" and in other comments throughout his book. Other useful writings on water law include Robert J. Hinton, "Water Laws, Past and Future," *Irrigation Age* 2 (November 1, 1891).

Understanding John Wesley Powell in western water history is important. Powell's report on the arid region is significant. Other works analyzing his contributions to western water policy are Wallace Stegner's *Beyond the Hundredth Meridian: John Wesley Powell and the Second Opening of the West*; Thomas G. Alexander's evaluation in "John Wesley Powell, the Irrigation Survey, and the Inauguration of the Second Phase of Irrigation Development in Utah"; and William Culp Darrah's, "John Wesley Powell and an Understanding of the West," both in *Utah Historical Quarterly* 37 (Spring 1969).

Works useful in understanding the reclamation policy and various projects in the West are many. Most useful in the study were Samuel P. Hays, *Conservation and the Gospel of Efficiency: The Progressive Conservation Movement, 1899–1920*, which treats the reclamation policies of the progressive era as a move towards centralized scientific policy. Donald Worster, *Rivers of Empire: Water Aridity and the Growth of the American West*, examines how federal water policy went astray and instead of creating a region of small farms as provided in the laws it led to an empire controlled by large agribusiness operations. Worster concludes that the reclamation policy had failed in its objective and had perhaps outlasted its usefulness. Worster's work is comprehensive, but it did give Utah some specific treatment. Worster saw the Utah system as an incipient one on a small enough scale that it did not build a despotic political system. This same conclusion is arrived at by Philip Fradkin in *A River No More: The Colorado River and the West*. Also important in federal policy considerations is William K. Wyant, *Westward in Eden: The Public Works and the Conservation Movement*, which deals mainly with public lands, but has some important references to water policy also. Samuel G. Houghton, in *A Trace of Desert Waters: The Great Basin Story*, shows the Utah water situation in relation to the Great Basin area.

Books that deal with the Bureau of Reclamation and its development are William E. Warne, *The Bureau of Reclamation*, and Michael C. Robinson, *Water for the West: The Bureau of Reclamation, 1902–1977*. Also helpful in understanding local projects is David Merrill's study, *An Historic Mitigation Study of the Strawberry Valley Project, Utah*, which has an important statement on Utah water growth and policy and treats specifically Utah's first federal project, the Strawberry Valley Project.

Finally, Karl A. Wittfogel's *Oriental Despotism: A Comparative Study of Total Power* deals with water control as a means of creating despotic government. Wittfogel makes some fleeting comments about Utah's water system, and the Mormons who did not develop a despotic system because of other important influences in their society—a wider nonhydraulic nexus which discouraged the formation of a despotic government.

Not all works give favorable accounts of Utah's water development. Two such works are Richard M. Alston's *Commercial Irrigation Enterprise, The Fear of Water Monopoly and the Genesis of Market Distortion in the Nineteenth Century American West*, and Keith D. Wilde's dissertation *Defining Efficient Water Resource Management in the Weber Drainage Basin, Utah*. Alston treats the economic failure of the speculative water projects in the nineteenth century and shows how they failed because of a lack of support from the farmers and the slow return of profits on the investments. Wilde's work contends that there is not a scarcity of water in the Weber Basin or of "storage and conveyance facilities. All are abundant, even redundant." He argues that any further building is unnecessary and special attention should be given to the distribution of the repayment burden of projects by the Weber Basin. Economists many times see profits as the only driving or acceptable outcome and do not give adequate attention to other factors such as aesthetics, traditions, sentimentality, or other factors that influence community decisions.[24]

The works cited above are some of the major works that impact importantly the theme of this work. Lastly, it goes without saying that the works cited in this study are significant as well in understanding the wide range of sources available on the subject.

Chapter 1
Water and a New Kingdom

Travel weary but seemingly free at last to build and maintain communities in accordance with their own beliefs and desires, the main body of Mormon pioneers arrived in the Salt Lake Valley July 24, 1847. These members of the Church of Jesus Christ of Latter-day Saints had crossed the Great Plains and Rocky Mountains under the leadership of church president Brigham Young to find new homes outside the boundaries of the United States, where they might be free of conflicts with their neighbors. They were pioneers in the isolated and arid high desert and mountain lands along the eastern rim of the Great Basin, and in adjusting to the region's demanding climate, they became innovators in the art and science of irrigation. Although Native Americans had practiced irrigation in parts of North America, including the Great Basin, European Americans had little need to artificially water their gardens until they began to settle the arid West. Through trial and error and experimental successes passed on by the ecclesiastical network from community to community, the Mormon settlers developed methods of irrigation that one day would be used the world over to develop desert resources, making them, as the Mormons described their efforts, "blossom as the rose" and provide crops to feed expanding settlements.

The very day of their arrival, the pioneers planted crops to sustain their lonely outpost. Wilford Woodruff recalled some years later that William Carter plowed the first furrows. "We arrived," Woodruff subsequently wrote, "at the encampment about 11:30 on the morning of July 24, 1847. The brethren had already turned out City Creek and irrigated the dry barren soil, this being the first irrigation ever performed by anyone in these mountains in this age."[1] The plot was located near what is now Fourth South and State streets in Salt Lake City.

Then, on July 26, William Clayton recorded that "some of the brethren have commenced making a garden about two miles to the southeast from the original encampment," near present-day Sugarhouse. By August 7 he also reported that "fifteen of the brethren commenced building a dam a little above the camp so as to bring the water around and inside the camp." After it was completed, there was "a pleasant little stream of cold water running on each side of the wagons all around," and many of the camp members were rebaptized in the water to reemphasize their religious commitment in the new land.[2]

Brigham Young was both ecclesiastical and secular leader of the new colony and played a major role in how the water, timber, and land were allocated. The broad policy decisions he made served as guidelines for local communities. How much of a role Brigham Young played in local decisions regarding such matters as water use is a topic that is still debated. Opinions of scholars range from that of Leonard J. Arrington, who portrays Brigham Young as playing a major part in making policy throughout the Mormon colonies, to the position of Charles S. Peterson, who tempers that view and sees more decision making on the local level. We conclude that Brigham Young had a great deal to do with making general policy and that local communities followed his direction as much as possible, but in many instances deviated to meet local needs.[3] In time the Salt Lake Stake of Zion was organized, with Young as president and a High Council composed of twelve members; this body continued to govern until the state of Deseret was established on March 10, 1849.

In August 1847, Orson Pratt and Henry G. Sherwood surveyed the city into 135 square blocks of ten acres each with 132-foot-wide streets and twenty-foot sidewalks. Irrigation ditches flowed on both sides of the streets. In 1852, Howard Stansbury reported in *An Expedition to the Great Salt Lake* that "an unfailing stream of pure sweet water" flowed "through the city itself . . . which, by an ingenious mode of irrigation, is made to traverse each side of the street, whence it is led into to every garden-spot, spreading life, verdure, and beauty over what was heretofore a barren waste. . . . The irrigating canals, which flow before every door, furnish abundance of water for the nourishment of shade trees. . . ." [4]

In time, complaints were made about contamination of this open water supply in the city, and in 1873 Alderman Felt "suggested the propriety of taking steps to prevent the *filthing* of water by horses tethered near the Temple block." The culinary water system was later improved by piping water from streams to houses, digging wells, and building covered water tanks.[5]

South of the city in an area called the Big Field a total of 5,133 acres were designated for agricultural usage. By 1856, a canal had been brought from Big Cottonwood Canyon to irrigate the Big Field under one of the first community irrigation projects.[6] This canal was later enlarged to extend its area of delivering water.

During these early years, irrigation standards evolved, including time intervals for irrigation, water quantities to be used, and other basic irrigation practices. This information was passed on among individuals and at public meetings held as part of the Sunday and weekday meetings of the Mormon church. Led by the ecclesiastical leader, the local Mormon bishop, the meetings provided an opportunity for disseminating and sharing farming experiences at a grass-roots level. Many meetings were devoted entirely to water and farming problems and irrigation techniques.

Church leaders made regular visits to outlying settlements to "spread the word" about farming innovations, and annual tours were made by Brigham Young and his advisors. A lengthy epistle written in 1849 summarized the gains made by the permanent settlement of Salt Lake and in the valleys extending twenty miles south and forty miles north of the city. The epistle talked of a hard winter and problems of crickets and that through experiment the pioneers had proved that "valuable crops may be raised in this valley by an attentive and judicious management." They had constructed a two-story council house and bridges across several creeks and the Jordan River. They had surveyed the Big Field to the south of the city and plotted this area into five- and ten- acre lots, and a church farm. These lots were distributed to the brethren "by casting lots," and every one who received property was "to help build a pole, ditch, or a stone fence" around the whole field. Also a canal on the east side was to be constructed for the purpose of irrigation. The letter also reported that the old fort was "rapidly breaking up, by the removal of the houses onto the city lots; and the city is already assuming the appearance of years, for any ordinary country; such is the industry and perseverance of the Saints."

Much of the success in the development of early Salt Lake City was attributed to the Big Cottonwood irrigation canal that had been completed on the east side. This canal was incorporated by the territorial legislature on January 15, 1855, and required several years to be completed. After visiting the digging of the canal for a whole week, Brigham Young spoke of its progress in a speech in the Bowery on June 8, 1856. He praised the bishops and brethren for their efforts by saying, "I think they have done extremely well. A great many men have labored on that canal during the past week, and had it not been for faith, or the Spirit of the Lord upon them, many might have sunk with fatigue, for they looked as though they would faint; but they have labored faithfully." On that occasion Brigham also talked of other canals that would bring water to Salt Lake City. He suggested a canal on the west side of the Jordan River, a canal to bring the water of the Provo River to Salt Lake and the waters of the Weber River be "brought to the Hot Springs, there to meet the waters from the south and empty into Jordan." He also talked of developing the waters of the Bear and Malad rivers. Hindsight from this position in time with the Weber Basin Project, the Central Utah Project, and consideration of a Bear River Project in prospect it seems that Brigham Young was prophetic in his ideas of water development in the Salt Lake area. [7]

As the population of the valley grew and demands for water increased, it became necessary to develop rules and regulations to control water resources. The first laws established by the church leaders came through the line of church authority—prophet (the church president) to apostles to bishops. In the spring of 1849, Salt Lake City was divided into nineteen wards (a ward is a basic ecclesiastical unit) with each ward under the direction of a bishop. Ward members participated in the planning and construction of projects, and ward units worked together cooperatively to develop timber, roads, gristmills, fences, and ditches.

At the first sabbath meeting in the Great Salt Lake Valley on July 25, 1847, Brigham Young declared, "No Man should buy or sell land. Every man should have his land measured off to him for city and farming purposes, what he could till. He might till it as he pleased, but he should be industrious and take care of it." Shortly after this it was also decreed that there would be no "private ownership of the water streams" and that wood and timber would also be regarded as community property.[8]

This land policy promoted small farms that could be managed by a family. The average Utah farm was about ten acres—one-fourth the national average. Church policy dictated that "each one was to have as much as he could till . . . and a polygamist [with several wives and families] could receive a sixty-acre tract."[9]

On August 22, 1847, Edison Whipple was appointed to oversee "the distribution of water over the plowed land,"[10] marking the beginning of the position of watermaster. When the wards were created in 1849, the bishops "became in practice watermasters, fence supervisors, and bridge builders." On April 5, 1849, instructions were given that the bishops of the nineteen wards were required to see "that ditches were cut around their respective wards and that bridges were provided where ditches crossed open streets."[11] On July 6, 1853, the position of watermaster as a separate office from the bishop was created by a Salt Lake City ordinance. The watermaster was to have responsibility for protecting the use of all waters flowing into the Salt Lake Valley.

Ward members were responsible for assisting in the construction and maintenance of ward ditches or canals to a degree based on the amount of property they worked and the amount of water they used. Each water user could work off his maintenance or construction assessment assigned by the watermaster by digging and maintaining a certain section of a proposed project. This close association of landowner as water user made the water appurtenant to the land, and it could not be transferred independently of it.

The construction of canals and ditches was crudely done at first. The survey of the ditch might be made using only a wooden triangle with a plumb dangling from its apex to determine the run of a canal. Makeshift levels were fashioned from wooden pipes and bottles filled with water. Canals were also laid out and leveled by "filling a broad pan with water and sighting along the surface." Several plows drawn by teams were used to dig canals, and some teams pulled an immense wedge-shaped device called a go-devil along the path of the canal. The go-devil was constructed of sharp, pointed logs about a rod long with a brace in the middle that gave it an A-shape. "The go-devil threw the earth upon the sides, and being very ponderous, it was drawn by a train of oxen. More frequently [the pioneers] divided the work into sections, and gave each man a rod of ditch to clean up with a shovel after each run of the plows over his rod." Depending on the number of workers, different construction techniques were used. If there were a hundred men to dig, the plows would run continuously through a hundred rods and return, thus running back and forth until the section was completed. "The ditch would probably be four to six feet at the bottom and twice as much at the top, with a depth corresponding with the bottom width, unless the unevenness of the ground required a broader cut upon the surface."[12]

Diversion dams were initially made by placing brush, willows, rocks, and earth in the main stream and diverting water into the "laterals" and ditches that would carry water to the individual farms. "Corrugates" or furrows were dug next to each row to carry the water to the crops. Each farmer was allowed a "turn" or a specified amount of time that he could use the stream or a portion of it. The early diversion dams were frequently washed out, but as time went on and materials and funds became available, stronger diversion dams were built and ditches were lined with masonry and concrete.[13]

In 1849 the state of Deseret was established by the Mormons as the governing body of the colony. In the constitution of the state of Deseret, no mention was made regarding the regulation of water rights; however, the legislature soon passed several laws controlling water usage. Between 1850 and 1851, special water rights were awarded to community leaders: Ezra T. Benson was given rights to Twin Springs and Rock Springs in Tooele County in 1850, and Brigham Young was given "sole control of City Creek, and Canyon" on payment of $500 to the public treasury. Historian Charles Brough wrote in *Irrigation in Utah* of these grants that the "exclusive control in these instances did not imply monopoly rights of use, but rather exclusive right to orderly regulation and apportionment of water usage in accordance with socially desired ends."[14] Apparently the legislative body resorted to the device of centralizing responsibility as a means of averting interminable controversies over this prime essential in food and crop production under arid conditions. This is verified in a grant to Heber C. Kimball on January 9, 1851, giving him the right to use two streams north of Salt Lake City for a sawmill, gristmill, and other power purposes on the condition that "nothing herein contained shall prevent the waters aforesaid, from being used, whenever and wherever it is necessary for irrigating."[15]

Among other legislative acts, the state of Deseret incorporated municipalities. On January 9, 1851, Great Salt Lake City was incorporated, and on February 6, 1851, Ogden, Provo, Manti, and Parowan were incorporated. Water jurisdiction was vested in each of the communities, and city councils had the authority to provide the cities with "water, to dig wells, lay pump logs, and pipes, and erect pumps in the streets for the extinguishment of fires, and convenience of the inhabitants."[16] A liberal interpretation of these laws gave city corporations the opportunity to exercise almost complete control over not only the culinary water supply, but also the irrigation streams and canals within their boundaries.

With the establishment of the territory of Utah in 1850, control of affairs in the community became more secular. The territorial government continued to act upon water problems, and the most influential law was enacted on February 4, 1852. Under this territorial act jurisdiction over water resources was transferred to the county courts:

> The County Court has the control of all timber, water privileges, or any water course or creek, to grant mill sites, and exercise such powers as in their judgment shall best preserve the timber, and subserve the interest of the settlements, in the distribution of water for irrigation or other purposes. All grants, held under legislative authority, shall not be interfered with.[17]

Until it was repealed in 1880, this law was basic to water control, with one important modification on February 20, 1865.

At that time, territorial legislators created the concept of water districts, which allowed landowners "in a county or part there of" power to organize as a controlling group. "The water district had the right to supply themselves water at the expense it cost to divert it from its natural channels and apply it to the lands."[18] In this way, all the people "to be benefitted may prosecute the work of appropriation and construction of the works, to the end that each acre of land requiring irrigation may have ample supply." Irrigation districts could be formed on petition to the county court by a majority of "any county or part thereof." All landowners in the district had the right to an equal share of the waters brought into the district if they paid their proportionate share of the construction and maintenance costs of the project.

Irrigation districts were also authorized to elect a board of trustees that had the contractual power to develop irrigation projects, employ and appoint agents to deliver and control the water, and establish bylaws, rules, and regulations to govern members and agents of the district. The district could acquire land and property and money from district members to support construction of projects. District members voted to approve the board of trustees and a treasurer and assessor, who carried out the district business. Members were given one vote for each acre of land held in the district.[19]

This law of 1865 transformed water development from a theocratic cooperative effort to a secular cooperative effort—a change that worked more effectively in the growing mixed Mormon and non-Mormon communities. By 1898 there were forty-one irrigation districts incorporated in Utah, most of them in northern Utah, the most populous part of the state where there was the most water development and where this type of organization could function best.

Another territorial law of March 3, 1852, regulated lands surveyed for cultivation and incorporated a significant aspect of water law—the concept of prior appropriation—when it stated: "If any surveyor shall survey land or lands for the purpose of cultivation, [and if] where to irrigate it would rob other previously cultivated lands of the needful portion of water, such last survey shall be void for cultivating purposes." Prior appropriation marked a break with the riparian laws that had governed water usage in the United States, and marked a shift in Utah policy.

The doctrine of riparian water rights had been a common law development that worked well in the eastern portion of the United States where water was abundant. In practice, landowners adjacent to a stream had the right to use the water if they did not diminish the flow of the stream. This was reasonable where water was plentiful and streams could be easily diverted for mill races to provide power for gristmills, sawmills, and the like, but in a region where water was scarce and needed for irrigation, the doctrine of prior appropriation worked better. Prior appropriation stipulated that the first claimants had the right to the water to irrigate their crops without returning water to the main stream. Eventually primary, secondary, and tertiary rights developed that allowed primary, secondary, and tertiary claims to be met as long as there was water, but in times of water shortage, the primary needs would be met first, then the secondary and tertiary, depending upon the amount of water available.

The earliest prior appropriation practices developed in the mining regions where water was used extensively in mining operations, and California was one of the first to recognize prior appropriation laws in relationship to the mining industry. Utah developed agriculturally, so the concept of prior appropriation in Utah became associated with agriculture. Donald J. Pisani writes,

> In Utah the Mormon Church initially doled out water and the state did not formally acknowledge appropriation until 1880 or 1897. There mining played no part in the formulation of water rights and water laws; irrigation took precedence from the beginning. Church control over natural resources and the dominance of agriculture in the state's economy permitted the attachment of water rights to the land.[20]

Seventeen states in the arid West adopted the principle of prior appropriation, but with variations that came to be known as the California doctrine and the Colorado doctrine. The California doctrine was a split doctrine. It applied riparian rights to private lands that had existed before prior appropriation was legally codified, but the doctrine of prior appropriation was applied to the lands that were public at the time of its legal recognition. States that followed the California style of water development included Washington, Oregon, North Dakota,

South Dakota, Nebraska, Kansas, Oklahoma, and Texas. The Colorado doctrine recognized only prior appropriation rights and states that followed this principle were Colorado, Montana, Idaho, Wyoming, Nevada, Arizona, New Mexico, and Utah. All "refused to recognize any right of water in any land owner merely because of land ownership. Beneficial use was declared to be the measure and the limit of the right and no rights were recognized as having been vested except from actual diversion and application to a beneficial use." There continues to be some disagreement over this classification by scholars who claim there are many exceptions, and Utah is currently perceived to be closer to the California position than at first thought.[21]

The effect of early communal practices gave prior appropriation a different application in Utah than in other western states. If strictly enforced, prior appropriation law dictates that first individual claims have first rights to the use of the water. As long as the supply of water is sufficient to meet all the claims, they all can use the water, but if the supply diminishes, the first claims get first use of the water and later claims have to give up their use of the water. In Utah there was a greater communal interest in water division. Under ecclesiastical leadership an attempt was made to carry out prior appropriation to meet the needs of the community in such a way as to give the "greatest benefit to the greatest number." This was satisfactory to most claimants for some time, and in part accounts for the small amount of water litigation in early Utah.

In 1898 Charles Brough wrote that under the Mormon communal organization "there was no tendency among the prior appropriations to abuse their monopoly. . . ." Roy Teele noted similarly in 1903 that

> the division of rights into classes is peculiar to Utah and is a wise compromise between absolute "priority" theory enforced in some states under which the first comer takes the full volume of water to which he is entitled regardless of the needs of his neighbors, and the utter disregard of priority, advocated by some, under which the lands of the early settlers can be robbed of their value by compelling him to divide his water supply with an ever-increasing number of irrigators.

Robert Dunbar observed that the Mormon "sense of ecclesiastical brotherhood" molded the Utah policy of prior appropriation in such a way that "none were cut off, unless the stream went dry. All the 'brothers' within each category were treated alike; their rights were correlative."[22]

The concept of prior appropriation was developed through legislative acts because land ownership in the early days of the West was often nebulous. When western lands came under federal government control after the Mexican War, there were no land offices in Utah where the ownership of land could be registered. The Homestead Act of 1862 and the establishment of the General Land Office in 1869 subsequently made it possible for Utah settlers to make official claims to land and water.

Little new legislation was passed from 1865 to 1880. After the federal land office was established in 1869 and up to 1880 "there had been petty trading, and buying and selling of water, but it was only an insignificant part of the whole and it was not based upon any legal right or established custom to do so. It was the rule for the water to pass with the land, and in the main it had been regarded as appurtenant." In 1880 the territorial legislature passed an act that separated water rights from land rights. It should be pointed out that this law was passed after Brigham Young died in 1877. This law was in total opposition to the policy of communal ownership, which Young had espoused during his era. In this law there was a general provision to make water "personal property and separate it from the land at the leisure of the owner." The owner, with water now considered private property, could transfer ownership of it, as in the sale of any property. The assignment of deeds for land, which was possible after the federal land office established private ownership of land, was also a necessary development.[23] This new concept of water rights under the 1880 law will be discussed in detail in a later chapter.

The first federal act to recognize prior appropriation was the General Mining Law enacted in 1866. In substance, it provided that the local rules and customs relating to mining recognized by the people were confirmed by that statute. No method was prescribed for appropriating water or initiating mining claims. The territorial legislature merely gave its sanction to the then well-established principles created by district rules and early court decisions. In 1872 a law was passed by the Utah Territory that prescribed methods to initiate a mining claim, but nothing was said about water. In 1884 the Utah Territorial Legislature "confirmed by statute local customs as Congress had done in 1866, but it was not until 1897 that a statute prescribing a method for appropriating water was enacted." This law was not an exclusive method and the courts held that "an appropriation could still proceed to initiate a water right by following the custom and usage of the area," and finally in 1903, the legislature of the state of Utah enacted a comprehensive water law, which has been "construed as prescribing the exclusive method for initiating water rights in Utah."[24]

The federal Desert Land Act of 1877 "severed the water from the land" on the public domain and cleared the way for each state to develop and pronounce its own system of water rights; this policy was upheld in federal court. Based on the Desert Land Act, the Utah Territorial Legislature developed the law of 1880 that separated water rights from land rights—a major setback, some thought, for community water development.

Private ownership was continued and substantiated in two laws in 1896, following Utah's admission to the union. One of these laws tried to establish a system of recording and defining appropriated rights, and the other created the office of the state water engineer to oversee the water rights recording program. There was no funding to support the program, so the attempt to organize the public water records was "virtually meaningless." In another legislative act in 1897, the Utah state engineer's office was given several technical duties to carry out. The state engineer was to survey and propose plans for the state reservoirs and approve the engineering safety of waterworks in the state. These tasks related to the development of 500,000 acres of public land that had been given to the new state under the Carey Act. Under this act the public land was to be sold to raise money to build federal irrigation works. The state engineer was also to conduct a complete hydrographic survey of the state's streams and make accurate measurements of the stream flow. With no money appropriated to carry this out, water data was not collected for a two-year period. In 1901 another law expanded the state engineer's authority and provided money to carry out the first survey of streams and canals, which promised to give order to and enforcement of prior appropriation water rights to a system that had been regulated for sometime by the state. In addition, the 1901 law ordered the county commissions to divide their counties into water districts and appoint water commissioners whose responsibility it was to measure the water and divide it among the recorded appropriations. The water commissioners were under the supervision of the state engineer. The counties were required to pay the salaries of the water commissioners. Several counties did not carry out this portion of the 1901 law.[25] It was not until 1903 that a comprehensive water law, which returned water ownership to the state, was enacted giving order to Utah water control.

Even before the community of Salt Lake City became firmly established and well before any water policy was codified, the outward push of colonization had begun. In August of 1847, explorations were carried out as far north as Cache Valley. Jesse C. Little led an exploring party of four men that left on August 9 and returned on August 14. Wilford Woodruff recorded in his journal that "the messenger brings a glorious report of Cache Valley and the country between us and there—that is rich soil and well watered, and well calculated for farming purposes." The party had called on Miles Goodyear at Fort Buenaventura on the Weber River and reported that he had a small garden of "corn and vegetables doing well."[26]

Weber County

UTAH

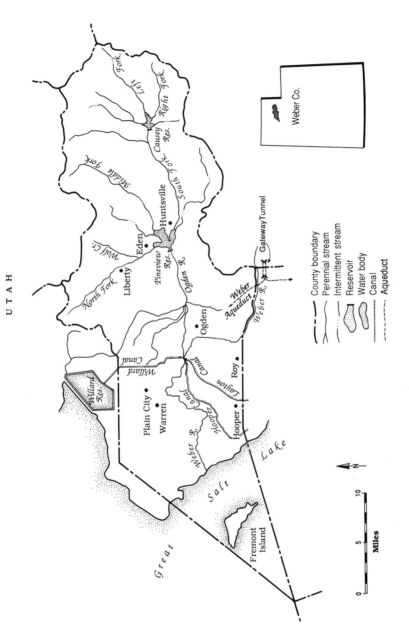

Weber Co.

County boundary
Perennial stream
Intermittent stream
Reservoir
Water body
Canal
Aqueduct

Left Fork

Right Fork

Causey Res.

Middle Fork

South Fork

Huntsville

Wolf Cr.

Eden

North Fork

Liberty

Pineview Res.

Ogden R.

Ogden

Gateway Tunnel

Weber Aqueduct

Weber R.

Willard Canal

Willard Res.

Plain City

Warren

Weber R.

Hooper Canal

Canal

Roy

Hooper

Layton

Great

Salt

Lake

Fremont
Island

N

0   5   10

Miles

Chapter 2

Early Water Development in Weber County

When Miles Goodyear built his picket fort on the banks of the Weber River from cottonwood logs in 1845, he named it Fort Buenaventura—Fort Good Fortune. The Weber and Ogden rivers, which join in present-day west Ogden, are the two major river systems that make up the Weber water basin area. Both rivers had been used for centuries as camping and hunting grounds for Indians moving into and out of the region, and later the mountain men trapped along the water courses in the Weber Basin from the 1820s to the 1840s seeking valuable beaver pelts. It was from these trappers that the rivers derived their names. Peter Skene Ogden trapped the upper portions of that river in 1825 in what became known as Ogden's Hole. The Weber River is believed to be named after John H. Weber, a member of the Andrew Henry and William H. Ashley fur company that trapped in the Wasatch Range in 1825.[1]

Goodyear established Fort Buenaventura as an "emigrant waystation" when the demand for beaver pelts began to decline. It was at Fort Buenaventura that the first irrigation in the area was actually practiced. It was the work of a Captain Wells to carry "water bucket by bucket from the Weber River to irrigate the beans, carrots—that reached one foot in length—cabbage, radishes and both Spanish and American corn."[2]

In November 1847, Goodyear sold his fort on the Weber River to the Mormons. Captain James Brown sent his two sons, Alexander and Jesse, to take possession of the property on January 12, 1848, and Captain Brown and his family, accompanied by other families, moved into the region and the new community, called Brownsville, began. The Brown family occupied the fort and the "other settlers scattered along the Weber River, some of them building two miles south from the fort and others locating as far north as the Ogden River and even beyond to Mound Fort."[3]

Most of the early settlements in the Weber Basin area were located along major canyon streams or on the lower plains surrounding the rivers where water was accessible and irrigation could be delivered to the flat lands with the least effort. There are eight major tributaries to the Weber River, and the Ogden River has four. This extensive water system also consists of thirty-eight small streams that run out of the Wasatch mountain-range canyons from North Ogden on the north to North Salt Lake on the south.[4]

The Weber River has its source in the high Uinta Mountains of eastern Utah, a major wilderness area even today. The Weber begins in a drainage basin lying north of Bald Mountain, where Reid's Peak and the Notch Mountains on the west form the perimeter of a basin and other peaks of the Hayden Range form the eastern edge. The river flows through Summit, Morgan, and Weber counties, and its waters are carried by canals and pipeline systems to Davis and Box Elder counties. The picturesque beauty of the Weber and Ogden rivers has long been known and publicized. The upper end of the Weber rushes from the high, snow-capped peaks of its source through pastoral settings, slows to a more placid pace through the farming valleys of lower Summit and upper Morgan counties, and provides thrilling views as it crashes wildly through Devil's Gate and the mouth of Weber Canyon. Leaving the mountains, it slows and winds its way through Ogden, then for twenty miles beyond makes a "big bend" before it enters the Great Salt Lake near Hooper. The Weber River picks up volume from various tributaries along the way and produces an annual average of 505,187 acre feet of water.

These rivers are great resources in northern Utah, and the growth and development that have centered around them have been extremely important to cultural and economic expansion in Utah. The Weber River Valley and Ogden River Valley are among the best naturally watered areas in the intermountain region, and extensive development evolved even with limited technological, financial, and institutional superstructures. A consequence of this was that Ogden, Utah's second city, and a prosperous and complex Weber County began to emerge almost at once. Water was the resource upon which the city and county were based, and water gave Ogden and Weber County a recognizable character.

At first it was a matter of subsistence farming for survival on Utah farms with few hands to help. Because of a lack of labor, the most convenient methods for getting water to the land were adopted. Initial settlement was in the lower valley floors with growth moving upward to higher areas and benches. Getting water to these less accessible places meant constructing high-line canals and ditches. Applying water to higher lands could cause soil saturation and subsequent flooding on the lower flood plain; soils could be leached of nutrients leaving only alkali and minerals behind and significantly reducing the productivity of the land. Irrigation projects that created problems had to be dealt with by the communities, and rules and regulations for water use had to be established.

Weber County, with Ogden as its focal point, evolved into an area of settled Mormon agricultural communities in the early pioneer period and remained fairly rural until the transcontinental railroad was completed in 1869. The evolution of Ogden as an important railroad city resulted in the immigration of non-

Mormons, creating a significant change in the composition of the population. As a result, there was a shift from ecclesiastical to secular control of the governing bodies and a shift from agriculture to commerce and industry. Water formerly used for agricultural purposes was shifted to municipal and industrial uses.

Brigham Young first visited the Mormon settlements on the Weber River in September of 1849 and, with other church officials, selected a site for the city of Ogden on the south side of Ogden's Fork "so that the waters from the Weber River and Ogden's Fork might be taken for irrigation and other purposes." Brigham Young gave James Brown permission "to build a bridge across the Weber River and another across Ogden's Fork and to collect toll from all parties crossing." Ezra Chase, who had moved to Ogden in 1848, commented that the land was very productive in grain, and added, "it will yield a hundred bushels of crickets to the acre and 50 bushels of mosquitoes."

Lorin Farr, Chase's son-in-law, moved to Ogden in 1850 and built Farr's Fort, and Brigham Young visited again in August 1850 to complete the business of laying out a city. The position of the town gave it "a most beautiful setting with several creeks and two major rivers for water. The President's party set the corner stake and gave a detailed plan for a modern, wide streeted city."[5] Brigham Young counseled community leaders to "move on to the city lots, build good houses, school houses, meeting houses, and other public buildings, tend their gardens and plant out fruit trees, that Ogden might be a permanent city and suitable headquarters for the northern country."[6]

The first systematic irrigation project used water in Canfield Creek, named for early settler Cyrus Canfield. The stream issued from Waterfall Canyon. In the spring of 1848, Jesse and Alexander Brown used a plow made of tire irons put together by blacksmith Artemus Sprague to plow the first furrow. Later they put a dam on the stream and turned water on the crops irrigating five acres of wheat, a patch of corn, some turnips, cabbage, potatoes, and a few watermelons grown from seeds brought from California by Captain Brown. In time Canfield Creek watered a tract of about ten acres.

In 1851 a number of small canals were dug, but the first major irrigation enterprise for the community was the Weber Canal, a seven-mile canal that brought water out of the Weber River near Riverdale to irrigate the lower part of Ogden. The Weber County Court, in its first session on April 24, 1852, ordered that two-thirds of the county revenues for the year be loaned to Ogden City to be applied to construction of the Weber Canal.[7]

The Weber Canal was an ambitious undertaking. It was fourteen feet wide and five feet deep and was taken out of the Weber River at the point where the river course turns from a westerly direction to flow northward (about one and one-half miles south of the spot where today in 1994 Riverdale Road crosses the railroad and the Weber River). Daniel Burch took advantage of the water in this ditch and built a gristmill in the Riverdale area in 1854.

Peter Boyle, a pioneer furniture maker, "built a dam across the Weber Canal which overflowed into an open ditch on 28th at Porter Avenue. By means of an overshot wheel, he turned a lathe to make furniture and, in the season, ground up cane to make sorgum molasses." At this point the water was diverted into a ditch and made its way to lower Ogden and eventually to 24th Street west of Wall Avenue, where it was used to power the molasses mill owned by James Brown. At this mill in 1863, Brown caught the sleeve of his coat in the cogs of the mill rollers and was pulled into the machinery. This first Mormon settler of Ogden died from the injuries he received, a victim of the water system he had helped create.[8]

The canal continued along the brow of the hill below Adams Avenue and flowed to 19th Street, where it joined Farr's mill race. As the canal waters moved through Ogden, several laterals were taken out. John Farr recalled how the canal functioned.

> Long before street paving started in Ogden, water ran down the dirt street gutters from the old Weber Bench canal. . . . Flash floods were common during rainy weather; otherwise, the water was clear, and thirsty throats were often quenched from the ditches. . . .

Farr said that one day while walking near 25th Street and Grant Avenue he saw an Indian woman "with her newly born baby, the baby sitting in the ditch, the mother holding the baby's head out of the water, while she gently gave it its first bath. Then, she tenderly wrapped the new-born in a warm blanket, and with comely dignity walked away."[9]

The canal was constructed with primitive tools, picks and shovels, and hand labor. Plows were used to loosen the ground. Water users along the canal were obligated to do the work based on how many acres they owned, and credit for work on the canal was allowed at $2 per day per person, $1 for use of oxen, and $.50 for the use of a plow. Laboring on the canal was tedious, and the citizens who participated did so at great sacrifice and privation; however, the project was worthwhile and provided water for gardens and homes in lower Ogden for many years, even up to the 1950s, when the water was piped to run underground to the river.[10]

Another significant canal for Ogden City was the Bench Canal, constructed in 1855 and 1856 by the Ogden Irrigating Company and built under the supervision of Isaac Newton Goodale, who was sent by Brigham Young in 1852 for the purpose of building canals and roads. Goodale recorded in his journal on September 3, 1855, that he went to Ogden City "to make arrangements about getine out of the watr of Ogdon Company [*sic*]" and was chosen director of the company. Construction began on the Ogden Bench Canal on September 5. The canal ran from the mouth of Ogden Canyon (at present-day Rainbow Gardens), and was approximately six to eight feet wide and eighteen inches deep. At the head of the canal, water was diverted out of the Ogden River and conducted by a wooden flume (an inclined channel for conveying water) for a short distance before flowing into the open ditch.

First used in May of 1856, the canal opened a large area to further settlement. The original cost was $22,000, but it was an investment that lasted until 1945 when it was replaced by the Pineview system. Portions of the old canal line can still be traced along the south side of the Ogden River ridges, a reminder of early pioneer ingenuity and commitment.[11]

In the early days of water development in Ogden, drinking water was taken out of streams, canals, or springs. "Water from irrigation ditches stood for hours before the silt sufficiently settled for water to be palatable."[12] Later wells were dug to supply individual dwellings with water.

In accordance with the laws of Utah Territory, passed in 1850 and 1852, city governments and irrigation companies could make claims on water rights. The citizens of Ogden proposed to the city council in 1879 that a reservoir be built on the bench area to collect waters from springs along the bench lands. On October 28, 1880, a group of Ogden citizens incorporated as the Ogden Water Company, to continue for thirty-six years, to construct and operate "water works" for the supplying of Ogden City and its inhabitants with water for fire, domestic, mechanical, and other purposes. "The capital stock of the corporation was $150,000—1,500 shares at $100 each."[13]

The Ogden Water Company was granted "the exclusive privilege and franchise for the term of 25 years, for providing and supplying the city and inhabitants with good, pure water" by the Ogden City Council on November 5, 1880. The city purchased a little over half the stock, which put the company on a firm financial basis. During 1881 and 1882, almost eleven miles of main lines were laid to bring water from a reservoir on 24th Street, located two hundred feet above Main Street near the old county court house. Pipes conducted water from the Ogden River and from Strong and Waterfall canyons to this reservoir that could hold a ten-month supply of water. In Ogden Canyon "about four miles from Main Street, was placed a receiving reservoir at an elevation of 350 feet above the level of the business part of town." Pipes were used to conduct the water from there to the distributor reservoir on 24th Street.

Water lines were run to most of the city from the reservoir; it also provided water for railroad operations. The mains took water along the eastern line of the town, and transversal pipes conducted the "indispensable liquid down as far as the depot, thus supplying all the aqueous fluid required in the inhabited part of town—for drinking, lavatory and mechanical purposes, as well as for protection from fires. . . ." In 1883 the city water supply serviced 250 consumers, including three railroad companies—the Central Pacific Railroad, Utah Central Railroad, and the Denver and Rio Grande Railway.[14]

The first water lines were built from wood logs hollowed out in the center.

James Macbeth, who worked for a long time as a plumber in Ogden, described the early water system.

> In the old days, Ogden had wooden water mains in the streets and they were constantly breaking and giving trouble. . . . When the city began to put in a good water system and install sewers, I had as many as twenty men working for me digging trenches and putting in pipes. People who had never seen any plumbing were quite amazed to see water come out of a pipe [by] merely turning a handle. [15]

The water system had a tremendous effect on the growth and appearance of the city. The 1883 *Directory of Ogden City* commented that the

> uses to which water is applied are too numerous to mention, but as the quality is very pure, free from lime or alumina, it is invaluable for manufacturing purposes. While the company does not desire to furnish water for irrigation, still by its sprinkling permit, many residences are surrounded by beautiful lawns heretofore an impossibility. [16]

The water system also provided hydrants, giving the city greater protection from fire.

Edward Tullidge, in his description of Ogden in 1881, wrote,

> Ogden offers a fine sight, as you view the lower western part from the bluff [bench], which rises in a smooth declevity towards the east. Your back toward the still snowclad mountain fortresses, you send your glances over a beautiful and fruitful country, rich in farms and fields, gardens and orchards, dotted and thriving settlements all over, as far as the alkaline shore of America's Dead Sea, whose wide and placid expanse glitters with silvery sheen at the foot of hazy hills, and the azure canopy of a cloudless sky. [17]

In later years the Ogden City water system went through some difficult times. In 1888 the city bought the water rights to Strong and Waterfall canyons to ensure adequate supplies for city development. In that same year a bond of $100,000 was issued, which included money for constructing a sewage system and extending the water system. In 1890 the city sold the water properties to the Bear Lake and River Water Company. This became a major political issue for the non-Mormon Liberal Party government. It was also an example of the new water laws of 1880, which allowed water rights to be sold to private companies for speculative purposes. This law was considered by many to be against community interests, and there was a great deal of opposition to selling the Ogden water rights to a private company.

In 1910 the city issued bonds for $100,000 and purchased the waterworks back. Since that time they have been operated as a municipal business. In 1914, the city developed the artesian wells in Ogden Valley as the main source of culinary water.[18]

Other significant water developments relating to the Ogden and Weber rivers occurred outside of Ogden City proper. At least twelve major water diversions were taken out of the Ogden River to deliver water to outlying areas. Each was planned by local leadership and constructed using community effort, a demonstration of the democratic nature of growth in the Weber River Basin.

The Chase ditch developed by Ezra Chase took water out of the Ogden River in 1849 for the area north of the river, and the Moore Ditch completed in the same year delivered water to the Mound Fort area. It was the Moore diversion that Lorin Farr utilized part way to get the water to his gristmill. Farr's diversion was known later as Mill Creek, and it became the main source for water to the communities on the lower and northern side of the Ogden River. This stream still provides most of the irrigation water for those communities.

These ditches were followed by others named after local citizens—Enoch Farr's Ditch or the Stone Ditch (named for Amos P. Stone) or the Tracy-Shaw Ditch (named after Moses Tracy and Ambrose Shaw). These were subsequently listed in other ways—the Stone Ditch became Mound Fort Irrigation #2 and #3 and the Tracy-Shaw Ditch became Mound Fort Ditch #5.

The Lynne Ditch, developed by Isaac Goodale, took water out of Mill Creek 350 yards west of Farr's Mill and conveyed it for about three miles to the Bingham Fort area. The ditch is still in use and serves a large area of Ogden all the way to the lower Second Street region.[19]

The Dinsdale and Marriott ditches were on the north side of Ogden and were named for Jeffrey Dinsdale and John Marriott. The Dinsdale ditch was opened in 1851 and the Marriott ditch in 1855.

On the south side of the Ogden River a ditch came to be known as the Shupe-Middleton Ditch, after John W. Shupe and Charles A. Middleton, major developers of that region; it was used as late as the 1950s. Charles Zeimer and John Broom, prominent community members, dug a canal in 1854, known as the Zeimer-Broom Canal, which went from Mill Creek to an area on the north side of lower 12th Street called Broom's Bench.[20]

Two other canals were dug from the Ogden River to carry water to remote areas—the North Ogden Canal and the Harrisville Canal. Many of the outlying towns had developed their water systems using mountain tributaries or drawing water from the Ogden and Weber rivers through extended canals. Some communities that had been settled independently were later incorporated into Ogden City (Mound Fort, Lynne [earlier Bingham's Fort], and Marriott). Others grew and created their own governments and water systems and followed the same general pattern of progression—early settlement by pioneer families organized as ecclesiastical branches of the Mormon church and then a move to a secular government format. Water systems also evolved from a few individuals diverting water from a convenient stream to more complex water systems as population grew. As time went by, water was taken out higher on streams and was run longer distances to lands developed on higher ground and farther from original settlements.

In 1850 several communities along the Ogden and Weber rivers were established and went through the water improvement sequence. The communities of 1850 were North Ogden, Harrisville, Slaterville, Riverdale, and Uintah. In that year North Ogden, mistakenly called Ogden's Hole in some early histories, was first settled. The community cleared "one large field, fenced it with poles and willow limbs, and planted potatoes, vegetables, wheat, and other grains," Isaac Riddle stated in his journal. "This was our first experience with irrigation and our first crops raised in Utah were good."

Water for the community was initially taken out of Rice Creek, which had its source in springs at the foot of the mountain west of Ben Lomond Peak and northwest of North Ogden Canyon. After the North Ogden Canal was put into use in 1857, Rice Creek waters were used below the canal. Rice Creek water ran as far west as Pleasant View.[21]

Other water was taken out of Cold Water Creek, which flowed from the mountains to the east. Randall Springs, developed by John Riddle, and Montgomery Springs were also early water sources. Robert Berrett, who located his family in a log cabin in 1855 on the west side of Pole Patch Road (4200 North 500 West in present-day Pleasant View) helped bring water to that area. The Berrett family dug an irrigation ditch from Berrett's Canyon just east of the base of Ben Lomond Peak. Another stream out of Ben Lomond is Alder Creek, and Simeon Cragun and John and Henry Mower are credited with exploiting it for later use by Pleasant View.

In 1856 it was decided that a canal should be constructed from the Ogden River to North Ogden. Jesse W. Fox was brought from Salt Lake City, and he surveyed the six-mile route from the Ogden River to the Hot Springs at the point of the mountain at the west end of Ben Lomond. Most of the male citizens participated in digging the canal, and one woman, Mrs. John Cardon, used a pick and shovel to help with the work.[22] Workers received shares in the canal or an appropriation of water on the basis of the amount of labor or money they expended. It was also necessary to make "as far as feasible" an exchange for waters that were taken into the canal system from the Cold Water Creek, Rice Creek, and Ogden Hole owners so that more of the lands might be served with water.

The actual digging of the canal started on October 9, 1856, and continued through 1857. In some places the banks had to be strengthened and near the mouth of Ogden Canyon an old rock wall still stands, built in pioneer times to hold the bank. William Hill was the first farmer to use the canal water on his land in 1857. In 1860 the canal was completed at a cost of $48,000, and it served the settlers as far as the Utah Hot Springs.

When the canal was extended, some dissatisfaction occurred among the water users and owners of the canal, which subsequently led to the end of the exchange system in 1856. On January 5, 1872, the North Ogden Irrigation District was organized and operated, somewhat "without much system," until April 23, 1873, when the old canal company was merged into the North Ogden Irrigation Company. The new corporation finished the canal to its full size, which gave it twice the former volume of water.

In later times additional water sources were developed from Willow Spring, and other private sources of water were established by tunneling into the mountains where there was seepage. Also scores of artesian wells were drilled to a depth of 150 feet for domestic and irrigation purposes to meet the needs of early North Ogden residents.[23]

Harrisville, six miles northwest of Ogden City, was first settled in the spring of 1850. Martin Harris, after whom the community is named, erected a house there and sowed ten acres of grain in 1851. A poor harvest led him to move to higher ground the next year.

The early water needs for Harrisville were met by using water from Four Mile Creek. In 1856 residents organized as the Ogden Western Irrigating Company and petitioned the Ogden City Council for the right to construct a canal from the Ogden River to Harrisville. Within the next year, the canal was located, surveyed, and completed, and the new settlers were supplied with an abundance of water. "The prosperity of the town can be largely attributed to the completion of this ditch."[24]

The western portion of Harrisville, and area north of Slaterville to the Box Elder-Weber County line at the Hot Springs, separated and organized into the community of Farr West, named in honor of Lorin Farr and Chauncey W. West. Water for this area came by a continuation of the Western Irrigation Canal and from several springs and wells.[25]

The experience of Luman Shurtliff in Harrisville illustrates the problems of early irrigation. In November 1851 Shurtliff established a forty-acre farm that raised apples, grapes, "seed locus" and "coffy nuts." He also raised cattle, wheat, and corn. Shurtliff described how difficult it was to maintain a fenced area due to the poor quality of fence poles available; instead a fence was created by digging a "ditch three feet deep and three feet wide [and] throw[ing] the earth all in a bank on the inside, then stick[ing] up stakes in that bank three or four feet high, then weav[ing] in willows thick to the top of the stakes. . . ." Thus the ditches became boundary lines for the property sections as well as the means of carrying water to the crops.

In 1852 Shurtliff recorded, "this spring we had much trouble in making water ditches or sects. Nor having any experience in such work we built a levy and made it to [too] small and we could not ge[t] enough water through it without raising it so high that Lewis or I had to stay by it and watch it while the other water [for] our grain and sect[ion] would break once or twice a day and it would take sometimes a half day to mend it, thus we worked night and day to water our crop."[26] Shurtliff's comments indicate some of the ongoing problems of early-day maintenance and attention that irrigation required.

Shurtliff's journal also indicates that the communal water system was not always carried out in the public interest. This eventually led to regulation by secular agencies and development of the prior appropriation doctrine, which recognized first rights claims because there was not enough water to serve every claim. In July of 1856 Shurtliff was shorted by North Ogden on his water rights, and he responded in the following fashion:

> Wrote a letter to Bishop [Thomas] Dunn of North Ogden concerning . . . devision of water for irrigating my farm as well as theirs, which agreement the people of his ward had broaken and instead of letting my share down as they agreed to, had used it on their grain which agreement if they do not heed and let down the water will damage me at least four hundred dollars in the loss of produce of my farm. And hearing nothing from Bp [Bishop] Dun, on Tuesday I went to North Ogden, saw the bishop watermasters and they agreed to turn the water down to my farm as formerly, and the watermaster started to do it. I went immediately to Br Terry and told him to do [it] now and water the crop on my farm, which he said he would do and supposed he had untill it was too late to save the crop.[27]

In August Shurtliff went again to North Ogden and "got damage[s] for using our share of the water. . . ." As is the case in almost all of the early water problems, the bishops of the ecclesiastical wards settled most of the difficulties. Later the courts settled disputes. That unwarranted use of water was a problem is evident. Shurtliff mentions that during the Mormon religious "reformation" of the late 1850s, one of several items of a "catechism of sins," which were acknowledged and agreed to by the Latter-day Saints, had to do with illegal use of irrigation water.[28]

Another example is found in the story of J. Willard Marriott, successful restaurant and hotel entrepreneur. Marriott started his life in the area of Marriott and Slaterville settlements. His family was descended from settlement founder John Marriott. They struggled through the hardships of early Utah farming and raised grain, potatoes, sugar beets, alfalfa, grass, cattle, and sheep. For the Marriotts, the river was always part of family life. It was the lifeblood of the irrigation canals and led to the sections of farm crops; it was a swimming hole for the kids; it was a place to catch frogs and trap muskrats; it was the source of water for the bucket brigade that tried to put out the fire that destroyed the family farmhouse; and it was the canal that was dragged for young J. Willard Marriott's body when it was thought he had been drowned in its waters.[29]

The abundance of water, which produced grass and meadows for pasture-lands, induced "the first settlers to locate on the spot where Slaterville now stands" in 1850. As the population grew and more land was put into production, water was brought from Mill Creek to irrigate the farms. The first canal extended three miles and cost $3,000. Later "the South Slaterville Ditch, coming out of the Ogden River from Marriott, ran along the southern part of the settlement, and the Warren Canal coming from Ogden River lower down stream ran along the north-ern part of Slaterville. These ditches were dug primarily with spades." Slaterville, located on the flat delta lands of the Ogden River, was susceptible to frequent flooding. In 1861 and 1862, flooding on the Ogden River ruined much of the Slaterville farming areas and forced many families to locate on higher ground.[30]

In 1850 two communities were organized along the Weber River, Riverdale and Uintah. Riverdale spread out along the river running from 33rd Street in Og-den south to the bend of the Weber River where it turns north. In the early days the region was principally sheep-herding ground. In 1854 Daniel Burch built his flour mill on the east side of the Weber River and later added a sawmill at the site. Burch developed a part of the Weber Canal that was approved and funded by the Weber County Court in 1852. The Ogden City Council records show some controversy as to how much of the water Daniel Burch was entitled to out of the new ditch and what he should be paid for early improvements. Later Burch was allowed $500 for work done. In May 1853 a tax was levied for use of Weber Ca-nal water; the city council approved a tax of $1 per acre for farm plots and $3 for every Ogden City lot.

The Riverdale settlers also took water out of the west side of the Weber River near where the river bends to the north. The soil was rich and produced "good crops of hay, potatoes, vegetables, fruits, berries, and, later, sugar beets." Also in this area was Burch Creek, named for Daniel Burch and his sons Robert and James, which flowed into the Weber River and provided irrigation water.[31]

Farther south and east, Uintah was located in the steep-sided valley west of where the Weber River breaks out of the Wasatch mountain range. Uintah was considered that "part of the country lying between the Weber River on the south and the hill, or bluff, on the north." This settlement was known first as East We-ber; then Easton; then, after the coming of the railroad in 1869, Deseret; and as of 1877, Uintah.

The first water was brought to the settlement in 1850 or 1851 when the Pio-neer Canal or ditch was constructed; water for this canal was taken out of Spring

Creek, which ran into the Weber River from the north near the mouth of the canyon. In 1852–53 River Ditch (now known as the Uintah Central Canal) was dug from the Weber River. The ditch cost $2,500 and provided water for the production of fruit trees, ground cherries, watermelons, cantaloupes, and other crops.[32]

West Weber organized as a community in 1851 on the south side of the "big bend" of the Weber River west of Ogden and below the confluence of the Weber and Ogden rivers. The first settlers brought their cattle there to graze during the summer. The first permanent settlers arrived in 1857.

In 1859 Archibald McFarland dug a canal in West Weber from one of the sloughs bordering the river and was able to irrigate part of his land, and in 1859 a canal was begun that put water on the land by 1860. The eight miles of ditch had to be "dug by hand over sun-baked plains, through sage brush hollows and mounds of sand." In 1860–61 it cost the settlers of West Weber $2,500 to irrigate ten small farms. There were disappointments, however, when the ditch broke from "want of sufficient fall" to carry the water. It was rebuilt, but in 1862 the Weber River rose higher than ever before and washed away the dams and destroyed a greater portion of the canals and ditches. The ditches and dams were rebuilt and in 1864 the main canal was enlarged and extended one and one-half miles to irrigate a newly surveyed parcel of land. In 1865 West Weber organized into an irrigation company to promote building a canal to the lower end of the settlement.

Many times flooding, not scarcity, continued to be a problem. In 1866 floods washed out the dams and ditches and ruined most of the nine hundred new acres of crop land. The community rebuilt the canal. On March 4, 1872, Hans Petterson and forty-three other citizens of West Weber were granted a charter for the West Weber Irrigation District. Chartering of companies did not stop the flooding, however. In 1873 the levee of the canal washed out and brought havoc to the community. During the winter of 1873–74 the people erected a new dam. The settlers purchased a pile driver and, using timber brought down from the mountains, drove piles into the river and created a strong levee with brush and gravel.

On July 22, 1875, James Robb, the watermaster, his son John, and Edward Fuller were drowned at the new dam while trying to repair a gap in it. These three represent only a few of the many deaths suffered through the years on Weber Basin waters. In 1878, the West Weber Diversion washed out again, and it was necessary for the water company to petition the Hooper City Irrigation Company to use their lateral to divert water on to the West Weber lands. After this both companies mutually agreed to blend their interests in the canal.

The development of the water in West Weber was difficult and fraught with problems. From 1859 to 1880, $75,000 had been expended to complete the system. The low farmlands on the Weber River had been a constant test for community effort.[33]

Other communities established in that area during the same time period were

Taylor in 1851, West Warren in 1853, Wilson in 1853, and Plain City in 1859. They too went through similar experiences in developing their water systems.

Taylor was a farming community, and some of the first crops were grain, alfalfa, sugar cane, and potatoes. In later years sugar beets, tomatoes, and peas were the main crops. Apples, peaches, and wild plums predominated in the early years, and pears, plums, apricots, and grapes predominated in later years. Kitchen gardens included peas, beans, rhubarb, berries, and native currants. To grow these crops, irrigation was essential, and water development in Taylor was associated with water development in West Weber. Taylor's main water source was the South Branch of the West Weber Canal.[34]

The Wilson settlement also played a significant role in the development of water on the Weber River. Until 1860, water for Wilson was taken "from high water which frequently backed up from the Weber River and over flowed its banks." In 1860 John Staker and the Wilson brothers constructed the first irrigation ditch. Wilson ditch was later enlarged until it was six feet wide and two feet deep and had sufficient water to irrigate four hundred to five hundred acres of land on Wilson Lane.

In 1879 a new canal was dug by the Wilson Irrigation Company. The canal headed near the old Wilson ditch and later branched, one branch going south and west into Kanesville and the other going to Hooper, where it later joined the Hooper Canal. The Wilson Canal was twelve miles long and watered four thousand acres.

One of the problems on the Wilson Canal was construction and maintenance of a flume on the north branch to elevate the water to the ridgeland areas. In 1888 the first wooden flume was built, but it was too low, and in 1893 the flume had to be rebuilt; it was repaired each year thereafter until 1919. By that time it "had reached a ripe old age. One morning late in August the farmers found the flume all lying on the ground. Not one post remained. . . ." In the summer of 1920 it was rebuilt again—this time out of concrete at a cost of $29,263.23.[35] This cement flume was finally removed in 1991 as the water routes had changed.

Plain City deviated from the pattern of communities moving outward from the Ogden hub in that most of its settlers came directly from another area without first being settlers in Ogden or Weber County. Located ten miles northwest of Ogden City, it was remote from the earlier settlements. Most of the first residents came with an exploring party from Lehi in 1858 searching for an area with water. After exploring the region thoroughly, including the Hot Springs on the northeast and other parts of Weber County, the party selected a site for the proposed settlement before returning to Lehi for their families. They liked the flat sagebrush plain and gave the town the name of Plain City. In March 1859 the settlers moved onto the land, and in May 1859 the work of digging a seven-mile canal commenced. Before the summer ended that year, the ditch was completed to Four Mile Creek, and water was brought to Plain City.

Over the years the cost of maintaining the Plain City Canal has been high because there have "been so many washouts on the big levee, and so many law suits with neighboring villages over water rights." When the big levee broke or was threatening to break from year to year it "caused a lot of excitement and men were kept there night and day to watch it."[36]

Cooperative water projects with other communities started as early as 1860 when "a proposition was made by the inhabitants of Marriotsville for the people of Plain City to join them, and operate together on one large ditch to bring out the water from a certain distance and then divide the stream at a point convenient for each settlement . . . that the volume of supply might not diminish." Through some misunderstanding, the agreement failed to have the desired effect; one-half of the crops in Plain City were lost as a result. Later the Plain City settlers brought water from the Ogden River above the Marriotsville Dam and obtained a fair supply of water, which gave a better crop the next year.

Again cooperative efforts were worked out with the people of Marriott, but in July of 1869, "a scarcity of water for irrigation purposes began to manifest itself," and the inhabitants of Plain City, in order to save their orchards and water their grain fields, went to Ogden and by "throwing up a dam in the Weber River" and putting in head gates, and by purchasing the right of way at a cost of $2,000, obtained partial relief.

In 1873 an additional eighty-two-rod-long canal was built to connect the Ogden River with the Weber River to bring more water to the Plain City Canal. Construction of a dam and head gates cost $2,000. On August 8, 1874, the Plain City Irrigation Company was organized, but even with new improvements, water was still scarce during dry years.

In 1889, Samuel Wayment drilled a 249-foot artesian well. Artesian wells proved to be an important source of water; however, it was not until the Plain City Irrigation Company subscribed for water in Echo Dam in 1924 that an adequate water supply was provided.

The first culinary water in Plain City was taken from springs on the west side of town and the first houses were built near those springs. Open wells were dug at first by tapping into the supply of underground water. These were followed by square-boarded wells with a covered top and bucket used to draw up water. Later hand pumps attached to pipes running deep into the ground were placed outside the houses, usually near the kitchen door. Later they were placed inside the kitchen at the sink, and, in more recent times, electric pumps were used to pump the water.[37]

Water development in Warren and West Warren was associated with Plain City development. This was a farmland area on the delta of the Weber River where silt from the canyons and mountain slopes had been deposited and had formed a rich alluvial fan.

Because of frequent flooding in early times, the surface of the land was very rough. Over the years the land was leveled out and small and large farms established. Like other Weber County towns, the supply of water for irrigation was vital to the economic prosperity of the community. The first effort in Warren after 1873 was construction of a high-line ditch from the Weber River, which served a few farmers in the early part of the growing season. Soon after another ditch was dug from Four Mile Creek, but this proved insufficient. In 1895, another project was undertaken; the Electric Power and Light Company acquired rights of way and water rights and spent $85,000 building head gates at Marriott and putting in an extensive distribution system. Much new land was to be opened up by this project, but when the system could not deliver water in the latter part of the season, the enterprise failed financially.

A committee was organized representing the water users and acquired the canal and water rights for $25,000 on easy terms without interest. From this group, the Warren Irrigation Company was established. It installed a pumping plant on the Weber River and purchased stock in Echo Reservoir to supplement water supplies. In later years the entire district of Warren and West Warren has been "quite amply supplied with irrigation water." With this increase, and the culinary water supply supplemented by artesian wells, the community was able to take on a larger number of farms and families.[38]

Hooper was the farthest extension of settlement in the southwest part of Weber County—in fact, the community is divided by the county line and the northern section lies in Weber County, the southern section in Davis County. The community is situated three miles east of the Great Salt Lake and is the last community on the Weber River before it empties into the Great Salt Lake.

Initially Hooper was a herd ground on Muskrat Springs, the main water source. In 1854 William H. Hooper built an adobe house there for herdsmen marking the beginning of settlement, but more settlers did not move in until 1867 and 1868.

Hyrum Belnap described the area and his move to Hooper in 1868:

> In the early Spring of 1868, my parents were called to build up Hooper, then called Musk Rat.
>
> My brother Gilbert R. Belnap had been working on the Southern Pacific Railway grade. One of the horses had taken sick from saleratus dust. I was placed on this ill-fated animal and taken to the Hooper ditch, the head of which began where the present Sugar Factory is built, and told to follow the same until I reached the end, where I would see a covered wagon, which was to be our future home. . . .
>
> On the east side of what is now called Hooper, I recall the round Hooper pond, about 200 by 300 feet, on the northwest side of which was a cold fresh water spring, from which later we obtained our

drinking water and supply for the family use and laundry. There was not a growing tree or willow after leaving Weber River, only some flags and rushes in Musk Rat Spring. Just as the sun was setting, I observed a small dugout, on the top of which was standing an Indian girl. . . . The shadows of the sun reflected over the wide spreading plain or vast prairie, and I saw two houses looking like tall trees or objects about two miles south, which were the only signs of a residence as far as the eye could reach. I finally reached the end of the ditch, which had never yet been filled with water. There was the wagon box, about ten and one-half feet long with a cover over it, where the family slept with our head to the ends and our feet overlapping each other. I lay near the opening at the end. Soon all were sound asleep, but I looked out and listened to the howling wolves. They cam[e] closer and closer to the wagon until the dog drove them away a short distance, then they in turn would drive her back. This pleasant music continued until the wee hours of the night.

The village of Musk Rat sprang up like mush room.[39]

Initially springs and later wells met the water needs in Hooper. John Hooper recalled:

Many years we carried all our drinking water from Everett's, a distance from their flowing well of close to 900 feet. Other water we used we got from our own surface well in our own yard, which we pulled up with a bucket, a common practice in our day. Washing was done with our own well water, which was hard and salty. Later, when Stones got a well, we carried from there, a distance of 700 feet or more. Even after we moved to the city blocks we did not get an artesian well of our own for about fifteen years, but carried water from "uncle John" Manning's across the street, a distance of nearly 300 feet. It was about 1906 before father finally drilled a well of his own, which was a big improvement and was a good flow. The well is still in use.[40]

Construction began on the largest canal in the county in 1866; when finished, it was seventeen-and-one-half-miles long and ran from Ogden to Syracuse in Davis County. It was eighteen feet wide and three feet deep when completed, had a fall of twenty-nine inches to the mile, and watered five thousand acres of farmland. The Hooper Canal, as it was known, had two branches—the North Fork and the South Fork. The canal was completed in 1875 at a cost of $75,432.53.

In 1916 a concrete diversion for the Hooper Canal was put in place on the Weber River, replacing the old brush and earth dam that required almost annual restoration and improvement, at considerable hazard to those who worked on it. The Hooper Irrigation Company passed a resolution in 1926 supporting construction of Echo Dam, and purchased seven thousand shares of stock in the Weber River Water Users Association to have claim on waters from the reservoir. Later the share amount was increased to 9,600. Water from the reservoir was first delivered in 1934 and finally guaranteed sufficient water for Hooper even in periods of heavy demand.[41]

Early settlement in Roy began between 1870 and 1873 when several families moved into the area, described as a "forbidden, forsaken piece of land without a name or settler. The acreage was mostly blowing sand covered with sandburrs, prickly pears, rabbit brush, sage brush, and bisquit root."[42]

During the first few years the farmers grew only wheat because there was insufficient water. The settlers had to haul water for culinary purposes from Muskrat Springs three miles northwest of Roy. Later William Baker dug a fifty-foot well and secured good drinking water, the only well between Ogden and Kaysville at the time.

Water development in Roy was associated with the building of the Davis and Weber Canal. This canal is more aligned with Davis County water history and will be treated in more detail in a later chapter. Although Roy had water from the Davis and Weber Canal as early as 1884 when it was completed, there was not sufficient water because prior appropriation of water claims on the Weber River took water out of the canal about June 10 of each year. This meant that Roy could not irrigate late-maturing crops. This problem was not solved until East Canyon Reservoir was completed and tied into the canal system after 1896. Then the farmers of Roy could plant orchards, small fruits, tomatoes, and vegetables.[43]

Another aspect of Weber County water history that deserves special consideration is the history of the Ogden Valley or Upper Ogden River communities, including Eden, established in 1859, Huntsville in 1860, and Liberty in 1892. These communities lie in a valley at a five-thousand-foot elevation completely surrounded by mountains, and the area was first referred to as a "hole" in mountain man language. The Ogden Valley is well watered, and several clear mountain streams issue from nearby canyons and ultimately find their way into the North, South, and Middle forks of the Ogden River.

Peter Skene Ogden, expedition leader for the British Hudson's Bay Company, brought his trappers into this valley in May 1825, but he was soon driven out of the area by the American trappers of the Ashley-Henry Company. Beginning in 1856, Ogden Valley was used by the Mormons as a cattle range in the summers. A few small cabins were built for summer occupancy, but it was not until 1860 that Captain Jefferson Hunt, his sons Joseph, Hyrum, and Marshall, and Joseph Wood, Nathan Coffin, and his mother, Abigail, went into the valley with the intention of making a permanent settlement. Others followed, and seven families made up the settling group. They located on the river bottoms of what is now Huntsville town proper.

After a severe winter, the creek overflowed its banks and the meadows became marshy. "Realizing it was an unsatisfactory site to build permanent homes, Jefferson Hunt moved his little colony up on the bench."[44]

An irrigation company was also organized in 1861 under the supervision of Jefferson Hunt. A ditch was dug out of the South Fork about two and one-half miles above the center of Huntsville, and it brought water onto the bench. "This same pioneer irrigation ditch still meanders from east to west through Huntsville," over 120 years after it was first constructed.[45]

With improvements being made, a larger population could be accommodated in the valley and it became necessary to enlarge the townsite. The Mountain Canal, which takes its water from the South Fork approximately one mile above the earlier ditch, was constructed at a cost of $3,500. By 1870 the population had grown to over one thousand people, with two hundred families who earned their livelihoods on approximately 1,700 acres of tillable land growing barley, oats, potatoes, grasses, and hay as the main crops and on hundreds of additional acres of pastureland.[46]

Eden sits between the North and Middle forks of the Ogden River. The first settlers were herders, and in 1859 Joseph Grover and John Riddle located on the North Fork northwest of Eden. Others soon came over North Ogden Pass to settle there.

Irrigation was initiated in Eden in 1861 by Richard Ballantyne. He took water from Wolfe Creek following the "most convenient route" to the town Later other farmers joined in constructing two main canals, laid out by using "a level made of boards."[47]

When demand increased, the Wolfe Creek waters proved insufficient. In 1867 David Jenkins surveyed a canal from North Fork and developers organized into the Eden Irrigation Company in 1871 to control water use. In 1889 James Ririe added to the water supply by drilling the first well in the valley. In years to come, the artesian wells became a main source of water for the Ogden City water supply.[48]

Eden, like other communities, had its share of problems over water rights. The Ogden River had claims on its water from one end to the other, and in drought years those holding prior rights insisted that latecomers give up use of the water. LaVerna B. Newey has written that

> there were more contentions and bitterness over water and fencing rights in the early days [in Ogden Valley] than anything else. Before the days of the civil courts, water troubles were often taken to the ward bishop to solve. This put the bishop in a serious predicament. Which ever ruling he made, he was bound to lose face with a mem-

ber of his congregation. Sometimes it involved members of his own bishopric, and then it was doubly hard to make a decision.

In one case a watermaster told a farmer he wasn't getting his share of water and he should put some rocks in the ditch to equal it out. The farmer was then hauled by another farmer into a bishop's court for doing this. When confronted, the watermaster denied his involvement in it. Who was to believe whom? . . . Gradually civil courts took over.[49]

The first settlers of Liberty in Ogden Valley came in 1859 and remained in the north end of the valley. These early settlers had access to individual springs, such as Fisher Springs, but these did not provide enough water for the settlement. Irrigation ditches and wells were needed to bring water to the farms, and an irrigation ditch was established in June 1864 by taking water out of the North Fork and from Spring Creek in Liberty.

Mary Chard McKee recalled:

Our settlers located near streams, creeks, or springs, but as more people moved in and the water was taken out of the original stream to be put into irrigation canals to irrigate the crops, the people at the north end of the streams found that their water became polluted from the farm animals and people who lived further up the heads of the streams.

I remember in my early childhood days of being called early in the mornings to go fill all the containers possible with water, so we would have enough water to last the day, or to do the washing, because the water was to be turned out of the ditch or canal into another ditch for someone's water turn. Many times we would go to the water hole with a tin cup and dip enough water up to use for culinary purposes. The drinking water was guarded like gold, because we would have to go a mile or so to get good water from some spring. This drinking water was put in earthen jars and put in a cold cellar and often had to last several days.[50]

To make improvements, the community was organized into the Liberty Irrigation Company in 1899. In 1910 William H. Chard, John Shaw, and O. A. Penrod signed a $2,500 note with the Ogden State Bank to provide financing for the Liberty system. An intake for the system was built on the North Fork and two and one-half miles of six-inch wooden pipe was laid. This early pipe system contributed to the Liberty water supply for many years.[51]

These Ogden Valley communities played significant parts in the early water development of the upper Ogden River, but there is another facet to this pioneer water history. One of the important pioneer industries that relied on water power was lumber milling. Lorin Farr had established the first mill site in Ogden and timber was brought out of the canyon by sliding logs into the river and floating them to the sawmill. In the 1850s Samuel Ferrin and his son, Josiah M., ran a sawmill near the mouth of Ogden Canyon, but it was washed out in 1862 by flooding on the river, and they moved to Ogden Valley to continue the business. In 1862 (some say 1858) Levi Wheeler and Chauncey W. West constructed a mill at Wheeler Canyon just below the present Pineview Dam, and they built a small dam that backed up the water from the Ogden River and Wheeler Canyon. This flutter-type mill was washed out and not rebuilt. In 1869 Francis M. Shurtliff and C. B. Hancock built a mill east of Lewis's Grove that operated on steam power. In 1871 they sold out to David Eccles, one of the largest lumber entrepreneurs in the area. The largest enduring sawmill in the canyon was the Wilson lumber mill and Wheeler camp established in 1873 by Billie Wilson, which was at the "Old Hermitage" site. A small dam and mill race delivered water to power the mill, and this mill ran longer than any others in the canyon.

In Ogden Valley the first sawmill was erected in 1861 by Thomas Bingham. It was a hand-driven shingle mill located on the North Fork. In 1862 Samuel and Josiah Ferrin put up a power-drive sawmill near the same site; later they moved to Wolfe Creek. LaVerna Newey summarizes:

> Enoch and Henry Fuller, just young men, started a lathe and picket mill in the Wolf Creek area too. Oscar Shaw and Joseph Southwick put up a mill on the North Fork in 1878. Steam mills came into being, and Levi Wheeler, who had a sawmill in Wheeler Canyon, took his mill into the Wheeler Creek on Monte Cristo. In the late 1870's Thomas Bingham began a shingle mill in Sugar Pine Creek. David Eccles operated a sawmill at the head of Sugar Pine at about the same time. In 1881, Stephen and Ephraim Nye ran Bingham's shingle mill, and Bingham moved to the head of Bear Gulch.[52]

The river also provided recreation. Several hotels and resorts were built near it that took advantage of the river's beauty and the abundance of fish and wildlife. Pioneer resorts included Glenwood Park with Jones Grove and Farr's Grove on the Ogden River at present-day Lorin Farr Park. The Ogden Canyon Hot Springs Resort at the mouth of the canyon was in use as early as 1890. The Winslow family established a hotel and eating establishment on a grassy flat near the riverbank close to where the Thiokol Center now stands. The dam built by the Utah Power and Light Company in 1889 above the Winslow property and below Wheeler Canyon made a small reservoir that provided a scenic view and boating.

The Hermitage Hotel was established in 1905, and the dam for the Wilson sawmill was used for a boating area at the resort. It was here that two of Billie Wilson's children drowned in a boating accident. The hotel was long known for its trout and chicken dinners, well served at a moderate price.

Clearly water played a significant role in many aspects of development in Weber County. As early settlements expanded, the people in various communities carried out final planning and construction of water projects.

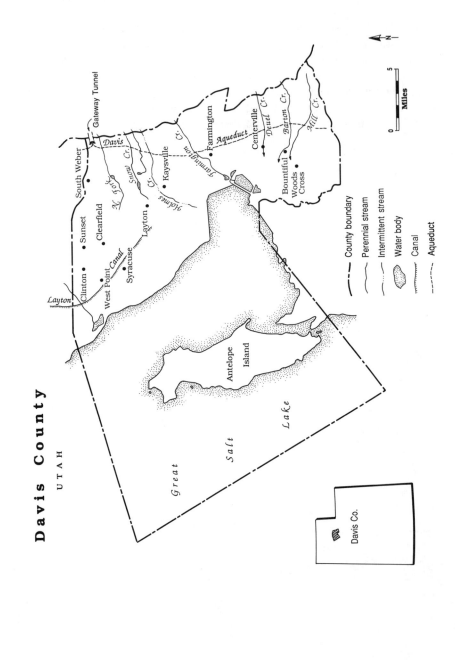

Davis County

UTAH

Gateway Tunnel

South Weber

Davis

Sunset

Clearfield

Clinton

Layton

West Point Canal

Syracuse

Layton

Kaysville

N. Fork

Snow Cr.

Holmes

Farmington

Davis Cr.

Farmington Aqueduct

Centerville

Deuel Cr.

Barton Cr.

Mill Cr.

Bountiful

Woods Cross

Antelope Island

Great Salt Lake

County boundary
Perennial stream
Intermittent stream
Water body
Canal
Aqueduct

N

5

0

Miles

Davis Co.

Chapter 3
Davis County Water Developments

Water development in Davis County was considerably less complicated than in Weber County, where early water usage was intimately related to the Ogden and Weber rivers. This was also a result of the lay of the land and a fewer number of communities making claim to the same water source. In Davis County, streams flowed directly from the mountains to the Great Salt Lake, and most communities claimed the water of each stream. Water development still followed a similar pattern—development first by individuals, then by the Mormon church, then by county governments creating irrigation and ditch companies, and, more recently, by the Weber Basin Project. In Davis County, cooperative efforts crossed county lines with construction of East Canyon Dam and the Davis and Weber Canal, but it was in the Weber Basin Project that Davis County joined forces with the greatest number of other counties to become a major recipient of the water resources the project generated.

Davis County was created by the legislature of the provisional state of Deseret on October 5, 1850. The county derived its name from one of its first and most notable pioneers, Daniel C. Davis, who settled in Farmington in 1849 and was revered for his service as a captain in the Mormon Battalion of volunteers who enlisted to fight in the Mexican War of 1846–48. Davis County is the smallest county in Utah. Its accessible and arable ribbon of lands comprise only one-sixth of the county's area, and they lie between the mountains and the Great Salt Lake in one of the most productive agricultural districts in the state. Today this region is rapidly shifting to an industrial economic base, and only a very small portion of Davis County land remains devoted to agricultural production.

The Wasatch range, with a maximum height in the county of 9,480 feet at Francis Peak, is the depository for the snow and moisture that feeds the county's streams and springs. The canyons of this section are "steeper and shorter than is usual to the range, the average length being only five to six miles in the southern portion of the county." Looking east in Davis County from the lake shore toward the mountains, a series of terraces can be seen rising from the barren alkaline flats to a saltweed line, then to saltsage and meadows, and then to various sagebrush covered upper belts that have been freed from the lake salts by drainage. In the county "the transition between well drained lands and wet bottom lands is very sudden and most of the lower terraces below the sagebrush line are almost indistinguishable."[1] The terrain then rises to benchlands that have distinct horizontal cuts marking the shorelines of ancient Lake Bonneville.

Still higher stand the mountains, rocky and patched with brush, cut only by narrow, steep canyons. These, too, display a response to altitude and rainfall in a gradual trend from scrub oak and deciduous trees along the creek levels to evergreens on the protected southern slopes. . . . It is here on this narrow inhabitable strip that the casual settlement of the county occurred.[2]

In February 1841 trapper Osborne Russell was among the first to notice the area's potential. While journeying through, he commented in his journal that "with little labor and expense the water could be turned to agricultural uses."[3] In 1846 several emigrant trains traversed Davis County on their way to California. Frederick Lienhard noted "the land extends from the mountains down to the lake in a splendid inclined plane broken only by the fresh water running down from ever-flowing streams above. The soil is rich, deep black sand, composition doubtless capable of producing good crops."[4]

In 1852 a government report, prepared by Lieutenant John W. Gunnison as part of the Stansbury expedition that surveyed the Great Salt Lake, said the area promised to be productive and able to sustain a large population.[5] This assessment came four years after Mormon settlers had made their move into the region. Some of the first settlers in the Davis County area struggled, and some even turned back to Salt Lake City or moved on after their initial experiences.

In order of development, the towns settled in the county were Bountiful (including West Bountiful and South Bountiful), Farmington, Centerville, Kaysville, Layton, Clearfield, West Point, Syracuse, Clinton, and Sunset. Later incorporation would occur in Woods Cross (1932), South Weber (1938), Fruit Heights (1939), North Salt Lake (1946), and Val Verde (1967). In order to participate in and develop water projects, communities were required to be officially incorporated municipal entities. Centerville, for example, incorporated in 1915 to be eligible to issue bonds for the installation of a town water system. Woods Cross incorporated in 1932 in order to participate in the Bonneville Canal proposal, and North Salt Lake incorporated in 1946 to gain control of the spring waters for the city.

On September 29, 1847, Perrigrine Sessions and his family, Samuel Brown, John Perry, and Hector C. Haight came to Davis County to make winter homes. Sessions and the others had been selected to care for five thousand head of cattle that had been moved to Bountiful to graze so they would not destroy unfenced crops in Salt Lake City. Sessions carved a dugout shelter into an embankment on Barton Creek to live in during that first winter.

In the spring of 1848, other families followed with some settling at the mouth of North Canyon (the first canyon north of Salt Lake), three and one-half miles south and east of the first Sessions encampment. This entire district was designated as North Mill Canyon Ward in 1849, and in 1855 the name of Bountiful was adopted. At first few settlers came to the Bountiful area because of the scarcity of water, although some did arrive in 1856 and 1860. In time the settlements of South Bountiful and West Bountiful grew near the Sessions encampment.

Early South Bountiful settlers claimed considerable land, creating a problem for later settlers seeking suitable acreage. Brigham Young strongly disapproved of this practice, and during 1848–55 Jesse W. Fox made a survey in the area, and lands were allotted to new settlers at a dollar per acre.

West Bountiful was developed on terrain that "was not very encouraging as the ground was all covered with willows, swampy places and small creeks." In 1848 and 1849, the first settlers cleared a small tract and planted it. In time scattered farms developed.

In 1849 Daniel Wood settled in the southwest end of Bountiful, and it is after him that Woods Cross was named. Farther south North Salt Lake stretched to the Salt Lake County boundary.

One of the early and most important concerns in Bountiful was development of a reliable water supply and establishment of an agricultural base for survival. The streams that ran from the mountains to the east were—from north to south— Dry Creek, which dried up before it reached the community; Stone Creek, which came out of Ward Canyon; Barton Creek, which came out of Holbrook Canyon; Mill Creek, which came out of Mill Creek Canyon; and North Canyon Creek, which came out of North Canyon, and whose insubstantial waters also vanished before they reached the lake.

Perrigrine Sessions and Jezreel Shoemaker plowed the first furrows in Bountiful in 1848. At the time this was fertile bottomland, but a good portion on the north and west of the settlement was badly cracked from a prolonged drought with ground "so hard that it was dangerous for the horses to walk." Injuries could easily result from a misstep into one of the large fissures. This was difficult ground to break, especially with a homemade plow.[6]

Other early farms in the Bountiful area were located near the Sessions homestead. Each settler claimed his water right on the basis of prior appropriation as designated by the ecclesiastical authority and then later by the civil authority of the county court, as supervised by the local watermasters. Each water user had the responsibility for constructing and maintaining a portion of the ditch according to the amount of water he had claim to.

As time went on and more settlers arrived in Bountiful, a need to apportion water fairly came about. In 1852 Aldonus D. L. Buckland was made watermaster of Barton Creek by the Davis County Court, and Chester Loveland was appointed to Stone Creek. More watermasters were appointed to newly created districts in 1853. These district organizations were established prior to the Utah Territorial District law of 1865 and did not have the power of the later district organizations. In Bountiful the water districts had the same boundaries as the school districts and road districts. In 1854 the districts were changed again, and by this time water was scarce. "Water supervisors drew the greatest amount of criticism of any public officials because of decisions they were forced to make. With all the problems over water it is no wonder the watermasters changed each year."[7] In June 1855 Judge Joseph Holbrook ordered the several Mormon bishops of Davis County to resolve the water problems, but complaints regarding water rights continued to be a problem. They became so numerous that county officials and Bishop Stoker appointed a committee to divide the water of Mill Creek in 1856 "according to the number of acres of land to be irrigated from the water of the said stream on each side."[8]

Most water problems resulted from one farmer taking too much water from the stream or keeping the water diverted to his ditch too long on his "turn." However, in 1856 Newton Tuttle was told by Jude Allen that he could not have water for his garden because he was not growing wheat—one of the products needed by the community.[9]

In the spring there would be an adequate supply of water, but "the water ran off to the lake too early," and it became evident that "other methods for water must be found."[10] In the 1880s artesian or flowing wells were developed to the west. Woodrow Barlow, who farmed portions of the Barlow land and Anson Call land for many years in West Bountiful, stated that in the early days Stone Creek water had been used, but the increased demand for water and the convenience and reliability of the wells brought almost exclusive use of the wells to the area west of 200 West Street in Bountiful.[11]

Charles R. Mabey wrote that the first wells were dug using a "16 to 18 foot length of 1 1/3-inch pipe perforated about half way, [which] was driven down with a sledge hammer and then washed out, the drillers extending the hole about fifteen feet beyond the casing [by using another section of pipe]." At any spot on the east side of Mabey's farm they could get a flow of from two to four gallons per minute at a depth of thirty feet, and at one time they had a dozen such wells scattered over the place. Mabey said that "these were small surface wells; they were often filled with sand and dirt and needed cleaning out frequently to keep up the flow, and, as other deeper ones were drilled, they failed entirely." As time went on professional well diggers started drilling numerous wells. Later wells were

> sunk down to sixty feet, later to 165 and still later over 200 feet. It had been learned that larger flows could be secured by using 2-inch casing, and this became the common rule for a number of years. We had a good flow for culinary purposes near the house, still another at

the southeast corner of the field and a real large gusher of about 150 gallons per minute. . . . [12]

In the winter when the wells were not used they were plugged with a wooden plug wrapped with burlap to prevent leakage.

In the next generation of users the well casings were increased in size and wells were driven from 180- to 250-foot depths. The water would gush from these wells under greater force and volume and they required a gate valve, which caused problems in the winter when it would freeze and break. To remedy this, a tiny niche was filed in the valve to cause a small leak so that the water would not freeze. Some wells on the extreme western edge of the area were allowed to run all year to keep them from freezing. Houses were provided with water from wells that usually used two-inch pipes that were drilled near each home, and these provided good-quality culinary water.

> In the early days there was little control of wells, but the men grew careless, largely through ignorance of what lay beneath the ground. The water table was gradually lowered, the older wells ceased to flow or gave out a mere dribble and their owners were forced to bore deeper for this, the most essential stuff of human existence. Farmers were often careless and left their wells uncapped the year round and this resulted in such waste that the State government stepped in with appropriate laws to control and regulate practices in this field. [13]

Up to 1944 water rights had been filed on the wells, but after this state control began to be more rigid.

Some cisterns and reservoirs were built to store water. In Bountiful cisterns were described as

> a large rocked up room below the earth; some were fourteen to twenty or twenty-five feet square. They were plastered and made to hold water. When a man had his water turn, he would put part of it in this cistern; later in the week he could either pump it or pull it up with a bucket for house use; although it was not fit to drink or to use for cooking it was used for washing, etc. [14]

Another account describes the original cistern built in Woods Cross by Steven H. Ellis, Sr., Jens Nelson, and Stearns Hatch. "At first the water was drawn up with a wooden bucket lowered on a long rope at the end of a well sweep. Then a pump was installed which required 175 strokes to fill one bucket to the top. Finally, an automatic pump was used to supply 41 homes with water." [15]

Water to West Bountiful came entirely from Mill Creek and from springs on the various farms until later years when artesian wells were driven. Mill Creek emerged from what is now called Mueller Park Canyon and dispersed into several indistinct channels. The Jordan River also flowed through the extreme western edge of the community and disgorged into the Great Salt Lake west of Centerville. It was the site of some early settlement; however, the area proved to be too wet and boggy for sustained community and for farming, and most settlers moved to higher ground. Spring flooding from the canyon creeks was always a problem for the lower ground of West Bountiful. As late as the 1930s these floods created serious incidents. The Mill Creek water was used mostly on higher ground and in the south part of West Bountiful. The western and northern areas relied upon springs and artesian wells for water for the farms.

Early records indicate that ditches were dug to bring the water to the various farms. James Waite had ditches on the edges of his field, and William Muir had a pond on his farm that was used for bathing and baptizing young candidates into the Mormon church. Mr. Muir "planted peach trees along the ditch banks from house to house which provided the community with a great deal of this rich fruit."[16]

By 1892 the population that had settled in the West Bountiful area was overtaxing water resources and it was necessary to find more equitable distribution. In that year Joseph T. Mabey and 151 others, including residents of the various suburbs of Bountiful, incorporated as Bountiful City. In March of 1893 meetings in the Barton-Stone Creek area, the Mill Creek area, and the North Canyon area were held to control water rights on these creeks. In October of that year, L. M. Grant and 144 other South and West Bountiful citizens petitioned for disincorporation, but their petition was disallowed. In early 1894, water equity split the residents of these areas into separately incorporated water companies. The water needs of the West Bountiful region were not significantly improved, however, until the Bonneville canal was built in the 1930s (which brought more water, but also extreme financial difficulties) and the Weber Basin Project developed greater water resources.[17]

In South Bountiful (which in early days occupied the extreme southern end of Davis County and today is encompassed by North Salt Lake, Val Verda, and Woods Cross) the land slopes gently in a northwesterly direction from the foothills on the east to the Jordan River on the west. At the extreme south end along the upper part for about one and one-half miles there is an abrupt rise that separates the benchland from the lower bottomlands.

> The mountain streams in their natural course had found their way to the lower lands where they had created swamps overgrown with willows, with rosebushes, bulrushes, and wheatgrass. There were also a few springs surrounded by a growth of willows and grass dotting the intervening country, which were soon selected as the location points of the pioneers, as they furnished a supply of good culinary water.

Depiction of the damming of City Creek by the Utah Pioneers in 1847. Irrigation was the essential element for settlement in the Weber Basin area. (Photograph, gift from the *Salt Lake Tribune* to the Utah State Historical Society.)

View of Main Street, Salt Lake City, about 1862. Photograph shows the irrigation water flowing through the ditches on both sides of the street. This pattern of water delivery down the streets to the plots of ground and residences was a common practice in the early Mormon towns. (Photograph from the Utah State Historical Society.)

Water wheel driven by waters of City Creek Canyon. Scene located at State Street and North Temple in Salt Lake City about 1880. This picture demonstrates the important role that water from the mountain streams played in giving power to drive the early pioneer mills and shops. (Photograph from the Lloyd I. and Zelma Alvord Collection, Utah State Historical Society.)

Etching showing how irrigation ditches were used to divide property boundaries and to provide water for farm animals. (Photograph from the Utah State Historical Society.)

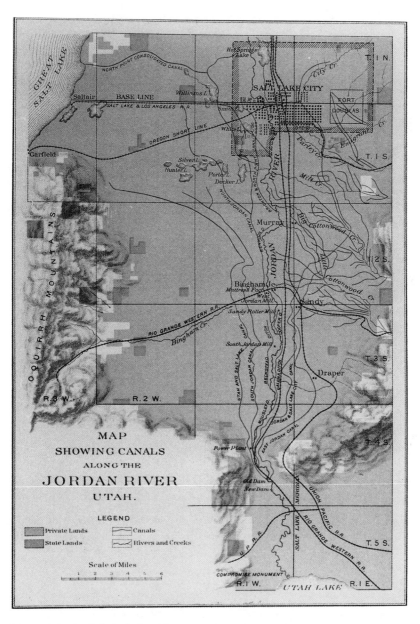

Map made by the U. S. Agricultural Department in a report of 1903 showing the main canals of Salt Lake County. (From Elwood Mead, *Report of Irrigation Investigations in Utah* [Washington: Government Printing Office, 1903].)

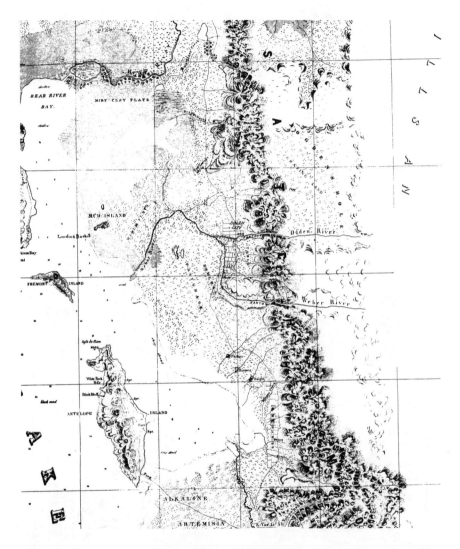

Howard Stansbury map surveyed during 1849 and 1850 showing the Weber and Ogden rivers. Map shown here is a portion of a larger map done by Stansbury. (From the Utah State Historical Society.)

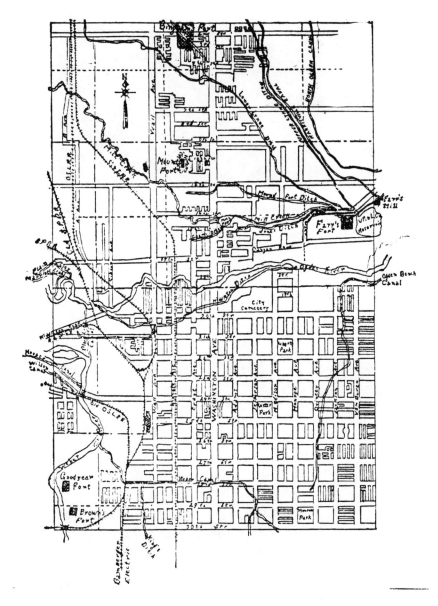

Early map of Ogden showing the various streams and canals in Ogden City area. (From Milton R. Hunter, comp. and ed., *Beneath Ben Lomond's Peak* [Salt Lake City: Deseret New Press, 1945].)

Weber Canal (in right foreground near 25th Street and Adams Avenue) flowed from the Weber River in Riverdale along the brow of the hill below Adams Avenue until it dropped into the Ogden River near 17th Street. (Photograph from the Utah State Historical Society.)

Pioneer Electric Power Plant in the distance, built in 1896 near the mouth of Ogden Canyon and the Mill Creek diversion. (Photograph from the Utah State Historical Society.)

Early twentieth-century photo showing auto travel on the left and Ogden Rapid Transit rails on the right of the Ogden River. (Photograph from the Utah State Historical Society.)

Automobile and horse-drawn vehicle traveling up Ogden Canyon in early 1900s. Water spilling from the Pioneer Electric Power Company pipeline creates a man-made waterfall. (Photograph from the J. S. Lewis Collection, Weber State University. Courtesy Jean Binnie.)

1896 construction of Pioneer Electric Power Company pipeline down Ogden Canyon. (Photograph from the J. S. Lewis Collection, Weber State University. Courtesy Jean Binnie.)

Workmen lowering into position the closing section of 75-inch diameter steel pipe of the Ogden Canyon Conduit during the rebuilding phase of the Ogden River Project of 1950. (Photograph by R. S. Billing. Courtesy United States Bureau of Reclamation.)

Ogden Canyon Conduit, which conducts water from Pineview Dam five and one-half miles through rugged Ogden Canyon to the South Ogden Highline Canal in Ogden, the Utah Power Pioneer Plant in Ogden, and the Ogden-Brigham City Canal. (1951 photograph by F. H. Anderson. Courtesy United States Bureau of Reclamation.)

Siphon of the Ogden Canyon Conduit that carries water from the north side of the canyon to the South Ogden Canal. (1951 photograph by F. H. Anderson. Courtesy United States Bureau of Reclamation.)

Pineview Dam and Reservoir as it existed after 1937 and before the 1956 enlargement of the Weber Basin Project. (Photograph courtesy United States Bureau of Reclamation.)

Weber Basin Project enlargement of Pineview Dam nearing completion, 1957. Dam height was raised from sixty to ninety feet, which increased the storage capacity of the reservoir from 44,200 acre-feet to 110,000 acre-feet. (Photograph by Stan Rasmussen. Courtesy United States Bureau of Reclamation.)

View looking upstream of section of South Ogden Highline Canal under construction. This was the Ogden River Project which was later made part of the Weber Basin Project. (1951 photograph by H. B. Wilcox. Courtesy United States Bureau of Reclamation.)

1961 photograph looking upstream on part of Huntsville and Mountain Canal near Huntsville in Weber County. This canal would later become the Ogden Valley Canal of the Weber Basin Project. (Photograph by C. S. Bolingbroke. Courtesy United States Bureau of Reclamation.)

Headgate and diversion dam for the Wilson Canal on the Weber River at 24th Street in Ogden in 1903. (Photograph from Elwood Mead, *Report of Irrigation Investigations in Utah* [Washington: Government Printing Office, 1903].)

Typical section of Wilson Canal looking downstream along hillside. (May 1951 photograph by J. S. Robinson. Courtesy United States Bureau of Reclamation.)

Warren Canal where it entered Four-Mile Creek Reservoir before the Weber Basin Project altered the stream flow. (1951 photograph by G. K. Wallace. Courtesy United States Bureau of Reclamation.)

Warren Canal Wasteway, which was located below Four-Mile Creek Reservoir and delivered waste water into the Weber River prior to the development of the Weber Basin Project. (1951 photograph by G. K. Wallace. Courtesy United States Bureau of Reclamation.)

Fall irrigation of a strawberry patch in Weber County under the Weber River Project prior to the Weber Basin development. (1946 photograph by P. E. Norine. Courtesy United States Bureau of Reclamation.)

Photograph taken near Plain City looking northeast toward Ben Lomond Peak—a 9,712-foot-high peak that catches snowfall, which becomes part of the water source of the Ogden River, which delivers water to the Plain City area. (1948 photograph by F. H. Anderson. Courtesy United States Bureau of Reclamation.)

Looking southeast from a point north of Roy, Utah, toward the mouth of Weber Canyon. Hot-capped tomatoes are representative of the extensive farming areas that were fed by waters of the Weber River in the southwest part of Weber County. In recent years these farm lands have been encroached upon by housing and industrial developments, and farming in any significant fashion is nearly extinct. (1948 photograph by F. H. Anderson. Courtesy United States Bureau of Reclamation.)

Early photo of Farmington, Utah, a typical Mormon village. Photograph taken from the Wasatch Mountains in 1896. (Photograph from the Utah State Historical Society.)

Early map of Kaysville and Layton, Davis County, area showing early settlement patterns along the main streams and the location of early reservoirs. (Kaysville map by Joseph Barton with updates by Noel Barton. Layton map from Dan Carlsruh and Eve Carlsruh, eds., *Layton, Utah: Historic Viewpoints* [Salt Lake City: Moench Printing, 1985].)

In the upper eastern portion of South Bountiful a great deal of the land was highly impregnated with alkali and saleratus. About 1851 precipitation increased, snows piled up in the mountains, and the spring and summer runoff made the lower land swampy. "By 1853 the springs had raised until the children could go swimming in them and the Indians had to hunt the high knolls on which to camp."[18] From that time on it became necessary to dig drains through which water could be diverted from the lower lands. The wet cycle continued until the lake waters reached a high point in 1876.

It seems certain that water was first brought to South Bountiful from nearby canyons by Bishop William Brown, Jesse N. Perkins, and John Perry in the 1850s. Their ditch remained practically unchanged through the years except for the last third of a mile. It was known as the Lower Mill Creek Ditch or South Fork Ditch.

Before 1870, the water claimants of Mill Creek drew up a charter claiming all waters from that canyon. In that year this group directed the cleaning and development of several springs. In 1871–72 the water claimants created the Upper Ditch to water farms on the east side. Water for land lying between the benchland and the South Fork ditch was provided by a third ditch that ran between the southwest corner of the Kimball Mill Pond turning southwest for about one and one-half miles.

These three ditches were jointly operated by a group of officers until February 26, 1894, when the Upper Ditch was incorporated as the independent Bountiful Irrigation Company.

In 1896 three other companies were incorporated—the West Bountiful Mill Creek Irrigation Company, the Bountiful Mill Creek Irrigation Company, and the Bountiful City Mill Creek Irrigation Company. When these companies were created, the waters of the stream were diverted to the various claimants by using a lumber divider placed in the main channel, which gave "proper measuring" of the water. In February 1896 the Mill Pond Ditch and South Fork Ditch incorporated under one set of officers. Mill Creek Fork had operated since 1876 with William Page as watermaster. The South Fork levied the water users ten cents per acre to pay the watermaster and ten cents per acre to cover ditch repairs and other operating expenses.

In South Bountiful early settlers obtained sufficient water for culinary use from readily available sources until the size of the community increased. Surface wells from sixty- to one-hundred-feet deep were dug into natural water basins and were rocked up and curbed. The water was drawn to the surface by a windlass, a pulley, or a hand pump.

As early as 1878 the artesian well became popular. Pipes driven into the earth by a large iron or wooden weight, lowered and lifted through a pulley attached to a derrick, tapped the underground water, which at that time had pressure enough to force a good stream for home and garden use. Bishop William Brown and William N. Atkinson were the first to have artesian wells. Atkinson was a blacksmith, and he welded an iron shoe to the end of the iron pipe to make it penetrate the ground more easily.

As the demand for well driving increased, professional well drivers came into the community and numerous wells were driven throughout the district. Most of these professionally dug wells "were of large bore and furnished large streams for irrigation, thus the underground water soon became depleted, the wells on the upper land ceased flowing, and a new problem now faced the people."[19] This led to the search for other sources of water, including development of the Bonneville Canal in the 1930s, which was fraught with many problems, and the Weber Basin Project in the 1950s, which solved most of the water needs of the area.

Even though the structure of water use changed from one period to the other, the actual methods of irrigating Bountiful farms changed very little. Scientific irrigation, which had an important impact on large farm operations in the agricultural regions of the West and other parts of the world, had little effect on the small truck garden farms that operated into the 1950s (and some small operations which function even today), as attested to by Richard C. Roberts's experience irrigating crops on the Henry Naylor farm on the east bench of Bountiful as late as 1951.

Charles R. Mabey wrote of those irrigation methods:

> There is no thrill sweeter than that which comes from watching the thirsty soil drink in the smooth-flowing water as it softly slithers down between two lines of growing vegetation. Irrigation is an art as well as a science and there are some who never learn its intricacies. Many a time have I seen a child of eight do this work to perfection where an unskilled grown man would botch the job beyond belief. It requires patience, solicitude and undeviating attention. He who knows turns down each carefully made furrow the right sized stream that will run to the end quickly without flooding and then just enough is turned off to leave a trickle that will keep the soil soaking all the way down the row until rootlets have had their full share of the life-giving fluid. Flooding will cake the ground and stunt the crop; too much water will stop the growth; too little is like too little food, it may keep body and soul together, but the subject will not thrive. Water-soaked land brings weeds, sours the soil and robs it of its fertility. What prettier sight on a sunny day than the tiny streams flowing down from ten to twenty rows of sturdy plants, the liquid glittering like ten thousand jewels! What more satisfying than to direct the water down a patch of young onions at night, telling by the reflection from the faint light of the stars how far the rivulet has trickled down the field! What greater enjoyment than to do this work

in the small hours under a full moon, the pale orb turning water and vegetation, willow brake and distant mountain, into silver! And yet they say there is not poetry on the farm and that it requires brains to be a husbandman![20]

Farmington, the second settlement in Davis County, was established the same year as Bountiful, 1847. It was the grazing of church cattle—as it was in Bountiful—that started the history of North Cottonwood settlement. The name Farmington was attached to the community in 1852 when the town was officially surveyed and plotted as a community.

In the fall of 1847, Hector C. Haight left his family in Salt Lake and took a herd of church cattle past where Perrigrine Sessions had camped to the lush grasses at the mouth of Farmington Canyon on North Cottonwood Creek. Haight lived in a tent at this original site for several months and herded cattle until he moved northwest about three miles in 1848 to Haight Creek, where he established a farm known as "Blooming Grove." This farm straddled the Farmington and Kaysville boundary, and thus Hector Haight is credited with being the founder of both communities.

Other venturesome pioneers moved into the area as well. By 1849, the population had increased to approximately twenty-five families, and by "January 1851 there were 280 person (about 65 families) living in the Farmington area (from Haight Creek to Lund Lane)."[21]

Farmington developed as an "unplanned, but not wholly unsupervised overflow of settlers from Salt Lake."[22] Foremost in its appeal as a place of settlement in the early days was the availability of land and a "reputation for its plentiful water supply," with one of the few streams large enough to drive mills.[23]

Near the north city boundary of Farmington today is Shepard Canyon. Next is Farmington Canyon, and south of it is a "slight crease in the steep front of the range" known as Rudd Canyon. Steed Canyon is next with steep Davis Canyon close to the Farmington southern boundary.

In the first years a smattering of farms scattered "within a two or three-mile radius of the major canyon" made up the North Cottonwood settlement. On February 18, 1852, the territorial legislature passed an act that designated the community as the county seat for Davis County and officially named it Farmington. The town of Farmington was not incorporated as a city until 1892, however.

The town remained in a scattered pattern for some time. As population increased, the Mormon bishop played an important role in distributing land. Within a few years there was no land left in Farmington for new settlers or for children of established families, and many had to go elsewhere to find satisfactory land.

With increased pressure on land came greater regulation of fencing and water resources. Water was needed for crops, mainly wheat and grain prior to 1890, that were shipped to outside markets. Water for culinary purposes in the first years was taken directly out of springs or streams. When irrigation developed, it was based on the "ward ditch," a community project. The system carried water from North Cottonwood Creek to town garden plots and the fields below town.

The Farmington water claims adopted the prior appropriation concept with a watermaster appointed to supervise distribution. Glen Leonard has given an excellent description of the way water was regulated in Farmington through integration of the ecclesiastical and court systems:

> In Farmington, as in every Mormon town, the bishop was directly responsible during the early years of settlement for the distribution of irrigation water in his ward. Even after the Davis County court assumed civil direction in 1852, the bishop, assisted by water masters, continued to manage local streams. Local irrigation waters were regulated through a cooperative arrangement between the county court and the ward bishop.
>
> The method for appointing water masters changed from time to time. From 1852 until 1861 the court appointed a separate water master for each stream. These men were nominated by the ward bishop, who no doubt consulted with his counselors, the ward teachers, and perhaps others in selecting water masters. . . .
>
> The water master regulated water turns so that each claimant received his rightful share. Construction of new ditches, alterations to the stream channel, and the cleaning and repairing of water ditches were also supervised by him. To accomplish this work, the water master would appoint a time and invite a general turnout or call on the required number of men. It was generally conceded that water users, rather than landowners, were responsible for keeping the ditches in repair. This, for example, freed widows from the responsibility of cleaning ditches through their city lots.
>
> Minor complaints concerning water rights were handled by the local water masters, according to mutually beneficial irrigation laws. This was the procedure recommended by the bishop, especially in later years. However, the water master was almost always himself a claimant to the stream he regulated, so the bishop and ward teachers were sometimes called in to arbitrate.
>
> The county court refused to accept jurisdiction in cases of irrigation water damaging property. Complaints of this nature were referred back to the water master or the local justice of the peace. The priesthood quorums were occasionally involved in settling complaints of the irrigation ditches over-running their banks. James Millard, watermaster for the town water ditch for many years, once apologized for discussing irrigation problems in priesthood meeting,

which, he felt, should be reserved for spiritual matters. But Bishop John W. Hess justified the discussion with a characteristic Mormon reply that temporal and spiritual matters were so closely related that they couldn't be separated.[24]

Farmington as county seat played an important role in the governing of Davis County. It was at the county court that much of the county water policy was made. One example of water policy making occurred in 1856 at Farmington when Brigham Young and church leaders met to plan the Davis and Weber Canal.

On January 12, 1856, the legislative assembly of the territory of Utah approved an act incorporating the Davis Canal Company. July 16, 1856, Brigham Young and Heber C. Kimball, who were the main proponents of the canal, attended a meeting in Farmington with "all of the prominent citizens of Davis County" to plan for a canal that would conduct water from the Weber River across the county to the southern boundary at the Hot Springs.[25] The canal would never be built to the planned termination, and it would not be finished to any part of Davis County prior to 1871 because of several frustrations.

When speaking to the bishops of the wards in Davis County in July 1856 in the courthouse in Farmington, Brigham Young stated the objective of the canal building program was to irrigate "a great section of land now lying virtually useless through lack of water." At the meeting surveyors who had been appointed to the company project reported how tunneling, fluming, and excavation of the canal would be necessary for its construction. The people in attendance understood the advantages to be gained by the project and "voted enthusiastically to begin the work at once, using all means available for its speedy completion."

Work on the canal started in early 1857, but the Utah War interrupted construction. In 1859 work began again, but a sand ridge at the mouth of Weber Canyon was not able to be penetrated. By 1860, it was obvious to the contractor that "since the hill was composed completely of sand, and not clay as they had supposed, tunneling operations were not going to be successful."[26] A canal on the lower bench was completed in 1855 and brought water to Davis County eventually as far as Kaysville. Not until the 1950s when the Weber Basin Project was completed was the dream of bringing water to the Hot Springs fulfilled.

In the spring of 1848 Thomas Grover and family brought church cattle to graze them at what would become Centerville south of Farmington. Shortly after arriving the Grovers were joined by William and Osmyn M. Deuel, after whom the main stream, Deuel Creek, was named. Grover found excellent meadowlands that were formed by spring flooding of the bottomlands. Grover and the Deuels settled on the south side of the stream and planted a small crop of wheat, corn, and vegetables. In 1852 the settlement was organized into an ecclesiastical ward with Sanford Porter as the first bishop.

Other settlers took positions on streams in Centerville. In 1848 Samuel Parrish, Sr., moved on to the stream one-half mile north of Deuel Creek, which was named Parrish Creek. In the winter of 1848–49 Parrish built a gristmill on the stream. "It was a crude structure, but it served to grind grain, and it is said to have been the first mill erected in Davis County."

The residents of the Centerville community worked together cooperatively to develop water sources. "The lands were watered by little ditches and canals that had been dug from the main canyon streams that flowed from the mountains," but Centerville had no elaborate water canals in the early years.[27]

Over the years several reservoirs for storage and controlled use of water were built in Centerville. Two mill ponds were also constructed at the mouth of Deuel Creek, and water was used to run a gristmill at that location up to 1905. These ponds also served as recreation sites in summer and winter as well as a place to baptize or rebaptize Mormon church members.[28]

The first water system in Centerville was run by the ecclesiastical government of the Centerville Ward. When the ward split in 1872, Centerville First Ward controlled the waters of Deuel Creek and the Second Ward controlled Parrish, Bernard, and Ricks creek waters. Open ditches were sufficient in the early days, but it required a great effort of community labor to maintain them. After 1900, when the Centerville Deuel Creek Irrigation Company was organized, the company carried out much of the maintenance and improvements of the water system.[29]

Improvements in the early 1900s included construction of ditches, weirs, gates, flumes, pipe from the canyon to the main weir, and pipe from the mill ponds. In January 1916 the directors of the Deuel Creek Irrigation Company approved a new irrigation system that was to be installed by watermaster Thomas H. Harris. In 1939 a new system was approved by the irrigation company under the direction of President George J. Miles. This project moved the weir higher in the canyon and ran new pipe. This was essentially the status of the irrigation system, except for some 1930 improvements, until the Weber Basin Project was installed.

In 1915 the town of Centerville was incorporated so $15,000 in bonds could be generated to install a culinary water system. Prior to this, "citizens above 300 east had to dip water out of the open irrigation ditch when their watering turn came, and fill barrels with it for drinking, cooking, and bathing."[30]

Kaysville, one of the first settlements in Davis County, included the present-day communities of Kaysville and Layton and all of northern Davis County. These communities played a significant part in early water development in the region. The first settler to come to that area was Hector C. Haight, who settled on Haight Creek.

In 1849 Samuel Oliver Holmes built a log cabin, which was referred to as the "herd house," at the junction of the North and South Forks of Holmes Creek. It was in this same region that the James Webb family established themselves, and Webb Canyon was named after them. In the spring of 1850, two men, John H. Green and Edward Phillips, joined by William Kay and their families, moved beyond Holmes Creek and settled farther north on Sandy Creek, now known as Kay's Creek, in present-day Layton. William Kay built a log house and later an adobe home on the stream that took his name.

The early 1850s brought an influx of people into the Kaysville area, most of whom came of their own accord and without encouragement or direction from Mormon church leaders. They took up land wherever it was available and wherever it suited them. Ezra Taft Benson commented in 1854 that he "found the people living in a scattered condition building in Gentile fashion." Emily Barnes described her first impressions of the new community as being made up of dugouts situated on the north bank of the creeks to benefit from the sunshine. On January 27, 1851, the community was organized into an ecclesiastical division known as Kay's Ward with William Kay as bishop and Edward Phillips and John Green as counsellors. Under the ward organization the community developed a fort as Brigham Young had counselled for protection from the Shoshoni Indians, who had been somewhat threatening to the northern communities.

Kay's Ward lasted until 1877 when South Hooper Ward separated and later Syracuse and Layton wards made their divisions. In 1858 Christopher Layton introduced a bill in the territorial legislature to incorporate Kaysville, and this was done on March 15, 1868, making it the first town incorporated in Davis County. The boundaries at that time embraced an area approximately five miles square, essentially today's city boundaries.

The first irrigation in Kaysville was believed to have been done by William Kay with water from Kay's Creek. Farming was conducted both by irrigation and dry farming. Joseph Barton stated that the settlers on Kay's Creek "joined hands and constructed the necessary ditches to convey water from the creek proper to their several claims, each doing his share of the work in proportion to the amount of land to be irrigated."[31]

In the early years water was not plentiful. Henry Blood records that "the stream coming through town became so small that for three months at a time no water crossed Main Street." The intermittent streams gave out during the dry months and during dry years the water sank into the alluvium a short distance from the mouths of the canyons. At first the people in Kaysville settled close to streams or near the lake shore so they could graze their cattle on the salt grass year round. This lower land was also comparatively easy to irrigate, but as more people came, water became more difficult to obtain. Later settlers moved higher up on the streams and onto the benchlands closer to the mountains when the lack of water in the lower areas became "a major factor in the immigration of the center of the settlement eastward." However, on the higher terraces the streams were confined to deeper channels making it difficult to bring water out onto the land.

As the demand for water rights increased, water companies were organized to allot water shares fairly. In June 1873 Haight's Creek Irrigation Company was the first to be organized; it was incorporated by the state on February 28, 1899. This company still functions today. Another company that was formed on April 19, 1921, with George Swan, Frank Muir, Thomas Bone, L. Henry Jacobs, and Joshua Conrad as officers, merged with Haight's Creek Irrigation Company and incorporated under a new charter on January 30, 1967. On February 25, 1897, the Holmes Creek Irrigation Company incorporated at $13,000 stated value, and two months later, on April 30, 1897, Kay's Creek Irrigation Company incorporated. These companies still function today under the Weber Basin Water Project.

Improvements along laterals and the main stream, storage of water in reservoirs, and the use of wells that tapped underground sources increased supplies for the community. Only by making improvements and using methods to store or extend water sources was Kaysville able to meet the needs of its growing population. Thirteen artesian wells were the "life-giving springs flowing out of the land."

The Davis and Weber Canal was another important development for the communities from the Weber River to Kaysville. The canal was proposed by Brigham Young as early as 1856. A few years later, on June 13, 1864, the *Deseret News* quoted Brigham Young as saying that "Davis is the richest county for grain and fruit that we have, and if a portion of the Weber were brought out thousands of acres of good land, now on the open prairie might be brought into cultivation." As mentioned previously, meetings were held in Farmington to plan and develop that project. In 1881 when the project was organized as the Central Canal Company, later named the Davis and Weber Canal Company, eighteen miles of canal were finally built to bring water to thirty thousand acres in the two counties.

On September 3, 1930, a contract was made to construct a high-line canal as an extension of the Davis and Weber Canal at a cost of $40,000. The extension was completed in the spring of 1931 and it followed the contour of land from Layton to Kaysville, where it went across Center Street and continued south to connect with Haight's Creek. The southern section was later abandoned, but the first eighteen miles is still in use with most of it now a cemented canal.

The Davis and Weber Canal had a difficult time getting enough water out of the Weber River. During dry years and in the late summer all the user claims could not be met. To develop more water for the canal, the company constructed East Canyon Dam in 1896 to store water to be used when the river level was low. Over the years the East Canyon Dam was raised in height to increase storage capacity. Also in 1929 to 1931 the Echo Dam was constructed to give increased storage capability. These developments will be detailed in a later chapter.

Kaysville had four major mills using water power from the mountain streams. In 1854, John Weinel, a German immigrant, established a gristmill on the South

Fork of Holmes Creek. There was a problem supplying enough water to run the mill, so a ditch was dug to bring water from Baer (Bair) Canyon to the mill. In 1866, Christopher Layton and William Jennings built a gristmill at the cost of $30,000 on the northeast corner of Main Street and 200 North in Layton. In 1855 John Bair built a sawmill at the mouth of Bair Canyon, and this became the main lumber producer for the community. When Bair moved away, William Beesley took over the operation. Beesley also built a gristmill on the South Fork of Holmes Creek east of the Weinel mill site. Later another mill was operated by H. J. Sheffield and Lamber Blamires near 325 Oak Lane that also used the water of the South Fork of Holmes Creek.

On March 15, 1868, the Utah Territorial Legislature approved a bill incorporating Kaysville as a city—the first incorporated in Davis County. The new city limits left much of Kaysville out of city jurisdiction and this, among other factors, led to the creation of other communities in northern Davis County. Under the city government on August 17, 1908, an ordinance passed charging Heber J. Steiner, a contractor, with construction, maintenance, and operation of a system to furnish Kaysville with water for "culinary, irrigation, and other purposes." This system was a piped system and greatly improved the water supply for city inhabitants.[32]

In 1877 some major ecclesiastical divisions occurred in northern Davis County that led to the establishment of new communities, among them Layton. On September 1, 1889, the Kaysville Second Ward was created and in 1892 that ward was named Layton Ward in honor of Christopher Layton, one of its prominent citizens. Much of Layton's history has been related in the history of Kaysville, but there are some unique aspects of water history in Layton.

In the early fall of 1850, Elias Adams settled at the mouth of the canyon, which was named Adams Canyon after him. Settlement along the east end of Holmes Creek and Kay's Creek was essential for community growth. In 1856 John Thornley went out north and east on Kay's Creek and established a farm where he raised irrigated crops. Other families went to the northwest section of the community on the North and Center forks of Kay's Creek.

Other pioneers built homesteads near the streams along the mountain road running from the Weber River and joining the Kaysville and Layton road at the North Farmington Junction. David Bybee maintained a toll bridge across the Weber River for this route, and Christopher Layton did the same across Kay's Creek.

Residents made significant contributions to water development in the community. Calvin Miles dug a ditch from the mountain to irrigate his 160-acre farm near the sand ridge and the Morris Town Ridge. Alexander Dawson settled and worked in Dawson Hollow. And in 1863 Elias Adams, Jr., built in Adams Hollow, formed by Snow Creek and Holmes Creek, and others located near him. Land in this area was plentiful, but water was extremely scarce. During the summer months water in Kay's Creek would never reach Layton, and it would be necessary for the settlers to drive their livestock to nearby Dawson's Hollow to find water. "It was a familiar sight to see the early settlers of Clearfield coming a distance of five miles for water, with oxen-drawn wagons filled with barrels. Dust from the unimproved roads could be seen long before the wagons and oxen were visible."[33]

In 1852 Elias Adams and his sons built a dam on the North Fork on Holmes Creek, the first reservoir for storing water for irrigation in western America. He realized that "mountain streams could never furnish enough water during the summer months to irrigate the vast acreage as the country developed." The dam was constructed by Adams and his sons using shovels and manual labor and was forty feet long and four feet high. Ward members used shovels and wheelbarrows in the winter of 1863 to raise the dam eventually to fifteen feet in height. However, the project failed when pressure from heavy spring run-off caused the dam to break. "This discouraged the builders, and they abandoned the project. Elias Adams and his sons, however, still had confidence in the project and continued to make improvements on the dam. . . . Eventually they built the dam to a height of seventy feet by 1930."[34] By then there were three other dams in the vicinity: Hobbs Reservoir, completed in 1919; Holmes Reservoir, operating before 1900 but completed in 1936; and the Kaysville Irrigation Reservoir.

West Layton established itself apart from Layton and Kaysville. Most of the settlers there "were sons and daughters of people who had settled in Kaysville and other towns of the county."[35] Most of the contact was with Layton and Kaysville and the main road was Gentile Street, which comes out of the center of Layton. Joseph (Cap) Hill was the first to come to the area in 1851 when he built a log cabin at the mouth of Kay's Creek near the Great Salt Lake. When the West Layton Ward was organized on February 22, 1895, there were thirty-three families residing there.

Settlement in West Layton was complicated by a lack of water, and most of the farms were dry farms and cattle ranches. Wheat and hay were the main products from the dry farms. As time went on, surface wells were developed, some of which were one hundred feet deep.

It was the Davis and Weber Canal and construction of East Canyon Dam that brought sufficient water to West Layton to aid in increased production on the farms. "Soon all the farmers in West Layton were increasing production of alfalfa, barley, potatoes, tomatoes, and sugar beets." Dairying also became a profitable enterprise in the region. Irrigation water came from the canal, but most of the culinary water came from artesian wells driven near homes. James Hill and George V. and Richard Stevenson drove the wells—many of which are still producing today.[36]

East Layton also has a significant water history that has had a contemporary effect. As the population grew, there were many demands for water and demands of areas lower down on streams complicated the situation. In 1936 the water problems in East Layton led the city to provide its own water supply separate from Kaysville and Layton. In that year a group of citizens found there was money available from the federal government Works Progress Administration to develop water for incorporated towns, forcing East Layton to incorporate. Still, under WPA guidelines, part of the funding had to come from local sources.

Funds were found, and approximately fifty-five men from the WPA worked digging the lines, "and nearly all the work was done by hand in order to create more jobs," a situation that suggests boondoggling, a charge often levied against the WPA. Much of the digging had to be done through oak brush and rocky areas making it very difficult and leading many to think the project would never be completed. With the help of the WPA workers and many unpaid community members, water was finally brought to each home.[37]

South Weber, to the west of the mouth of Weber Canyon and on the south side of the Weber River obtained water for irrigation and household purposes by constructing the South Weber Canal in the spring of 1852. The present Bambrough Canal is the old South Weber Canal channel. The Dunn Canal in South Weber was built in 1869 and irrigated 261 acres of land.[38]

The communities of Syracuse, Clinton, and West Point developed out of South Hooper. By 1867 enough settlement had occurred in the Hooper area to warrant division into a Weber County section and a Davis County section. The Davis County section was called South Hooper, and the people were organized into a Mormon ward on June 26, 1877, with Henry B. Williams as the first bishop. In 1895 the South Hooper Ward was divided and the south ward was called Syracuse. In 1896 the east section became the Clinton Ward.

In May of 1910 the east section of South Hooper was renamed West Point by a unanimous vote of the members of the ward. The town itself was not legally incorporated until 1935, when that became necessary for installation of a culinary water system. Much of the trenching for pipelines for the system was dug by WPA workers.[39]

Dry farming predominated in Clinton in the early years until the Davis and Weber Canal was completed and East Canyon Dam was installed. Wells were eventually dug to supply water, but they needed to be deep and some were salty.

Chauncy Hadlock, who worked on the Davis and Weber Canal, noted that "water only came a little at a time until the East Canyon dam was built in 1890 [1896]. Before completion of the dam the canal would carry water in the Spring of the year to the farms along the way, but the source would stop as the hot weather period lowered the size of the Weber River."[40]

After initial settlement of Syracuse in 1876, there followed a decade of growth. In the west part of the settlement the only drinking water available was a small spring of water on William H. Miller's farm, and most drinking water was hauled from there and stock was driven there daily for water. Later, artesian wells were driven west of Syracuse and by 1875 the Hooper Canal was extended to the community.

In 1890 the Davis and Weber Canal brought water to Syracuse making it possible to create some large acreage farms. Syracuse developed as a community and became noted for a resort on the lake established by D. C. Adams of Salt Lake and Fred J. Kiesel of Ogden which, before 1900, was the largest resort in Utah.[41]

Clearfield was first settled by Richard Hamblin, his wife, and two sons in 1878. At that time the nearest water was in Kay's Creek in Layton. Around 1880 Hamblin developed a source of water from a well after several attempts had produced no result. He built a "small concrete-lined reservoir" to which he attached a windmill to pump water from the well to irrigate his garden and small orchard. From the beginning Hamblin grew strawberries, and when the Davis and Weber Canal reached the area, it permitted him to harvest a large crop for the Ogden market.[42]

Sunset occupied a fertile area between two forks of the sand ridge, and the area was called the Basin by early settlers. All the land was dry and covered with sagebrush. There was no water for the parched earth, and even drinking water had to be hauled a distance of five miles. Eventually Sunset also benefitted from the Davis and Weber Canal, as did all of northern Davis County. Bringing water to Sunset allowed "this arid soil" to be "transformed into irrigated farms. The prospect of such a canal brought other settlers into the Basin and by diligent cooperative and incessant labor the canal was finally completed. . . ."[43] The canal was not without its problems, however, and it was not until East Canyon Dam was constructed that there was sufficient water to irrigate crops all summer long.

From this time on things changed for settlers of North Davis County and Roy, and the region became one of the most prosperous farming areas in Utah, especially for growing sugar beets for the sugar industry. The cooperative efforts experienced by the various communities in Davis County led to the improvement of the canals and ditches that carried water and expansion of water systems. These were small steps along the way to the Weber Basin Project, which would provide water for even larger populations.

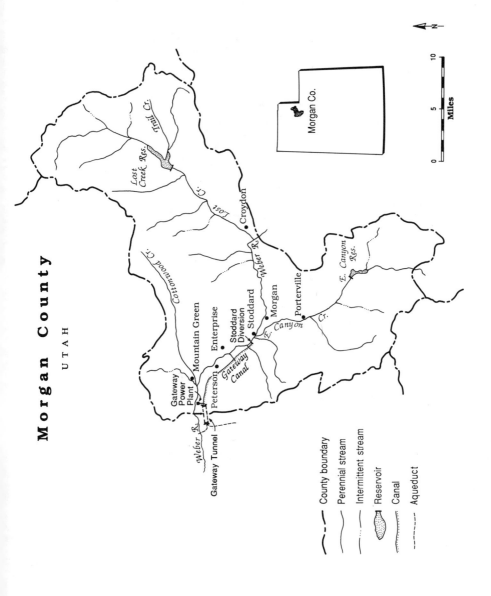

Morgan County

U T A H

Trail Cr.

Lost Creek Res.

Lost Cr.

Croydon

Cottonwood Cr.

Lost Cr.

Weber R.

Mountain Green

Enterprise

Stoddard Diversion

Stoddard

Morgan

E. Canyon Cr.

Porterville

E. Canyon Res.

Gateway Power Plant

Peterson

Gateway Canal

Weber R.

Gateway Tunnel

Morgan Co.

Miles
0 5 10

N

— · — County boundary
——— Perennial stream
········· Intermittent stream
Reservoir
- - - - Canal
– – – Aqueduct

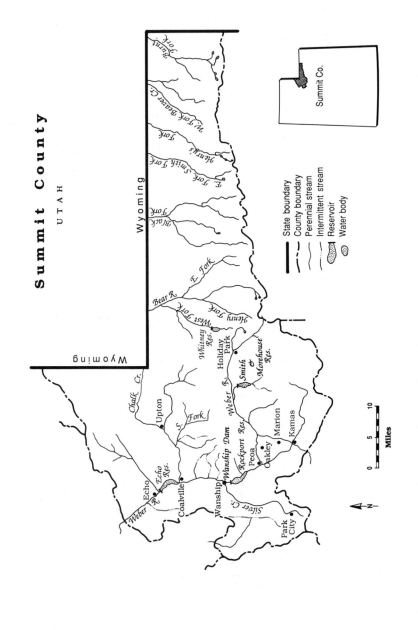

Summit County

UTAH

Wyoming

Wyoming

Burch Fork

W. York Beaver Cr.

Henry's Fork

E. Fork Smith Fork

Black Fork

Bear R.

E. Fork

West Fork

Henry Fork

Whitney Res.

Holiday Park

Smith & Morehouse Res.

Chalk Cr.

Upton

S. Fork

Weber R.

Wanship Dam

Rockport Res.

Peoa

Oakley

Marion

Kamas

Echo

Echo Res.

Coalville

Wanship

Weber R.

Silver Cr.

Park City

N

State boundary
County boundary
Perennial stream
Intermittent stream
Reservoir
Water body

Summit Co.

0   5   10
Miles

Chapter 4

Water Development in Box Elder, Morgan, and Summit Counties

More rural, less populated, and less developed than Weber and Davis counties, Box Elder, Morgan, and Summit counties nevertheless have played important roles in the Weber Basin Project. Only three communities in Box Elder County—Willard, Perry, and Brigham City—use water from the project, but in the southern part of the county, Willard Bay (Arthur V. Watkins Reservoir), adjacent to the Great Salt Lake, stores water that can be pumped in or out of the system. Sparsely populated and mountainous Morgan and Summit counties provide water storage sites and much of the water supply for the project. One hundred and fifty miles of the Weber River and its tributaries, the major portion of the watershed, delivering 359,520 of the river's annual 505,187 acre feet of water, lie within these two counties, as do most of the system's major dams. Since residents of these counties had less to gain from downstream use of the mountains' snowpack, they resisted more the development of the Weber Basin Project, and developers put forth a greater effort to convince the counties' citizens to participate and sacrifice for the "benefit of the whole." While appeals from secular and religious leaders to the cooperative nature residents had displayed in earlier church and county water control projects softened their opposition, it eventually took a vote from the entire Weber Basin Project area to override it.

BOX ELDER COUNTY

Box Elder County was created on January 5, 1856, from a portion of Weber County. It remained under the ecclesiastical jurisdiction of the Weber LDS stake until 1877. A lack of adequate records from the early period has confused the history of the county, but some historians credit William Davis with being, in 1849, the first settler in the Brigham City area.[1] He explored the region and returned in the spring of 1850 with a few other Mormon settlers, who established themselves near Box Elder Creek. In March 1851 another group settled along North Willow Creek in what is now Willard, while, in April, William D. Brooks and Thomas Pierce joined the Box Elder, later to become Brigham City, settlement. These infant communities' names reflected their Mormon orientation—Willard after church leader Willard Richards and Brigham City after Brigham Young.

Box Elder Creek and County took their names from the native trees growing along the stream's banks. The county encompasses the largest portion of Utah north of the Great Salt Lake, an area where agriculture historically has been the dominant way of life. The strip of land at the southern end of the county, between the east side of the Great Salt Lake and the foot of the mountains and running from the Weber County boundary north to Brigham City, comprised the area involved in development of the Weber Basin Project.

Traveling ahead of his company in 1850, Mathew W. Dalton described the sites of Brigham City, Perry, and Willard: "I passed alone and afoot through a country solitary and desolate, not a man nor a house along my line of march."[2] Later descriptions stated that the townsite of Brigham "was covered with a heavy growth of grass extended far up the mountainside and was bordered on the south by a sagebrush growth which extended south in more or less unbroken patches until Three Mile Creek," now Perry, was reached:

> Here the country was covered with a heavy growth of choke cherry, oak, and maple brush, there being only a narrow strip of tillable land between the brush covered section and the lake which extended east very near the western side of that little settlement.
>
> Three Mile Creek emerged from the canyon, wound in and out northward among the brush, and gradually turned to the west until it reached the swampy sections caused by the overflow of water from the Porter Springs and the chain of springs extending southward. The brush growth continued southward past Willow Creek. Along the banks of that crystal stream grew willows of every size and variety.[3]

To the south, past the site of Willard, the lake came up close to the mountains, and the area between the Hot Springs at the mountain point on the Weber County line and Willow Creek was desertlike. But when water was applied from the mountain streams to its very rich land, it became an area of "beautiful farms and orchards."

The main streams in the southern strip of the county, from south to north, are Willow Creek, flowing from Willard Canyon to Willard; Three Mile Creek, running through Perry; and Mathias Creek, flowing near the north boundary of Perry. A series of springs lying along the strip—Cold Spring near the county border, Marsh Spring, White's Spring to the northeast on the north end of Willard, and Porter Spring to the northwest at Perry—as well as Wright's Spring north of Brigham City provided additional sources of water. Dunn's Canyon, between Perry and Brigham City, was named for Simeon A. Dunn, an early settler at nearby Reeder's Grove. Box Elder Creek was the main source of water for

Brigham City. It occasionally overflowed and created bogs and marshes in the village. In time the channel was cleared and the low areas filled with dirt to make the area habitable. Later, irrigation ditches were added that emptied into ponds on each side of Forrest Street between Fifth and Sixth West streets.

In the fall of 1851, a fort was built on the clay soils of the lake plain beyond the Box Elder Creek alluvial fan. In 1852 Box Elder Fort replaced it. Water was taken out of Box Elder Creek at Bott's Mill (227 East Second North) and brought by a ditch down Second North to the fort. It ran from south to north through the fort, then to Third West, then south to Forrest Street and the Big Field to the west. In the spring of 1854, the townsite of Brigham City was surveyed. Streams flowed through town on either side of Main Street, shaded by trees planted to mark the boundary between the street and sidewalks, neither of which was graded in early times.

As population increased, the community spread out upon the land. In the initial survey, farms ranged from forty to eighty acres in size, but a second survey considerably reduced farm acreage. As in other Mormon communities, a Big Field was surveyed. Plotted northwest of Brigham City, the field's lots varied in size but averaged 4.5 acres. Larger farms at a greater distance from town ranged from 20 to 1,400 acres. A prohibition of individual private fences or irrigation ditches in the Big Field simplified getting water to the land. To protect crop land from livestock, the field was enclosed by a fence. All landowners were jointly responsible for constructing and maintaining the fence and irrigation canals.

Most early settlement in Brigham City was on the "lower sloping lands of the piedmont."[4] This allowed enough water to be brought to community lands by canal and still provided sufficient slope for drainage. The lower, flat lake plains to the west were avoided because of poor drainage, alkaline soils, and sloughs. The higher alluvial slopes on the east of the city were also avoided because they were "above the canal [from Box Elder Creek] and remained unirrigated until after the turn of the century." Varied needs for water led to division of the streams. The main division, on Box Elder Creek at the mill race, allowed water to run to both the north and south parts of town: five-sixteenths of this division went to the Big Field, seven-sixteenths to Brigham City, two-sixteenths to Perry, and two-sixteenths to a ten-acre field known as Lindsey Park.[5]

Mormon Apostle Lorenzo Snow's first visit to the community in 1853 demonstrated the role of church leaders in encouraging cooperative control of water:

> he found that all of the water coming from Box Elder Canyon was
> already appropriated and that the existing settlers considered it only
> sufficient to supply the needs of the inhabitants already there. Snow
> persuaded them to relinquish their right to the water and agree to a
> new division which would include new settlers that would be arriving.[6]

Because water was so precious, each resident was granted a share equal to the relative size of his lot. To ensure equal access, turns, including night irrigation, were organized. On June 6, 1859, the town appointed H. P. Jensen watermaster general to oversee the system.

Beginning in 1892 the Brigham City government developed a culinary water system to replace use of irrigation ditches and of open and pumped wells. Over the objections of a few bitterly opposed citizens, organized as the Safety Society, who failed in their attempt to obtain a court injunction, city voters approved bonds for $24,000, and construction commenced. The system consisted of a reservoir with a 160,000-gallon capacity and a network of eight-inch mains. It was finished by July 9, 1892, and a celebration was held to commemorate the event. The city later bonded for an additional $35,000 to install a new and larger pipeline to the reservoir and to extend mains throughout the city.

In 1890 or 1891 interested parties organized the Brigham City Electric Company to develop a water-driven electric plant near the old Oregon Short Line depot in Brigham City. The plant was later converted to steam power, which proved inadequate and too expensive, and the company built a new plant at the mouth of Box Elder Canyon. By 1903 its capacity was not sufficient to meet the city's growing needs. In the fall of 1903, the Brigham City Municipal Corporation began building a one-unit power plant of approximately 470-horsepower capacity to be financed by a voter-approved $30,000 bond. By 1920 the plant also provided electricity to the Perry Power and Light Company, and in August of that year, voters, by supporting a $200,000 bond, approved construction of a larger plant, including a concrete intake dam at the head of Box Elder Canyon and a thirty-inch water pipeline from the intake to the new power house. Completed on January 10, 1922, the new plant started operation with a 1200-horsepower capacity.[7]

As population increased and more farms and orchards were laid out, demands for water increased. To supply Brigham City and southern Box Elder County, a canal project was promoted, first in conjunction with the Bear River Canal in 1903 and later with the Pineview Dam project in 1937. The Pineview water was brought to Box Elder County by the Ogden-Brigham Canal, which was later incorporated into the Weber Basin Project.

The town of Willow Creek was established five miles south of Brigham City on the north side of North Willow creek on March 31, 1851. In 1853 Mormon church officials directed an additional 150 families to move to Box Elder County to strengthen its communities against Indian uprisings during the Walker War. Complying with an order by Brigham Young, the Willow Creek settlers built a fort in the western part of town. After Indian troubles passed, the community spread out, and the town was plotted on the eastern portion of farm lands. In 1857 the town honored Mormon Apostle Willard Richards by changing its name to Willard, and in February 1870 the territorial legislature granted Willard City a charter.

Water from Willow Creek and White's Spring provided both irrigation and culinary water until a piped water system was installed between 1910 and 1912. During that same period Willard City filed for water rights to establish its original electric power plant. By the 1920s, however, electric power was being furnished by Utah Power and Light.[8]

Perry, approximately three miles south of Brigham and north of Willard, was first known as Three Mile Creek. The first recorded permanent settler was William Plummer Tippetts, who arrived early in the spring of 1853. In May 1853, William Walker and Lorenzo Perry took up land north of the Tippetts farm. Their land extended almost as far north as Porter Springs—a major water source named for Porter Rockwell, who, in 1857, owned the land surrounding the spring. While these people were building the settlement on Three Mile Creek, a group of Welsh families were locating farther north in what came to be known as the Welsh Settlement. These settlers plowed the first ditch to carry water from Box Elder Creek to the Welsh Fields.

Like other settlers at Three Mile Creek, the Barnard White family struggled to develop sufficient water sources. Water for the White farm and for a few close neighbors came from Porter, White's, and other springs, but most of the community depended on the mountain creek near which they lived:

> Although there was an abundance of water when the spring run-off occurred, there was barely enough in the latter part of the summer to water the crops. The water from this creek went into three different streams: one went south to North Willard, one went north to what was called the Welsh Settlement, and the other stream went to the central part of the community. For a number of years the farmers of Three Mile Creek had talked of the advisability of building a dam in the canyon, as there was an ideal site for one.[9]

The town had little money to invest in a project that would provide a better system. Most of the work would have to be done by donated labor. Barnard White insisted on using a competent engineer to direct dam construction and suggested that the dam be made of concrete. Local leaders felt these plans were too costly and disregarded White's advice. In the fall of 1894 the community began building a reservoir and a dirt dam at the head of Three Mile Creek Canyon, but frost of the oncoming winter stopped work before the spillway was completed. Nothing was done the next year to complete the dam. In the fall of 1895 farmers once again worked on the dam and added about twenty feet to the top.

On a late-spring morning, Sunday, June 10, 1896, the dam burst and a forty-foot wall of water came pouring out of the canyon. The wall of water, clogged with debris, rolled down through the orchards. When it hit the Whites' spring and pond, more volume was added, and the cascade did not stop until it hit the railroad embankment to the west of town, which impounded a large body of water. Members of the White family successfully released the water by opening up culverts beneath the tracks.

Several families suffered in the flood. In the midst of it, there was great confusion. The Thorn home was caved in; the McDonald and Henderson homes were washed away. Great boulders were carried down to the mouth of the canyon. Many families fled to high ground in the foothills to escape and many possessions were lost. Barnard White toured the flood area on horseback, assisting victims after the water subsided.

> He made his way first to Thorn's farm where he found Rebecca Thorn . . . . [who], seeing no means of getting away from the flood, had crawled into a very small cherry tree, the only place she could reach to get away from the swirling waters. . . . [S]he had to stay in the same position all day. She was not a small woman and she had been in terror that the little tree would give away under her weight or the tree would be hit by debris. Barnard got her on the horse and they made their way to Barnard's home. Then he found Geneva Nelson, who had been caught in the flood as she ran across the yard to try to find a place of safety. She had rolled over and over in the mud and water and narrowly escaped being drowned. A log from Henderson's house came by. By hanging on to that she was able to get her breath. Then James Judson saw her and pulled her out of the deep water. Before Barnard found her she had made her way to a barn where there was a trunk containing some old clothes, none of them suitable for her, but she had dressed herself in a man's starched-bosomed shirt, which she put on backwards, and a red flannel petticoat. Barnard took her home, too. . . .
>
> [T]hroughout that fateful Sunday one person after another was accounted for until everyone had been found. Some had narrowly escaped but no one was killed or badly injured. . . .
>
> When Isaac Thorn complained to Barnard that their land was ruined, Barnard told him the flood was a blessing in disguise, as all the uneven places in their fields had been filled with silt and the land would be level and more productive than ever in the future. The immediate problem for Barnard was to get the spring cleaned properly so that they could have water. There were about 1200 gallons of water in the tank, but that would not last a large family very long. His problem was unusual, for no one else in Three Mile Creek had a water system. Even while the mud and slime of the flood was just

beginning to dry, he had men dredging out the spring. One of his faithful horses slipped in the mire and sank into the spring. . . .

After a couple of years Barnard's remarks to Isaac Thorn were proved to be true. For all they had lost was a crop, their farms had benefitted by the flood. The new soil filled the hollows and after it had been worked a while, it proved to be some of the best soil in Three Mile Creek. The ones who suffered were the ones on whose farms the boulders had been dropped by the flood. Their land was almost worthless. . . .

Besides excitement and hardship for some, which for many years kept it fresh in memory, the flood brought changes in Three Mile Creek farm operations. The settlers cleared away orchards coated with silt. They realigned irrigation ditches and changed the boundaries of their fields accordingly.[10]

In 1911 the town was renamed Perry, after its first bishop, O. A. Perry, and incorporated so that a municipal water system could be installed by means of community bonding. The municipal water first came "from a spring in the basin east of Perry just above the second bench where a storage reservoir was built," but it proved necessary to replace this source with a well drilled on a nearby side hill.[11]

In south Box Elder County, water sources drove mills too. In 1851–52 the first flour mill was erected on Willow Creek midway between Willard and the lake. In 1855 Brigham City built its first gristmill on Box Elder Creek, and a few years later Brigham City Roller Mills built a mill at the mouth of Box Elder Canyon. It operated continuously until it was destroyed by fire in 1927 and replaced by a chopping mill. In 1862 Ole Paulson used Box Elder Creek for his saw and molasses mills. Later he converted them into a gristmill.

Horticulture required a water system, and horticulture characterized south Box Elder County. Before cultivating fruits, settlers gathered choke cherries, service berries, ground cherries, and wild currants from the natural shrubs that grew along streams and in the ravines and canyons. In 1855, William Wrighton, encouraged by Brigham Young's advice to grow fruit trees around the farm, planted the area's first peach trees on his lot. He later added cherry, pear, apricot, prune, apple, hazelnut, gooseberry, currant, strawberry, and raspberry starts. In Willard, Matthew W. Dalton is credited with setting out the first orchard to contain a variety of fruit trees. At Three Mile Creek, Asael Thorn planted the first large orchard.[12]

In the late 1890s, Box Elder County developed commercial orchards, and Brigham City shipped considerable amounts of fruit to markets in Ogden and throughout the country. During two weeks in 1907, 110 cars of Elberta peaches were shipped from Brigham City railroad stations. Upwards of thirty additional carloads were hauled by trucks and teams to other counties. Two hundred and fifty thousand dollars was a conservative estimate of the value of Brigham City's peach crop. Since 1904, Brigham City has celebrated Peach Days each fall in recognition of the importance of the fruit to the community.[13]

Such developments demanded more water than the ditches and canals from local streams could provide. In the 1880s and 1890s, the water needs of southern Box Elder County and northern Weber County were included in planning for a canal system intended to bring water from Bear River to farms and communities north and east of the Great Salt Lake. (The company behind the plan ultimately failed, as is described in Chapter 5 below.) Construction on this Bear River Canal project was largely completed by June 1891.

The system was known by various names—the Bear Lake and River Water Works and Irrigation Company, the Bear River Irrigation and Ogden Water Works, the Bear River Water Company, and the Bothwell Canal, after its principal promoter. It included two main canals that carried water out of the Bear River Canyon. The East Side, or, later, Hammond, Canal ran to Collinston and Dewey-ville and was intended to extend south along the east bench. The West Side, or Bear River, Canal brought water to communities west of the river. The original plan included a survey through Brigham City and beyond to Ogden so that the system could be continued south. Israel Hunsaker, who was a major force in the Bear River Canal project and its successor, the Hammond Canal, expected the canal system would "cover the entire area from the mouth of Bear River Canyon, to a point west of Brigham City."[14] Conceived to provide water to the orchards and farms of southern Box Elder County, this plan was never realized, and the Ogden River-Pineview Dam system was developed in the 1930s to bring the needed additional water to south Box Elder County.

MORGAN COUNTY

Morgan County was organized from a part of Davis County and established by the Utah Territorial Legislature on January 17, 1862. Its boundary ran along the crest of the Wasatch range and bordered on Weber County in a way that allowed stream basins to remain largely as units. The Ogden River drainage went to Weber County and the middle Weber River drainage to Morgan County. Similar considerations applied to boundaries with Rich County, Summit County, Wasatch County, and Salt Lake County, allowing Morgan County the whole drainage areas of those streams—Lost Creek, East Canyon Creek, and Hardscrabble Creek—that entered the Weber River within the county boundaries. The southern Morgan County line is approximately one mile above the landmark Devil's Slide on the Weber River. From there, the river runs northwesterly for twenty-one miles to the western boundary of the county at Devil's Gate. There it makes a U-bend, which was a difficult obstacle to travel in early days.

Due to its extensive mountain ranges, narrow river valleys, relatively short growing seasons, and Indian presence, the Morgan County area was somewhat foreboding and, hence, was left unsettled for several years after pioneers arrived in Utah. The first continuous settlement came after Thomas Jefferson Thurston of Centerville and his sons observed Morgan Valley while logging in the mountains to the west. Accompanied by friends, he later explored the area and was delighted with the beauty there. "The only great obstacle was the unaccessibility. Entirely surrounded by mountains, there was only one narrow canyon through which the Weber River flowed."

In the winter of 1855 Thurston persuaded Charles Shreeve Peterson, his two sons, and his son-in-law, Roswell Stevens, to work their way through Weber Canyon into the high valley. Once there, the Petersons and Stevens located on the Weber River at what was first known as Weber City, later as Peterson. Thurston moved five miles upstream to what would become Milton. The Utah legislature on January 2, 1856, gave Jedediah Morgan Grant, his brother George Grant, and Thomas Thurston, a large section of land between Line Creek where Milton is located and Deep Creek where Littleton is located. Milton was first called Morganville, honoring Jedediah Morgan Grant, who as a member of the Mormon church presidency, supervised development of the area. In 1868 the name of Morgan was officially transferred to the centrally located town that became the county seat.[15]

Thomas Thurston served as presiding elder over the entire valley until 1863, when two ecclesiastical wards were organized. He then presided over Milton, Littleton, South Morgan, Richville, and Porterville; and Charles Peterson presided over Weber City, Mountain Green, Enterprise, North Morgan, and Round Valley.

Thurston built a house on Deep Creek, ran his cattle on a large tract of land nearby, and operated a sawmill in Hardscrabble Canyon. His daughter Cordelia described his early struggles in her journal:

> My Father took his wife, Elizabeth (and I went along for company) and settled farther up the valley (from Peterson) when with the help of Jedediah M. Grant's men and teams, he plowed some land and sowed about ten or twelve acres of wheat, which failed to mature for lack of water. They built a dam in Deep Creek in the spring of that year, and took out the water in ditches. But the stream soon failed, so they went on about two miles to Canyon Creek [East Canyon Creek], put in a dam, surveyed a ditch, dug it and brought the water to Deep Creek, where they turned it in above the dam. By this time there was not enough water to fill the creek bed or to wet the canal, which was about two miles long. So again their crops were a complete failure.
>
> Father had injured himself with the heavy work of lifting rocks in the canyon and before he had finished his work he became very ill

and came near dying. He probably never would have gotten out of the valley had not J. Morgan Grant come up with his carriage and brought him to Centerville, where he could be under the care of a physician. He got better, but not well, and took his wife, Elizabeth and baby and again went into the valley. He plowed the same land, and a little more, and again sowed it to grain. This, however, did not mature on account of frost.[16]

Line Creek provided water for the town of Milton. In 1860 ditches were dug by pick and shovel after plows had loosened the dirt. The engineering was done by eye. The head of the ditch was at a sawmill on Line Creek built by James Hansen, where "a very good measuring devise of hardwood with screw type head gates to control the water" was employed. Here the main ditch was split into north and south branches. The two-miles-long North Line Ditch carried two-thirds of the water to 320 acres of land. The South Line Ditch, carrying the other third, flowed south one and one-half miles to eighty-five acres of land. In 1928 the ditch was incorporated under the name of Line Creek Irrigation Company. The system was improved and upgraded in 1928, 1945, and 1976.[17]

Farther north, John Anderson took water out of Smith Creek at a diversion about one-half mile above the county road. Anderson dug three-fourths-mile-long ditches on both sides of the creek by hand. The south ditch watered about 21.5 acres, and the north ditch about 49 acres. Between Line and Smith creeks was the head of the Mecham-Madsen-Nelson Ditch. Started in 1858, it watered 62 acres, an area doubled by improvements in 1861. The water was taken out of the Weber River about one-quarter mile below the Milton Bridge at a diversion made in the river with trees and brush, which each year had to be replaced. The ditch was dug by hand, ran for one and one-half miles, and carried about five second feet of water. This original ditch was united with the Mecham-Ekstrom Ditch, dug, also by hand, in April 1938 in conjunction with the development of Echo Reservoir. It irrigated 59.3 acres. In 1955 and again in 1965, the diversion was moved upstream from the Milton Bridge.[18]

Littleton, on the west side of the Weber River, two miles south of Milton, was settled in 1856 by Jesse C. Little, whom Brigham Young had asked to go to Morgan Valley and assist in planning settlement. From his log cabin, Little directed a variety of developments, including installation on Deep Creek of a large water wheel and mill machinery that remained in service for several years. Another sawmill used the water of Line Creek for power.

Farther down the Weber, in 1860, settlers dug two ditches out of Dalton Creek. The first, from the south side of the creek, irrigated twenty-five acres of land on the Oscar Gorder farm east of the county road. The second, from the north side one-half mile west of the county road, was known as the North Dalton Ditch. In 1900 and 1902 the Mrs. James Whitear Ditch and Lewis Cobabe Ditch were taken from Dalton Creek to water additional acres to the north. In 1888 the Whitear-Gorder Ditch was diverted from Sandbar Slough and Springs just north of where Smith Creek runs into the Weber River. It provided water for the Whitear and Gorder farms in the Dalton Creek area. This was a difficult ditch to maintain along the steep embankment south of Dalton Creek.[19]

About a mile north of the mouth of Dalton Creek, the Anderson-Bohman Ditch was dug in 1861 to divert water from the Weber. Over the years it too was a difficult ditch to maintain because of beavers damming the ditch, the river changing its course, or spring flood waters washing out the head gates. It was not until the 1960s that these problems were worked out by combining the Anderson-Bohman Ditch with the Lower River Ditch and making one diversion about one hundred yards below the Dalton Creek confluence. Water was piped to the old ditches. Combined efforts of the water users, the Soil Conservation Service, and the Utah Water and Power Board accomplished the improvements for a total cost of $45,000.

During the first years after Charles Peterson and Roswell Stevens's settlement of Weber City, grasshoppers destroyed their crops in the summer and severe winters killed many of their cattle. Nevertheless, in 1861 several other families came to the spot and established squatter's claims. They surveyed a townsite and plotted one-fourth-acre lots on the flat. With "not much water available at this time for irrigation," Charles Peterson, as leader of the community, divided the water among the settlers, a practice "true to the pioneer custom of sharing."[20] Joshua Williams and Roswell Stevens built and operated a sawmill driven by water from Peterson Creek to provide building lumber for the community.

In 1860 Albert and Arthur Whitear watered 29.6 acres on the Peterson Bench and 40 acres under the bench from a ditch taken out one-half mile up the west side of the creek. This was the oldest water claim on Peterson Creek. In 1862 the Peterson Ditch was taken out of the west side of the Weber River. In 1868 the Carrigan brothers, James W. and Irvin, developed two ditches to water their 186-acre farm—one from the North Fork of Peterson Creek and another from Jacob Creek. Today the Weber Basin Canal runs through this ranch, which is still operated by the Carrigan family.

Culinary water for the town of Peterson came from wells and springs at first. In 1917 a new school was built on the east side of the Weber River, and a four-inch wooden pipe was installed to bring water to it from springs two and one-half miles to the east. A small reservoir was later built on the hill about one-quarter mile east of the school. Victor R. Bohman and Reinhardt Olsen got permission of the school board to run a pipe across the Weber River. Several families on the west side obtained culinary water from this source. In 1934, a drought year, the Peterson community obtained state permission to develop a culinary system out of Peterson Creek. At a point just below the fork of Peterson Creek, a sump was dug, and a pipe was laid to the town approximately two miles away, where a small reservoir was constructed. Lines to the homes were dug by picks and shovels.[21]

In 1854 Warriner and Sanford Porter, Jr., of Centerville, came over the mountains to Hardscrabble Canyon and built a sawmill on Beaver Creek (a tributary to Hardscrabble Creek), transporting the lumber back over the mountains to Centerville. The Hardscrabble stream was first called Mill Creek because of this sawmill. In the spring of 1861, Sanford Porter, Sr., arrived with his sons and families and settled permanently on the site of Porterville. The scene from their new homes was described as, to the west, "a dense growth of willows and wild gooseberry bushes. On the east of the valley was bunch grass, extending to the foothills. Along the creeks grew groves of huge cottonwood trees."[22]

Over the years ill feelings developed in the community, and East Canyon Creek became a dividing line between East and West Porterville. The original settlement of Porterville was located principally on the low ground east of the Creek, and the first settlers on this rich bottom land charged "a very high price for building lots." The land on the bench west of the creek could be acquired for next to nothing; it was fifty feet higher in elevation and had a heavier clay soil. After counsel came from Brigham Young to build a town on the bench, more people settled on the west side of the creek. The high costs of the east side land contributed to the separation of the community. The conflict was reflected in the 1877 division of ecclesiastical responsibilities between two wards, which were not reunited until 1897. Henry Florence and his sons operated a sawmill in Sheep Canyon, and it was moved in 1898 to Porterville to make lumber for the building of the new ward church house.[23]

The Porterville community dug a canal in 1861, loosening the soil with plows pulled by oxen and then shovelling by hand. The work was made difficult because of rock ledges, springs, and unstable soil along the canal route. The East Porterville Canal began about five hundred yards below White's Crossing on East Canyon Creek, followed the east side of the valley for about one-half mile, and then crossed west back to the creek. It irrigated 352 acres and was incorporated in 1926 as the East Porterville Canal Company. In 1865 the men of West Porterville dug a ditch that brought water to the west benchland.

East Canyon, sixteen miles in length, and its main tributary, Hardscrabble Creek, became a major milling area. The first mill in Morgan Valley was the one the Porters set up at the mouth of Beaver Creek in the spring of 1854. This mill operated until 1891 and provided a great deal of lumber for communities within and outside the county. Samuel Brough built a mill to make shingles on the tributary of Hardscrabble Creek known as Shingle Mill Creek. Lumber was a major product from 1860 until 1875, much of it used to make ties for the Union Pacific Railroad. Six mills—two steam and four water powered—operated in Hardscrabble Canyon at the time. One of the largest operations was that of William Ferrel, who had a sawmill at Ferrel's Fork (sometimes spelled Ferrell or even as Farrell's Creek on current maps) and a lumber yard in Richville. About 600 feet down Hardscrabble canyon, Holdman's sawmill made shingles and ties. Farther down was another water-powered mill, run by Bill Dickson. Shingle Canyon held both Sam Brough's mill and Joseph Taylor's at its mouth. At the mouth of Arthur's Fork, Jake Arthur ran a water-powered mill. At first, Bert and George Turner ran the last sawmill in the canyon, but they sold it to Joe Carpenter, who moved the machinery to the canyon's Black Grove. In the early 1900s, an electric plant was built in Hardscrabble Canyon at the mouth of Pole Hollow.[24]

Richville, between Porterville and Morgan, was settled in 1859 by David E. Henderson and Isaac Morris. In 1860 Brigham Young sent John H. Rich, Gillispie W. Waldron, and Jonathan Hemingway to join them, and in that same year, Thomas Rich, Solomon Conley, and John Wood also arrived. Located at Taggart Hollow on the East Canyon River, the town took its name from Thomas Rich, the first presiding elder there. It consisted of several families who resided in a sort of "stringtown lying on the west side of East Canyon Creek at the base of the hills" and extending about one and one-half miles north and south.[25] Only three families resided on the east side of the creek.

Gillispie Walter Waldron recalled that when he arrived at Richville on April 6, 1861 (may have been 1860), he, John Rich, and the others began work immediately on the ditch that afterwards was the basis of the Richville Irrigation and Canal Company. This ditch connected with East Canyon Creek about one mile south of the center of Richville and continued for about two miles north of the town. It, like most of its contemporaries, was leveled by eye and dug mostly by pick and shovel, with the help of a plow in some spots. In the first season of 1861, the canal delivered water for irrigation, livestock, and domestic use. Good crops of wheat, oats, and vegetables were raised that year, and about six families stayed the winter in Richville.

In 1863 the townsfolk added the East Richville Ditch, heading from the same dam, on the east side of East Canyon Creek. Later another ditch was taken out just below the confluence of East Canyon and Hardscrabble creeks to water the higher land on the east side. In 1901, the Richville Irrigation Canal Company was incorporated, based on the claims from 1861. The company provided water for culinary and domestic use and irrigated approximately 550 acres of land.

In 1863 George W. Taggart and Morgan and Henry Hinman began building a gristmill in Richville. They soon encountered difficulties. The irrigation ditch that they intended to use had to be enlarged into a mill race, but the race did not have enough fall to drive the water wheel. One account claims that Brigham Young, after inspecting the project, told Taggart, "it will not run. The draft is in dead water." To get sufficient water pressure, it proved necessary to enlarge the wheel and sink it deeper. Completed in 1866, the mill became an important asset to the town. It drew patrons from all over the valley, and as many as thirty teams pulling grain-laden wagons would line up for the milling.

Under the leadership of Bishop John H. Rose the small town built in 1912 the Richville Pipeline, 4.13 miles long, to bring culinary water from a spring to their homes:

> The work was done by the local shareholders with some assistance by the Wheelwright Plumbing Company of Ogden, in laying the pipe. Digging the trench was the great task. It was done with a plow and pick and shovel method, no modern machinery to do it. Each family who subscribed to stock put $300 in the bank to the company credit before the work began. Those who did not have the ready cash borrowed from the bank. Pipe was much cheaper at that time, about 20 cents per foot for 3-inch steel pipe. It was a big undertaking for a few families, and most of them worked remarkably well together to accomplish the task.[26]

The whole project cost approximately $8,000. In 1924 the town added a reservoir to the system.

In 1950 a new irrigation canal 6,600 feet long and 8 feet wide was constructed through the lower town of Richville to replace the old crooked and uneven east canal. In March of 1951 a new dam, at a cost of $6,353, was constructed across East Canyon Creek to divert water to the West Richville Irrigation and Canal Company. The old dam had washed out. The cost also included widening, straightening, and repairing the canal.[27]

Opening into Morgan Valley about five miles above Morgan City is a fertile little valley bisected by the Weber River and called Round Valley. At one time about forty people made their living on the two sides of the river known as South and North Round Valley. They began arriving in 1860. A Mr. Filshop (or Phil Shop) came into the valley around that time and settled at the mouth of the hollow which took his name. In 1863 Edward Geary moved in, and one of the hollows soon bore his name. Samuel Carter became the presiding elder of the community in 1864 and oversaw much of its development. In 1875 Edward Hunter purchased the upper part of the valley and began farming it. Olof B. Anderson worked on Hunter's farm and is credited with there introducing the first alfalfa in the Morgan Valley region. A ditch diverted from the Weber River in 1861 provided water to the south part of the valley. In 1928 this ditch was incorporated as the South Round Valley Canal.[28]

Downstream, across the river from Milton, Judson L. Stoddard of Farmington in 1860 purchased the property of Ben Simon, a trapper who had first settled at what would become the town of Stoddard. For a time a considerable number of settlers resided in the community, drawing on Stoddard Spring for most of their water needs, farming, and raising stock.

Mountain Green, north of the river at the lower end of Morgan Valley, was named for the green grasses in the valley and on the lower hills. In 1859 George Higley built a cabin there and spent the winter. Ben Simon claimed part of this area for herd grounds, but Higley got his consent to establish a dairy herd and in connection with it built a butter churn from water power provided by the Weber River. In the spring of 1860 several families settled in, including that of Gordon Beckstead, after whom Gordon Creek is named. Cottonwood Creek was another major water source for the area. On the downriver side of Mountain Green, a McLean family built a sawmill on the Weber River near the location of the Strawberry Creek bridge.

Enterprise, named for the enterprising character of its first inhabitants, is located on the east side of the Weber River between Stoddard and Mountain Green and about two and one-half miles southeast of Peterson. It was settled in 1861 by Roswell Stevens, who built a cabin at the mouth of what would be called Roswell Canyon.

Water for irrigating in Enterprise was taken from the Stoddard-Enterprise Slough at Enterprise Point in a ditch dug in 1860. The open ditch ran two miles and watered 281 acres. In 1863 residents dug a new ditch that tapped the Stoddard Ditch at a point higher up. Known as the Stoddard-Enterprise Ditch, it extended from Stoddard for about three and one-half miles along the hillside to Enterprise. It connected at Stoddard to the North Morgan extension ditch, whose diversion point was across the river from Como Springs. Built with plow and scrapers and horse and hand work, the Stoddard-Enterprise Ditch was abandoned after seepage and washouts made it impractical to maintain.

Other ditches established in the area included the Madsen Olsen Ditch, 1860; Heiner and Morris Ditch Company, 1861; Roy Heiner Ditch, 1862; Poulsen Ditch, 1865; Madsen Olsen Ditch Company, 1870; James Fronguer Ditch, 1870; Chris Johnson Ditch, 1885; George M. Robinson Ditch, 1888; Harriet Clausen Ditch, 1888; Clawson, Smith and Robinson Ditch, 1888; and Big Hollow Irrigation Company, 1889. Lack of water, however, was a continuing problem for Enterprise, and it retarded settlement. In 1928 when the Echo Reservoir was built, the Enterprise Field Ditch, which had no rights in the Weber River, bought stock in the Echo project to obtain water for the community.[29]

Croydon is located two miles east of Devil's Slide on the Weber, in a little valley on the bank of Lost Creek (also called Plumbar Creek). So named "because during the heat of summer it loses itself underground and comes to the surface again just before entering the Weber River."[30] Lost Creek heads at the Morgan-Rich county line and flows southwesterly for a distance of twenty-two miles. Croydon is named after a town in Surrey, England, from which some of the early settlers came.

In 1862 George Knight, George Shill, James Walker, and Levi Savage moved into the valley. Other families followed, and because of Indian problems, eight houses were built in a square-fort formation. The next year Jesse Fox surveyed and laid out a townsite. Early settlers who took up claims included Thomas Condie, W. H. Toone, and George and Charles Shill, after whom tributary canyons to Lost Creek were named. Irrigation ditches were dug in 1863 to bring water to the town lots and farms. The first ditch was known as Lower Water Ditch Company and irrigated 348 acres. In 1877 a portion of the unappropriated waters of Lost Creek were diverted higher up the creek into what was called the Upper Water Ditch Company, which watered 128 acres. On July 22, 1893, the two companies incorporated as the Croydon Irrigation Company.[31]

Morgan City, the major town in the county, was divided by the Weber River into pioneer South Morgan and North Morgan. The territory of Utah on February 13, 1868, defined a city boundary that united the two sections, which survived as the North and South Morgan wards of the Mormon church.

In September 1860, Richard Fry, Richard Norwood, and Daniel Bull were the first to settle in South Morgan. Fry in 1865 became the ecclesiastical leader. In the spring of 1862 several families moved from their original homes on the bottom lands to a "hollow between the base of the hills and East Canyon Creek" because they were afraid that the rapidly rising East Canyon Creek would inundate them. Their new location became known as Monday Town Hollow because Monday was the day they moved. It attracted other settlers but soon filled up, and newcomers thereafter made their homes in South Morgan, which became the main settlement.

Irrigation in South Morgan began in 1861 with a ditch first named after Richard Fry but now known as the City Ditch. It was constructed by Fry, Daniel Bull, and Richard Norwood. Originally the diversion of this ditch was in the city near where the present post office is located, from where it ran westerly to water the area called the Island between the Weber River and East Canyon Creek. A few years later the diversion was moved upriver to a point just below Como Springs. In 1957, the ditch, which ran along Young and State streets, was piped and covered over. Local history states that the early ditch was excavated by plowing and the use of a "ditcher" that consisted of the "large fork of a tree pulled by a head of oxen."[32]

In 1862 the Welch Field Ditch or East Richville Canal was taken out of the east side of East Canyon Creek approximately one-fourth mile below the Richville crossroad. This canal ran down stream through the bottom lands for about one and one-fourth miles. T. R. G. Welch was the largest stockholder, thus the canal's name. It irrigated approximately 160 acres of South Morgan and was incorporated in 1908 with a capital stock of $5,760. Later, the ditch was extended along the contour of the east hills for approximately five miles to the South Morgan Cemetery.[33]

In the spring of 1864, under the direction of Thomas J. Thurston and committee members Richard Fry, Charles Turner, and George Simmons, a diversion from the Weber River known as the Weber Ditch or Weber Canal was begun. The project took three years to complete under very difficult circumstances. After a few years the town of Milton built a new ditch from East Canyon Creek and the lower end of the Weber Ditch, which had supplied Milton, was abandoned. On March 20, 1893, the Weber Ditch was incorporated, and in 1901, the head of the canal was moved to the George Compton property to provide a better fall. Later the head was again moved downstream and the ditch line taken through Como Springs. In 1918 the canal company installed new head gates and divided the water into three sections: one at the head of the ditch, one at the head of the City Ditch branch, and one at the head of the middle ditch. A substantial amount of rock was put into the Weber River to make the diversion, but it was an annual task to rebuild the dam after the spring floods. In the 1930s a new diversion dam was put in, but it was 1952 before the company completed a permanent cement-core dam that raised the water level at the head gate and made the diversion reliable.[34]

Como Springs come out of a volcanic formation, and in the 1860s, flooding from the Weber River created a pool or lake that was warm because of the hot sulfuric springs. Even in early days the spot was used as a bathing pool, partly due to the supposed medicinal qualities of its mineral waters, and as a recreational area for fishing and camping. The springs were named by Dr. T. S. Wadsworth in honor of Mrs. Samuel Francis (Esther Charlotte Emily Weisbrodt), who was born in Northern Italy near Lake Como at the base of the Italian Alps.[35]

Settlers first came to North Morgan, across the river, in 1861. In the spring of 1864 this community laid out a channel from the Weber River on the east side of the Daniel Williams Meadows west of the Stoddard Meadow:

> This canal, since called Big Ditch, was nine feet wide and eighteen inches deep. Daniel Williams was employed to do the construction from the river to the town. From that point, each man was required to excavate the channel across his own land. It was completed by irrigation time which resulted in a bountiful crop of grain that year. Daniel Williams was paid 200 bushels of wheat for his work.[36]

Two springs known as the Bennett Springs and later as North Spring and East Spring emerge in the foothills east of North Morgan. In 1864, Martin Heiner and Daniel Robison obtained a right to a continuous stream of water for culinary use from the North Spring. The remaining water was used to irrigate town lots of one acre each. Turns were limited to eight hours, and it required eight days and fourteen hours to complete the round of use. In July 1869 the city created the office of watermaster, whose job was to regulate the turns and use of the water. Robert Hogge was the first to hold the office.

North Morgan shared East Spring with a number of users. In 1869 the Union Pacific Railroad obtained a right to a portion of the spring and built a reservoir and pipeline that supplied water to their depot and to steam engines of the trains. In 1900 the Morgan Canning Company also obtained a right to a portion of the spring. In 1902 the Heiner brothers built a reservoir with the help of the members of the North Morgan LDS ward and laid a pipe to the county road. From this source, each North Morgan homeowner obtained culinary water. The mayor and city council then negotiated for a right to a portion of East Spring for culinary use so that the city could also build a reservoir and lay a pipeline across the river into South Morgan. In 1925 the reservoir was enlarged to provide a more efficient water system, which still serves a large section of Morgan City.[37]

SUMMIT COUNTY

Although Summit County was designated as such by the Utah Territorial Legislature on January 13, 1854, it remained "attached to Great Salt Lake County for election, revenue and judicial purposes" until 1861 when it was granted a charter to organize independently. The county takes its name from the fact that it "occupies the Summit of the watershed between Green River drainage and Salt Lake Valley (Great Basin drainage)." It was not until December 1872 that the final boundaries of the county were fixed by Jesse W. Fox's survey. The county borders on the state of Wyoming and on the Utah counties of Rich, Morgan, Salt Lake, Wasatch, Duchesne, and Daggett. The Provo River and the crest of the Uinta Mountains, the highest peaks in Utah, form part of its southern boundary.

In these mountains, within Summit County, the Weber River heads. Rising in Reid's Meadow in the basin north of Bald Mountain and east of Reid's Peak and Notch Mountain, the main stream of the Weber River flows to the northwest. Approximately six miles downstream it is joined by other tributaries—Middle Fork, Gardner's Fork, and Dry Fork—in the area now called Holiday Park. At this point the river begins providing needed water to community enterprises. At Larrabee Creek the Weber turns and flows west to Oakley, and a few miles beyond that town, it turns again and flows north until it leaves Summit County three miles northwest of Henefer.

Summit County encloses about one-half of the length of the Weber River and several of its major tributaries. The main tributaries that enter the river in the county, from the headwaters on the southeast to the boundary of Morgan County on the north, are Smith and Morehouse Creek, entering from the south up the canyon above Oakley; Beaver Creek, arriving from the south after passing through Kamas; Crandall's Creek, entering from east of Rockport; Silver Creek, flowing from Park City to the west and entering the Weber at Wanship; Chalk Creek, joining from the east at Coalville; and Echo Canyon, which enters from the east between the towns of Coalville and Echo. Where these tributaries join the Weber River, spaces between the mountains allowed farming communities to develop. As in other areas of the Weber Basin, the original settlers came in small numbers to the places near the streams that were most easily cultivated and had the best access to water.

Summit County was, from its earliest history, a transitional area where the Weber River and its main tributaries provided passage routes to major population areas. It is thought that trappers in 1824 were at the headwaters of the Weber and Provo rivers. These mountain men included Etienne Provost, after whom the Provo River is named, and trappers for William Ashley and Andrew Henry's fur company. Among the latter was John H. Weber, whom most historians credit as the namesake of the river, although some suggest his contemporary Pauline Weaver. Jim Bridger also is credited with trapping the upper Weber.

In the 1840s the lower part of the Summit County section of the Weber became the main trail of migration into Utah. In 1846 several parties, convinced by Lansford W. Hastings that a route to California around the south end of the Great Salt Lake would cut 250–300 miles from their trip, traveled from Fort Bridger down Echo Canyon to the site of Henefer and either proceeded down Weber Canyon, as the Bryant-Russell, the Harlan-Young, and the Frederick Leinhard parties did, or went up Henefer's Main Canyon to East Canyon via Dixie Hollow and over Big Mountain and down Emigration Canyon, as the Donner-Reed Party did. In 1847 this Emigration Canyon route became for several years the trail of Mormon migration. Weber Canyon was the route of the transcontinental railroad line in 1869.

Over the years alternate routes through the region were explored and developed. In 1848 Parley Pratt discovered the Golden Pass up Parley's Canyon to Parley's Park and on down Silver Creek to the Weber, and in 1850 Captain Howard Stansbury visited the Kamas area, where he was impressed with the lushness of the valley. Stansbury wrote in his report that "it may be remarked here that the Camass Prairie consists of most excellent land and can be irrigated over its whole extent with comparatively little labor. Water for stock is abundant and timber for ordinary farming is plentiful and convenient." The report continued that "the Weber bottoms as far as the mouth of Red Fork [Echo Canyon], five miles beyond, presents many beautiful little prairies on either side of the stream fringed with belts of large cotton woods affording good location for many smart grain and stock farms."[38]

Another military report, that of the topographical engineer Lieutenant E. G. Beckwith, shed more light on the Summit County area. Beckwith was continuing the work of Captain John W. Gunnison, who had been killed earlier during the transcontinental railroad explorations. In April 1854, Beckwith made his way up Weber Canyon to Fort Bridger and found a new route up Chalk Creek. He described the trail through Weber Canyon: "followed a precipitous and rocky path leading over the retreating craggy sides of the canyon, so steep that a single misstep would have permitted both mule and rider to fall into the foaming torrent, hundreds of feet below us." The party proceeded to White Clay Creek (Chalk

Creek), where they turned upstream to the east. White Clay Creek was lined with cottonwood and willow in the lower part but destitute of timber higher up, though "while scattered cedars are seen on the nearest hills and pines, fir and aspen fill the ravines of the mountains." Beckwith also noted tracks and signs of grizzly bear and porcupines and that a fine silver-gray fox watched their progress up the valley.[39]

The familiarity with Summit County territory that such reports created soon led to settlement by those who took a liking to the setting and hoped for prosperous times there. Henefer was the first major settlement. Its location on "the west side of the Weber River, about four and a half miles below the mouth of Echo Canyon," was on the emigrant trail where it forked to go down Weber Canyon or up Main Canyon. This position made it an early place of interest to the Mormons. In 1853 Brigham Young directed two brothers, William and James Henefer, to start a settlement in this valley. They took their families and "claimed land where the town now stands, and built the first houses of logs near the river."[40]

Other early founders included Charles S. Appleby and William Bachelor, who settled at the mouth of Bachelor's Canyon on the southwest of Henefer. Southeast from Henefer is Leonard's Canyon, named after homesteader Leonard Richins. Next downstream is Fire Canyon, where an extensive fire burned during the Black Hawk Indian War. Next north is Bridge Canyon, which required a bridge to cross its stream. Here the Union Pacific Railroad constructed a water tank to supply their steam engines, and hence the canyon is sometimes called Tank Canyon. Farther north is Bald Rock Canyon, named after a white sandstone rock that protruded over the roadway until it was blasted away in 1956. Morgan and Dank's canyons, named after settlers, are next. Across the river from Henefer is Owens Canyon, named after William J. Owens, who lived near its mouth. This is the largest canyon on the east side of the river. Five small ravines complete the list of canyons on the east—Three, Willow, and Cottonwood hollows and Anderton and Harris canyons. Each of these canyons provided sources of water to Henefer Valley.

On the west range of mountains to the south of Henefer are Franklin Canyon, where Captain Thomas Job Franklin built his homestead; Bachelor Canyon; Main Canyon, the emigrant route from Henefer to Salt Lake City; Lone Tree Canyon, identified by a solitary cottonwood tree in early days; Hog's Back, whose ridge divided the waters flowing to Henefer and East Canyon; and farthest north in Henefer Valley, Bishop's Canyon, named after Bishop Charles Richins.

Up Main Canyon the settlers found a spring that became known as Jack Beard's Spring. There was a spring up Franklin's Canyon, but it was two miles away from the valley. There was also a big spring up Owens Canyon that provided water for culinary use and irrigation and to generate electric power for the Owens homestead. Another significant spring was located near the Narrows of the Weber River: "The water was delicious to the taste and the spring never did 'go dry' as some others did. In addition to refreshing drinking water this large spring provided a plentiful supply of succulent water cress."[41] A spring that was found to be 90 percent pure was developed in Bachelor's Canyon. Its water was piped into every house in Henefer in 1909, and it continued to have the same purity as late as 1958. Many small hollows branch from Main Canyon. Up them can be seen "water holes used by Pioneers for watering their cattle." Each played a part in providing water and other resources to Henefer.

Although the first water for Henefer came from them, "all of these springs were quite a distance away from the settlements so the people were forced to haul water from the river."[42] Some settlers hauled water from the springs, and later others developed wells near their homes. In 1860 citizens finished a canal to bring water from the canyon springs to their town on the bench. In 1861 the town plot was surveyed by Jesse Fox and a water ditch was tapped from the Weber River, but little work was done on it for several years. The original cost of bringing water from the Weber River to the bench was $4.50 per acre. This was too costly for the settlers, and most depended on the small supply provided by the Main Canyon stream; Bachelor's Creek, a mile to the south; and Franklin Creek, another quarter of a mile south.

The community made a concerted effort in 1867 to obtain irrigation water. The Henefer Irrigation Company, which would bring the first water from the Weber River to the Henefer bench, was organized. To construct a canal from the Weber River, the builders "used horses and scrapers and men with shovels. It took many men, and many hours, but soon the river was converted [diverted] into the canal. Farmers dug ditches on their farm lands and the little valley began to blossom."[43]

In 1909 a pipeline association, incorporated with thirty-seven stockholders, installed, from the spring in Bachelor's Canyon, the first piped water system for the community. In February 1957, the same Henefer Pipe Line Company determined that the old system had "become inadequate to carry enough water to supply the needs of the town and the over-flow water was running to waste." The company decided to build a new reservoir and pipelines through the town. To pay for the improvements, stockholders and water users were assessed according to the number of shares each held. Construction started on August 5, 1957, and water was turned into the new concrete reservoir, which is still in use, on June 24, 1958. Forty feet in diameter and ten feet high, it had a capacity of 95,000 gallons, a five-times increase over the old reservoir.

Water was not always beneficial; it sometimes threatened Henefer. In the late summer of 1892, a cloud burst in Bachelor's Canyon sent a wave of flood water down the canyon. It crashed through the Charles Brewer home, almost washing it away. The Nephi Bond family, who had taken refuge there, fled and took safe positions on the side of the hill. The hay in the field below the potato patch had just been raked and was ready to haul. "The flood washed the hay and rake along with it way across the fields and into the 'Big Ditch.' The potato patch was covered a foot deep with silt, dirt and sage brush, roots and rock."[44]

The Weber River, which for the most part was "a fast-flowing but quiet stream," was an annual threat to the small farms bordering the river banks. Prior to the construction of Echo and Wanship dams by the Weber Basin Project, the spring run-off annually overflowed the river banks and carried away the fertile soil and planted crops on which the community depended. "The flooded bottom ground, after the swollen river had receded, would be covered with fish, and the Henefer settlers would gather them in tubs and buckets."[45]

Water caused human, sometimes humorous, problems as well. On September 12, 1895, a complaint of assault was brought before the Summit County Court by Caroline Phillips against George Baker. On September 18, 1895, defendant Baker pleaded guilty to having put the complainant in a ditch, making the following statement: "On the 10th day of September, 1895, I wanted some water for my farm, went to turn it into my ditch. Mrs. Phillips forbade me from touching the ditch and placed herself in the way, and said she wanted the water and was going to have it. I told her she should have it, and thereupon, I laid her down in the ditch." The prosecuting attorney, H. S. Townsend, said there was no need of taking testimony and asked that "a fine sufficient for the gravity of the offense be imposed; to warn the defendant that hereafter he must respect the law." Based on that recommendation, Robert A. Jones, justice of the peace, imposed a fine of $15 and a court cost of $9.

Echo, four miles upstream from Henefer and at the mouth of Echo Canyon, was first settled in 1854 as a freight and stage station on the main trail through Utah. James Bromley, who was sent to Echo to take charge of the station, was the first settler to have a major canyon entering Weber Canyon from the west named after him. The land on which the station stood was sold in 1868 and became the Echo townsite.

The farmers who settled Echo had an abundance of water from Echo Creek and the Weber River. Some farmed and ranched west of the Weber River, and during the spring, high water made it difficult to communicate across the river. One family, William Stevensen, received butter and eggs they ordered in a basket "pulled across the river on a trolley."[46] When Echo later became a railroad town, it ceased to be a quiet farming village.

The next town settled in Summit County, Peoa, took its name from an Indian word that the first settlers found carved on a log. They believed it was the name of a chief or a tribe. The choice of the site is credited to W. W. Phelps, a prominent Mormon leader who came to Peoa in 1857 with a group sent by Brigham Young and who laid claim to land at the mouth of Fort Creek on the Weber, about two miles above where the river bends to the north at its confluence with Beaver Creek. In May of 1859 other settlers arrived to make the community permanent.

In 1861 settlers moved south into the area referred to as Sage Bottom Flat or Woodenshoe. In that same year two canals were constructed to bring water to the Sage Bottoms. Parts of these canals are still used today. One canal "flows in a northwesterly direction irrigating the land west of the county road," and "the other flows north and is used to irrigate the land east of the county road. At first the canal on the east side went as far as Peoa (called the North Bench Canal), and was used to irrigate everything as far as Fort Creek, but [it] since has been abandoned for that purpose."[47]

In 1868 Peoa expanded east approximately three miles and took in Oak Creek, now known as Oakley. This area was designated as the New Field, and in 1868, to bring water to it, the citizens constructed a canal that headed at the mouth of Weber Canyon, where the river emerges from the Uinta Mountains above Oakley. By 1881 more land was needed, and through the efforts of Peoa's citizens "the north bench canal [was] joined with the New Field making one larger canal."[48]

By 1887 Peoa had three thousand acres of fenced land, "for which there was an abundance of water." In that year the town of 450 inhabitants produced 24,000 bushels of small grain, 4,000 bushels of potatoes, and 1,600 tons of hay. Water for these crops came from the canals and from little Fort Creek, which drains the north and south benches above the town. There are also numerous springs along the bluffs immediately east of Peoa and in the bottom lands between the town and the river. The springs and some wells provided culinary water. As of 1903, Fort Creek and the various springs provided water for four hundred irrigated acres in and around Peoa.[49]

Wanship took its name from a prominent Ute Indian leader whom Mormon pioneers met upon their arrival in Salt Lake Valley. The first major community in Summit County, Wanship served as county seat until 1872; by then it was overshadowed by more populous Coalville, which became the seat.

Steven Nixon, his daughter Margaret, and young Henry Roper settled at the site of Wanship after emigrating from Provo through Three Mile Canyon into the Weber Valley in September 1859. From where they entered the valley, they moved about four miles north to the juncture of Silver Creek and the Weber. Here they built a log cabin and spent the winter. In the next year, Thomas Nixon, a son of Steven, and his family moved to the area, and that spring they broke six acres of ground and "raised a light crop." By 1861 a road had been completed from Salt Lake City to Parley's Park and down Silver Creek to Wanship, where it connected with the Echo Canyon and Weber River trails. The new accessibility to the upper Weber Valley stimulated more settlement, and in the 1860s Wanship acquired the characteristics of a frontier Mormon town. In 1867 a school and a gristmill were built. The mill was built by Henry Alexander and George S. Snyder.

The first water project, the West Wanship Ditch, was begun in 1860. It originally came out of Silver Creek, but pollutants put into the creek by Park City mining made the water unfit for irrigation. After 1893, the diversion for the ditch was moved south of town to tap water in the Weber River above the pollution. Later this ditch was known as the West Wanship Ditch No. 2. In 1861 the East Wanship Ditch No. 1, heading on the Weber about two miles above town, was surveyed. This ditch has continued to serve the farmers on the east side of the river until the present day, giving over 140 years of service.

In 1887, 180 Wanship inhabitants had 900 acres under cultivation, which produced 8,000 bushels of grain, 900 tons of hay, and a considerable quantity of potatoes and other hardy vegetables. A report on the Weber River made in 1903 stated that "East Wanship Ditch No. 1 . . . supplies water to a little more than 400 acres. Wanship Ditch No. 2 supplies water to the town of Wanship and . . . for the irrigation of about 170 acres of farming land."[50]

Coalville, established permanently in 1859, is located five miles south of where Echo Canyon enters the Weber River Valley, so at first it was out of the regular traffic down that canyon toward Salt Lake City. The early town's setting was three-fourths of a mile east of the Weber on a "low bench formed by alluvial deposits from Chalk Creek Canyon." The Weber River at this point sweeps around the foothills on the west of the valley and creates a rich bottom of alluvial soils that are excellent for agriculture. In 1889 the banks of the Weber were "heavily fringed with cottonwoods and willows which, with the adjoining fields, give pleasing variety to a landscape that would otherwise be monotonous." The mountains around were rugged and broken and "more majestic than pleasing, they afford in spots a scrubby growth of cedars, supplemented with white pine and balsam on the higher peaks in the distance." The mountains and foothills provided areas for stock-raising and in the 1880s were estimated "to be utilized to their utmost capacity."[51]

Edward Tullidge wrote that "trifling circumstances often produce important results. This was exemplified in the settlement of Coalville."[52] Tullidge referred to an incident in which William H. Smith, an early pioneer, spotted some mature wheat growing along the Weber and determined that if wheat that was apparently dropped by accident could mature to its fullest, then cultivated grain would be all the more productive, making the upper valley fit for settlement. On April 26, 1859, Smith returned to the site of Coalville with Alanson Norton and Andrew Williams.

At the time, their nearest neighbors lived at Snyder's Mill in Parley's Park, twenty-five miles to the southwest; at Peoa, five miles up the Weber; and at the Echo Canyon mail station. At Coalville the new settlers "found the ground bare, and hardy vegetation putting on the green verdure of spring." They set to work cultivating "a field of four or five acres, the same ground being now occupied by a part of the town of Coalville, including the Stake house." Even though the planting was not completed until June 8, 1859, "a fair crop of wheat and vegetables was raised."[53]

At about the time of Coalville's settlement—some believe even before settlement—coal outcroppings were discovered nearby. It was the development of the coal mines that "subsequently built up the town of Coalville."[54] The community was first called Chalk Creek, but the importance of the coal activity inspired a change of names on May 9, 1866. Besides serving as county seat, Coalville became the LDS stake center for the upper Weber Valley.

George Beard, a prominent artist and a long-time citizen and leader of Coalville, remembered some of the town's water problems during the 1860s:

> a very large slough covered all the ground lying between the ledge and the bench on the south on which Coalville is built. Main Street was surveyed commencing at the west end of the ledge and running through the slough to the Copley farm. Rocks had been hauled into the slough to make a narrow road and it was barely wide enough for two wagons to pass each other.[55]

An 1880 account described Main Street "with its shade trees, neat public buildings and private residences" as "a pleasant avenue through the town" and noted its location relative to Chalk Creek:

> At the north end it crosses a substantial bridge over Chalk Creek from which it extends about a mile south in a direct line and then bends to the right. The town extends on the east a few blocks up Chalk Creek, and this makes its greatest width east and west. Chalk Creek is a rapid mountain stream with an average width of about forty feet and a depth of twelve inches. It runs along the north side of the town into the river.[56]

Coalville takes most of its water from the east out of Chalk Creek. By 1903 several ditches and canals from the creek divided its waters to irrigate approximately 2,000 total acres in upper and lower areas of Coalville. In the upper section, about 780 acres were irrigated "from private ditches along the upper course of Chalk Creek and tributaries in the upper valley." In the lower valley, about 875 acres of land were irrigated, including 384 acres on the south side of the creek watered by the Upper Robinson, the Lower Robinson, the South Chalk Creek, and the Coalville City ditches. On the north side of Chalk Creek were 533 acres watered by the North Narrows, the City Cemetery, the Chalk Creek, and the Middle Chalk Creek ditches. The North Chalk Creek Ditch furnished water for 256 acres in the Weber Valley below Coalville.

Other ditches were constructed below Coalville. In the six miles between

Coalville and Echo, 150 acres were watered by the Weber River in 1903. Ninety acres on the west bank were watered through the Calderwood Ditch, and sixty acres on the east bank by the Asper and Brian ditches. About midway between Coalville and Echo, Grass Creek joins the Weber. It furnished water only during the early part of the season for about sixty acres of land, through the Lawson and Williams ditches.

In 1906 a group of Coalville citizens began to develop a culinary water system for the city. The water would come from a large spring. In 1911 the City acquired the farm where the spring was located. Several city meetings were held in 1912, and with the help of Harry S. Joseph of Salt Lake City, a water works system with a reservoir and an electric light plant was designed. On August 5, 1912, a bond election was approved by a vote of 115 to 16. The system used eight-inch and six-inch wooden pipes to conduct water from the reservoir to town and two-inch galvanized pipe for the pipelines in town. Wooden pipe was used because it was less expensive. The town could bond up to a limit of $15,000, and the cost for freight alone to ship the iron pipe from the manufacturing plants in the East would have cost almost the entire amount of the bond. "It was a case either of using wood pipe or having no water system."[57]

Work on the system started as soon as the bond was approved, and was completed before winter of that same year. A power house was completed and the lights turned on for the first time on New Year's night, 1906. A few years later the plant was sold to Utah Power and Light Company, which became the supplier of electricity thereafter.

Hoytsville, located south of Coalville and north of Wanship, was first settled in 1859 by Thomas Bradberry and his wife, who built a cottonwood log house on the west side of the Weber River. Nearby Bradberry (spelling is according to local histories, but current maps have spelled it Bradbury) Canyon carries their name. Soon others arrived. Brigham Young sent Samuel P. Hoyt in 1861 to establish a gristmill, which in the winter of 1863, began operating as the first mill in Summit County. The mill was closed in 1867 because of insufficient fall in the river and mill race to drive the machinery and because of difficulty in obtaining a satisfactory water right. It was moved to Oakley, where it operated for many years.

Several canals provided water for Hoytsville: the Coalville-Hoytsville Irrigation Canal, the Hoytsville Irrigation Company No. 1 Canal, the West Hoytsville Irrigation Company Canal, the Elkhorn Ditch Company Canal, and the Pipeline Canal. The Coalville-Hoytsville Canal irrigated 100 acres of ground until 1883, when it was improved and enlarged. In recent years the canal still functioned, watering approximately 250 acres. The Hoytsville Irrigation Company No. 1 was constructed in 1864, irrigating in early years 350 acres and later approximately 500 acres. Over time the company added more water by purchasing claims to storage water from the Chalk Creek-Hoytsville Water Users' Corporation. The

West Hoytsville Irrigation Company was incorporated in 1865 and watered 16 acres, which was later increased to 320 acres.

The Elkhorn Ditch Company irrigated approximately 400 acres of land. The canal system transferred water from Elkhorn Creek, a tributary of Chalk Creek, to Spring Canyon, where it could run to the Hoytsville farms. Reservoirs and lakes in the mountains to the east were built or enlarged to store the water above Spring Canyon. The lakes are known as the Sargent Lakes after Nephi and Amos Sargent, early developers of the system. Construction started in 1883, and the first water was turned into the ditches on June 27, 1884.

Nephi Sargent was also instrumental in developing Hoytsville's culinary water supply. In 1907, with Nephi as chairman and C. H. West as secretary, a group of citizens "worked diligently to develop a community reservoir and pipeline system so that the towns people might have pure spring water in their homes." The group "insisted that it be a community project and not privately owned."[58]

The town of Kamas, on Beaver Creek above its confluence with the Weber River at Peoa, was a product of 1859 colonization. Thomas Rhoads (there is disagreement on the spelling of Rhoads's name and the year that he came to Kamas Valley), a Mormon entrepreneur, first came into what was then named Rhoads Valley in 1855 to promote mining, trapping, and ranching. The name Kamas was applied to the valley after several years of settlement. It derived from the Mootka Indian word *chanas*, sometimes written as *camass*, which identified several plants whose bulbs were widely used as a food by western American Indians. The word also designated "a small grassy plain among the hills," which fits well the Kamas Valley. Twenty settlers came to the area in 1859 in response to a "call" from Brigham Young. These families lived in a fort near a spring on the east side of the valley. The extensive meadow land was an ideal summer range for cattle.[59]

Francis and Marion are small communities that splintered off from Kamas. Francis, two miles to the south, is oriented toward the Provo River, which just west of town collects water that is transferred from the Weber River via the Weber-Provo Canal. Marion is a small agricultural community north of Kamas. Thomas Rhoads was the first to settle there. After the Mormon Kamas Ward was organized, Marion was known as North Kamas. On May 1, 1909, it was organized as a separate ward by Mormon presiding bishop Francis Marion Lyman, whose middle name was thereafter attached to it. (Another account claims the name came when the town applied for a post office and that it derived from Marion Sorensen, the leading woman in the community).

In 1861 Samuel P. Hoyt moved his family to Marion and built his home at the mouth of the canyon that is named after him. For culinary and stock water he used the spring that would later furnish water for the Marion water system. Abundant water from lakes, rivers, and reservoirs was potentially available to

Marion. Several companies and systems were organized to develop that water supply: the Marion Upper Ditch Company, the Marion Lower Ditch Company, the Boulderville Ditch, Gibbon's Ditch, Hoyt Ditch, and the Smith and Morehouse Reservoir. Getting the water into these systems was slow and difficult work. Many of the ditches were completed by "stent work," which provided that "the distance of the ditch was divided into sections and each man drew for a certain section which was his responsibility to build."

These ditches provided enough water in the spring and early summer, but additional water was needed for late summer. Because of that need a group of Marion and Oakley farmers planned a reservoir for Smith and Morehouse Creek. A. Z. Richards, an engineer from Salt Lake City, contracted to survey the reservoir site in 1915,[60] but after the survey was made, the reservoir plan was abandoned for five years. "The project seemed too expensive and the money to pay the bills already accumulated could not be raised. Finally Albrey Gibbons, Joseph D. Hoyt, and Joseph William Myrick gave the money which was in excess of $300, the bills were paid and in 1920 work again started on the project."[61] At first the reservoir had a 400-acre-feet capacity, but later the size was increased to store 1,048 acre feet.

The work was done by hand with teams. No power machinery was available. For several years the farmers of the region spent the fall working on the dam, which was made of dirt with rock facings. Their work was threatened in 1937 by high water that almost washed the spillway out. Miners from Park City came to the aid of the valley farmers, and by working night and day, they saved the dam. The cost of the dam had exceeded $60,000, but it was completely paid for by 1944.

In 1931, under the supervision of William Myrick, first president of the water company for Marion, a culinary water system was installed using water piped from the spring at the mouth of Hoyt's Canyon. For a cost of $800 to each family, it was thought that the "the health of the people was improved and their surroundings made attractive by their new water system."

Rockport, settled in 1860, was located approximately halfway between Wanship and Peoa, at the mouth of Three Mile Canyon, more recently known as Crandall Creek. After constructing log cabins, the settlers, under the supervision of Jesse Fox, laid out the townsite and surveyed a water ditch from the Weber River. The town was first called Enock City, probably after the prophet Enoch who in Mormon belief built a City of God. In 1866–67 fear of Indians led townspeople to build a fort with walls eight feet high and two feet thick. The village was renamed and "became known as Rockport in commemoration of the rock fort."[62]

The Weber Valley is narrow at Rockport, and consequently there was but one street through town. The Weber Basin Project eventually chose this constricted site for the Wanship Dam, and Rockport Reservoir inundated the pioneer town.

Located in Kamas Valley near where the upper Weber River emerges from the Uinta Mountains and then makes a big bend to the north, Oakley was an obvious place for settlement. The homesteading and cattle raising of Thomas Rhoads in 1855 first brought attention to the area, but the first permanent settlers did not arrive until 1868. Early settlers lived in dugouts on the banks of the Weber River at what was called, as an extension of Peoa, New Field or Oak Creek. The town name was changed to Oakley sometime between November 7, 1886, and May 15, 1887. The soil here in the river bottoms and on the bench is "fairly good" for farms and for pasture land, which was valuable to the dairy business. One section at the south end of Oakley is named Boulderville for a rocky area "where a great deal of water is necessary."

As usual in the history of these communities, the Mormon church organization became the first motivator of water development, encouraging the division of water shares and the digging of the first canals. The first in Oakley was the later-abandoned Hopkins Ditch. The two main irrigation companies that organized in Oakley were the New Field and North Bench Irrigation Company and the South Bench Irrigation Company. The New Field ditch, dug in 1876 (the Peoa account says it was dug in 1868), heads just below the dugway in Weber Canyon. The South Bench Ditch was taken out of the Weber River in 1880 about two miles below the north bench.[63]

To supplement the Oakley water supply, reservoirs were built in the headwater canyons of the Weber River above Holiday Park. Several were completed in the 1920s and 1930s by damming up the outlets of mountain lakes to give them greater holding capacity. The Fish Lake Reservoir Company, the Markson Extension Company, and Glen Gibbons and Joseph B. Andrus developed these projects. In 1920 Albrey Gibbons, Joseph Hoyt, and Joseph Myrick put up the money to start work on the first dam on the Smith and Morehouse reservoir. The farmers worked on this first dam through the years until it was finally completed in 1944. The Fish Lake Reservoir Company built dams on Fish Lake, Sand Lake, Lavinia Lake, Seymour Lake, and Cliff Lake. Gibbons and Andrus dammed Kamas Lake near Reid's Meadow. Some of these projects provided water for the Provo River side as well as the Weber River side. Forest Service maps have changed the names of some of these lakes in recent times. Without these reservoirs, the Oakley area could have never prospered.

Building the dams was accomplished with a great deal of difficulty because materials had to be taken in with pack horses over rugged mountain country. Access was achieved by going up the Weber River or the Provo River and down from Bald Mountain around Reid's Peak or Notch Mountain. Ralph A. Richards of Oakley remembered how he worked on these projects as a young man. Work usually began after the fall harvest and continued until the workers were driven out by deep snow. They worked for cash, for company stock, or for credit against their assessment for funds as company members.

Richards noted that "those early dams were built by wagons or rock boats and teams, picks and shovels. There was no large machinery at all used on any of them when they were originally built. Some of them were even built with pack horses—the gravel and sand that they used for what little cement work that they did was packed in pack bags to the dam site." A rock boat was a piece of tin or of heavy iron sheet metal about twelve to fifteen feet long and four feet wide. Rock or dirt could be shovelled on to it and a team could pull it. Generally a pole was attached as a runner on each of the long sides, as on a sleigh, to keep the boat from catching on rocks or stumps.[64]

The Weber-Provo Canal, which carries water from the Weber River to the Provo River, begins at Oakley. The water feeds, via the Provo, Deer Creek Reservoir, where it is stored for use in Utah and Salt Lake counties. Oakley thus plays an important role in two critical river and water supply systems.

The question of building reservoirs on the upper Weber River was raised during the early 1900s. By then water was becoming scarce in the lower Weber Basin. Commissioners from Morgan, Weber, and Davis counties came up the river seeking more water. "First they asked to have the water turned out of the canals of upper Weber River for ten days or two weeks in order that crops might mature." The commissioners then "demanded that water be turned down to them. They were determined they had rights which were prior to many of the Upper Weber." The citizens on the upper river believed that they also had rights which should be respected and that by "using water in early spring their land would act as a reservoir and the water used by them [would] come out in return flow." The two groups of claimants seriously considered lawsuits to settle their differences. Then, Joseph F. Smith, president of the Mormon church, called a meeting of the presidents of six church stakes whose ecclesiastical jurisdictions included Weber River water users: Joseph W. Hess and Joseph H. Grant of Davis Stake, Lewis W. Shurtleff of Weber Stake, Daniel Heiner from Morgan Stake, J. R. Murdock from Wasatch Stake, Stephen L. Chipman from Utah Stake, and Moses W. Taylor of Summit Stake. President Smith advised these men "to make a study of reservoir possibilities, not to spend their means and time in law suits, but to take care of the early high water. He told them there was enough water for all if it was stored so that it could be used when it was needed. He advised building reservoirs and storing the spring flood waters."

The stake presidents went back to their communities and used their positions of authority to encourage water storage. "There is no doubt a law suit was averted by their influence." In Summit County that influence went furthest, after President Moses W. Taylor assigned his first counselor, Thomas L. Allen, "to make a study of reservoirs sites." Allen made many excursions into the canyons which drained into the Weber River and studied the reservoir possibilities. His studies led to recommendation of sites at Roos Hill in Peoa, between Wanship and Rockport, and where Echo Dam was eventually located. Allen's work laid the groundwork for later efforts to determine dam sites.[65] This early cooperation by upper and lower Weber River water users set the precedent that later would be followed by the Weber Basin Project in the building of storage reservoirs. It was another example of grass roots democracy in the Weber River Basin.

Logging and mills have been major uses of water from the upper Weber River. In early days needs for lumber and fuel sent men up the Weber Canyon above Oakley to harvest logs and trees from the forested slopes of the Uinta Mountains. Railroads and mines in the surrounding region required ties, lumber, and cordwood. Many people on the upper Weber supplemented their incomes in the lumber business, and many participated in early spring "tie drives" to bring logs out of the canyon. Canyons such as Perdue Canyon, Frazier Gulch, and Neil Hollow were named after these early loggers.

At first, logs were floated down the Weber River during the spring surge of water. Men had to break up log jams to keep the timber moving. This was dangerous work, and many times the loggers were thrown into the cold, rushing waters. One unfortunate logger, George Cartio, drowned when he was thrown from a log jam near the Pines ranch in 1877. The logs and even rough-cut ties for the railroad were floated down the Weber to a boom in Wanship where they were taken out. The logs were then cut into lumber or ties. Cordwood was cut for fuel and hauled out by oxen or horse teams. Some of it was hauled to Snyderville, where it was burned to make charcoal for forges at the Park City mines. Cordwood and logs cut above Holiday Park and piled there in great stacks could not be floated out because the water was not deep enough at a point that high up the river. Sawmills were needed to prepare the wood nearer its origin points. As a result about twenty-five sawmills were built in Weber Canyon. The first sawmill was built near the mouth of South Fork in the canyon in 1870. John Taylor, an apostle and, later, president of the Mormon church, owned it. Many of the Weber's tributary canyons eventually had their own stream-powered sawmills.

George Wilde listed the following sawmill operators: Gardener at Gardener Fork, J. H. Stalling at Moffet Creek, Jack Marchant on the Weber before Holiday Park, George Wardell at Pine Ranch, Joseph Williams at Slater (maps show it as Slader) Basin in Smith and Morehouse Canyon, Dan Lambert in Smith and Morehouse, the Wilde Brothers and Vernon at Wilde ranch, Willis Stinson at his place, James Evans at his ranch, J. H. Stallings at Swift Creek, Roe at the Noblet Fork of South Fork, James Welsh on Pullum Creek up South Fork, John Maxwell on the Shingle Mill Creek South Fork, Jade Lambert over the Rock Slide, James Evans at the mouth of South Fork, Jones below Holiday Park, Butler below Holiday Park, Merrell on the fork of Smith and Morehouse, Stillman at the Stillman ranch, Ad Miles at the first mill up South Fork, and an unnamed shingle mill operator at Shingle Mill Canyon.[66]

Early pioneers throughout the Weber Basin labored mightily to put into place the irrigation and culinary water system that was essential for their survival and prosperity. Brigham Young to a great extent directed and encouraged development of water resources, but local church members followed his guidance selectively according to their needs. As leader of the LDS church, Young claimed the inspiration of a prophet, but over time he also learned from observing the successes and failures of various communities throughout the territory. He drew on this accumulated practical experience to counsel other settlers, passing his conclusions down through the church hierarchy to local presiding elders and bishops, who became the key figures in instigating water projects in local communities, which were organized as Mormon ecclesiastical divisions.

On the scene, water was delivered through democratic and cooperative effort. In the beginning, a few persons or families who had come to a place on a stream diverted water as quickly and conveniently as possible to nearby land in order to meet immediate survival needs. Then, as more settlers arrived at the settlement, more land was taken up further away from and higher on the streams. This required more effort and planning, but there were more hands to complete the work and an ecclesiastical system to organize it.

Church wards and quorums carried out the projects cooperatively. Thus, for example, Luman Shurtliff of Harrisville credited his seventies quorum, a unit of the Mormon priesthood, with assisting him with his water problems and claimed that without their aid he would have surely failed. All ward members, land holders, or water users became responsible for building and maintaining a portion of the canals or ditches that brought water from the main sources, nearby springs or streams, to plots of land. In early days these were either parts of a common big field or individual farms recognized or assigned by the ecclesiastical system. Wells sometimes supplemented the water supply, especially for individual households. The work was universally done by hand. Surveyed and leveled by eye and with primitive instruments, canals were dug with picks and shovels, often with the help of a plow or cruder tools to loosen the soil. Diversion dams were built of whatever material was locally available.

These pioneer developments had to overcome human problems as well as natural obstacles. Personal greed, misunderstandings, and jealousy interfered, as they do whenever a scarce and valuable resource is involved, but the system brought progress and a base of development on which diversified economies of trade and industry could build. Communities did not collapse or break up as a result of conflict, although a few divided. Rather, they generally settled their grievances through ecclesiastical channels and through county courts that were heavily influenced by the church. Some complainants undoubtedly remained dissatisfied, but for the most part the system benefitted overarching community interests.

All of what happened in the pioneer times was only a prologue for what was to follow. The Weber Basin area could not remain as an agricultural area only, and its resources could not remain under the control of Mormon institutions. The nature of the area began to change as early as the Utah War of 1857–58, when a new governor was appointed to replace Brigham Young. The arrival in 1862 of an army sent under Patrick Connor to make certain Mormons remained true to the Union further weakened church control. Most significantly, the completion of the transcontinental railroad in 1869 brought non-Mormons to Utah who did not fit into early patterns of doing things. Because of this new population, the early cooperative system had to change, but to a great extent it would have a lingering effect on subsequent water development.

In subsequent decades, water-user organizations, irrigation districts, private companies, and town governments took over much of the water development role church units had played. However, although new laws eventually codified private ownership of water, the democratic will of local water users and land-owners continued to be expressed through the new organizations, and well into the twentieth century, many projects were still completed by community-based planning and cooperation. Organizational and legal changes in Utah were re-sponses to changes in federal law, to a more diversified population, to more var-ied needs for water, and to greater needs for capital, but the organizations were made up of and controlled by local water users. Municipal governments gradu-ally took more responsibility, especially for providing culinary water; they and the bonding they increasingly used to finance the projects were also democratic responses to the needs and wishes of the local citizenry. Brigham Young was gone, but the impact of the early Mormon approach to developing water re-mained.

Chapter 5

 Transitions in Water Development from 1880 to 1940

O ut of the pioneer period in the Weber Basin came several important developments that shaped the modern era. The pioneers had commenced irrigating in a semiarid region with no experience and scant information about irrigation practices. What they learned was from trial and error and from directions which were spread by the Mormon hierarchy through the church system, sometimes directly from the church prophet to local leaders. Settlements usually began with a few settlers taking water from the natural flow of the streams and applying it to the neighboring land. As more people arrived, they constructed ditches and canals to deliver water to areas farther away and higher on mountain benches. To do this, organization was required, and the Mormon church and its leadership fitted the needs admirably well. Local bishops and priesthood quorums provided the authority and manpower that allowed communities to plan, control, and construct water projects in a cooperative way. A great amount of land was brought into production under this system, which met the needs of early residents.

WATER LAWS AND ADJUDICATION

Early laws supported this approach. The decrees of territorial governor and church president Brigham Young guided many of the initial irrigation efforts, and territorial laws of 1852 and 1865 gave an appearance of secular control by putting county courts in charge of regulating water use and claims. Although specific laws did not at first outline or implement it as a doctrine, practical experience and the lack of sufficient water to meet all claims led to the policy of determining water rights based on prior appropriation for beneficial use, the system which became common, with some variations, throughout the western United States. Until 1875 this was all well and good, but increases in Utah's population brought greater demand for water. A significant part of the increase consisted of Gentile, or non-Mormon, citizens, and pressure mounted to remove water control from the Mormon church and give it to a secular authority. The pioneer system was no longer sufficient, and something different had to be worked out.

Several water laws addressed this need: the law of 1880, a brief attempt to try an entrepreneurial, laissez-faire method of developing water; the law of 1903, which brought water rights back under community control; the 1902 federal legislation that created the Bureau of Reclamation and offered a way to finance major projects within the state; and the Utah water law of 1919, which called for a study of water claims and led to a court decree in 1937 that designated which claims would be recognized in the Weber River area. Under the umbrella of these laws, major dams—East Canyon, Echo, and Pineview—were built, as was the finale to the area's water development in the modern era, the Weber Basin Project, which, with its dams and canal system, brought coordination to water delivery in the entire region, meeting the demands of a growing population for irrigation, industrial, and culinary water.

The decades of 1880 to 1940 brought not only important new laws and new relationships with the federal government, especially those engendered by the reclamation law of 1902 and the projects it funded, but also many legal issues that had to be settled or adjusted by court action. Their settlement allowed, by 1940, the Davis and Weber Canal, East Canyon Dam, Echo Dam, and Pineview Dam to be constructed and various water user organizations to emerge to utilize the water the projects provided.

The territorial law of 1852 stated that "the County Court has control of all timber, water privileges, or any water course or creek, to gristmill sites, and exercise such powers as in their judgment shall best preserve the timber, and subserve the interests of the settlements, in the distribution of water for irrigation or other purposes." In 1865 this law was changed so that water districts could be organized "in a county or a part there of" and empowered "to supply themselves with water at the expense it cost to divert it from its natural channels and apply it to the lands." The water districts could tax their water users for the costs of projects. The new laws were implemented in accordance with the doctrine of prior appropriation rights, the custom established by trial and error in this arid region where all needs and wants for water could not be met. Under these laws, many mutual irrigation companies were established in the Weber Basin. By 1898 there were forty-one irrigation districts in Utah, most of them in the northern part of the state. Laws and practices that were sufficient for a society dominated and controlled by one church would no longer suffice once that control was under challenge from the influx of non-Mormons the transcontinental railroad brought after 1869 and from attempts of the national government in the 1870s and 1880s to halt Mormon polygamy and break the power of the church's leadership. In this same period these and other factors were also forcing Utah's economy to change from the communal system of the Mormons under Brigham Young toward mainstream American free-enterprise values and practices. After 1877 Brigham Young was not around to defend his views that had shaped the pioneer era.

Laws are made to shape public policy, and they change to meet the changing needs of society. The Utah water law of 1880, "An Act Providing for Recording Vested Rights to the Use of Water and Regulating their Exercise," enacted on February 20, 1880, smacked of an economic trend that placed laissez-faire principles over communal interests. It repealed the law of 1852 that had placed control and approval of water claims in the hands of county courts consisting of a probate judge and three county selectmen. Under the old law, the "common wealth was a party to every water claim or controversy."

Under the new law, authority over water passed into the hands of new selectmen who became ex-officio the water commissioners of a county. Also, "it was no longer the duty of the territory to enforce a beneficial and economical use of the public waters but merely to supply a means of adjudicating the difficulties which may arise between different appropriators and not concern itself as to whether the claims were excessive or not as long as each claimant was adequately protected." In issuing certificates to water users, the commissioners were inclined to grant full requests for water as long as no one protested, which resulted in "grants so large as to be absurdly excessive."[1] The law of 1880 in essence promoted the development of water claims as private rights subject to challenge only by other water users claiming the same rights. This was in line with the period's trend toward supporting individual property rights against the common interest. Individual claimants took their chances that no one would challenge them and they would get a favorable award.

The selectmen issued certificates for the water on receipt of "satisfactory proof" to claims for water, and it was their duty to distribute the water fairly to "each . . . corporation or persons" applying. In cases of dispute "as to the nature or extent of their rights to the use of water, or right of way or damages," the selectmen were "to hear and decide upon all such disputed rights and file a copy of their findings with the County Recorder, and distribute the water according to such findings." Any challenges to the determinations of a commission could be appealed to the district courts of Utah.

Water commissioners were also authorized to measure streams and record their flow. This part of the act was never carried out because of a lack of funding to accomplish it. It could have been a major contribution to water development in Utah and the Weber Basin and could have prevented much of the adjudication and water waste that persisted for several years.

The law vested rights to the use of water by stating that "a right to the use of water for any useful purpose such as for domestic purposes, irrigating lands, propelling machinery, working and sluicing areas and other like purposes is hereby recognized and acknowledged to have vested and accrued as a primary right to the extent of, and reasonable necessity for such use thereof." After open, peaceable, uninterrupted, and continuous use of water for a period of seven years, primary or first rights to the unappropriated waters were established. The law also defined secondary and waste water rights and made all rights saleable. Utah was the first territory to define classes of primary and secondary rights.

There previously had been limited sales of water rights in Utah, but for the most part they had been appurtenant to the land. In 1877 the federal Desert Land Act freed water from the land nationally, and the 1880 law did the same in Utah, legally establishing "the right of the user to declare his water rights personal property and dispose of it as such." The act further provided that "water companies could organize and conduct their business upon the corporation plan."

There was much fear that the law would lead to detrimental monopolization of water by individuals and corporations working against the public interest. George Thomas wrote that "it is doubtful, if in all the legislative history of irrigation a more retrograde piece of legislation was ever placed upon the statute books than the law of 1880." He felt that the legislature should have passed instead laws that protected public interests throughout the territory.[2]

In the Weber Basin the law created numerous challenges to water appropriations. It also led water claimants to seek protection of their claims by applying for certificates from their counties. Summit County records show that under the law of 1880, sixty-four primary rights and eight secondary rights were issued to the waters and tributaries of the Weber River and to springs in its watershed. Forty-six of these certificates were issued to individuals and the remainder to groups of individuals. In Morgan County, the Board of County Commissioners granted 247 water claim certificates, 220 for primary rights and 27 for secondary rights. These certificates covered irrigation of a total of 5,056 acres: 963 acres with water from the Weber, 665 acres from Lost Creek, 631 from Deep Creek, 263 from Line Creek, 278 from Peterson Creek, and the balance from springs and smaller streams. There were also claims for irrigation water from the towns of Croydon, Morgan, Porterville, and Peterson.

Davis County records show no claims under the law of 1880 for irrigation from the Weber River. The Weber County records are meager on certificates issued. They list six certificates issued by the Board of County Commissioners, granting to fifty-eight people primary rights to the Weber River for irrigating of 1,066 acres. Two certificates granted primary rights to the Hooper and Wilson canals for "2 by 8 feet of water" and "1 by 8 feet," respectively. Another grant gave a secondary right to the Wilson Irrigation Company of forty-seven square feet of Weber River water. Twenty certificates granted primary rights to the Weber to twenty people who had an aggregate of 670 acres to irrigate, and one certificate gave Charles de la Baume of Uintah the right to water twenty-three acres from an unspecified source.

Certificates issued under the 1880 law were ineffective because they generally did not define the amount of water that was appropriated and because no uniform measurements were made. A few grants specified the number of acres to be irrigated, "but no definite description of the land was given." In 1903 Jay D. Stannard, a field investigator who helped prepare Elwood Mead's *Report of Irrigation Investigation in Utah,* noted that "these records show that the matter of water rights in the Weber Valley is in a very confused state."[3]

From 1880 into the 1890s, county water commissioners were kept active adjudicating conflicting claims. The Box Elder commissioners, treating the county irrigation district and some of the companies as units, dealt with practically every stream and spring in the county. Their adjudications involved 307 claimants and 429 water certificates. In Davis County the water commissioners adjudicated fifteen streams and one spring and issued 306 water certificates after reviewing 633 applications. In Weber County the water commissioners adjudicated several streams and springs and issued 198 water certificates. Even though the commissioners kept busy, their work was considered ineffective overall.

George Thomas concluded that "if this same system had been built upon by allowing the county water commissioners to employ expert advice, protect the public interest, and to receive judicial approval for the work done it would have proven ideal." As the state grew in population and counties became less isolated, the county water commissions also could have been consolidated into one state commission. However, "the public doubts the wisdom of the past and looks afar for some new institution when a slight change in an old one would meet the situation better."[4]

The separation, under the 1880 law, of water rights from the land encouraged corporations to speculate in irrigation projects. The main speculative investment in Utah was the Bear River Canal. The first survey of the canal and dam was done in 1868, but the project was of such magnitude that the local promoters soon realized that they could not accomplish it without outside assistance. They first petitioned the United States Congress to support the project at that time, but the petition was denied, demonstrating that federal reclamation was not yet an accepted method of financing western water development.

To get the Bear River project going the Corinne Mill Canal and Stock Company in 1883 bought 45,000 acres of land in Bear River Valley from the Central Pacific Railroad. The company undertook new surveys and developed plans for a diversion dam, but soon realized that financing remained insufficient. In 1888, John R. Bothwell, an eastern entrepreneur, became convinced after a visit to Bear River Valley that the project would be profitable as a consolidated land and water enterprise.

Bothwell succeeded in getting funding from the Jarvis-Conklin Mortgage and Trust Company of Kansas City with $2 million in bonds. In 1889 the Bear Lake and River Water Works and Irrigation Company was incorporated and took over the contract. Although it was primarily an eastern manipulation, the corporation had several prominent Utah men as officers and directors. The purpose of the company was to "supply water for domestic, municipal and manufacturing uses to Ogden City, Corinne City, Brigham City, Bear River City and other cities and villages and their inhabitants and for irrigation of land and for all other useful and beneficial purposes." The water was to be supplied by the Ogden River, Bear Lake, and a major diversion dam on the Bear River just east of the Cache divide. The bonds of the company were sold, a large quantity in Great Britain. Samuel Fortier and Elwood Mead, prominent irrigation engineers, were hired for the project.

Two six-mile canals were constructed through the Bear River Canyon above Collinston. They were fifteen feet wide and ten feet deep. One canal went to the valley north and west of the canyon, and the other went south for the purpose of irrigating lands on the east side of the valley, in the original plan as far south as Ogden. Construction was largely completed by June 1891, and water was delivered to farmers in the spring of 1892. The south branch, however, was only completed as far as Deweyville until the Hammond canal was extended in 1903.

The project became engulfed in numerous difficulties and eventually failed, becoming a negative example of free enterprise attempts to build large-scale irrigation projects. Problems developed when William Garland, the construction contractor, could not get his money from the Bear Lake and River Water Works and Irrigation Company. Costs had been underestimated by an engineer. Garland won his court suit for additional money, but the canal company was in bad financial shape.

Sales of land separate from water rights had created distress: "The lands of the Corinne Mill Canal and Stock Company were not selling well. The public lands, in part at least, were held by speculators and they were not buying water rights, but simply holding their lands with a view of selling them at an enhanced value which resulted simply from the fact that water was available." Farmers who honestly felt that the annual fee of two dollars an acre for maintenance and operation of the canals was too high would not buy water. "Some few held off no doubt in the hope that if the company failed, they could buy the water cheaper."[5] And fail the company did, a victim of the speculative system created by separating water rights from the land.

On June 6, 1893, the Bear River Canal company was put in receivership, and in the same year, the Jarvis-Conklin Mortgage Trust Company went bankrupt. The partly finished canal system was reorganized in September 1894 as the Bear River Irrigation and Ogden Water Works Company, with W. H. Rowe of Salt Lake City as president. Litigation continued, and the new company was forced to sell out to the Utah-Idaho Sugar Company for $300,000 for the canal system and $150,000 for the land. In 1902, under the Utah-Idaho Sugar Company, 45,000 acres were irrigated and cultivated with a possibility of developing 100,000 acres later.

As part of the Bear River Canal project, the Bear River Irrigation and Ogden

Water Works Company purchased the Ogden City Water Works with the intention of selling water rights. This created fear in Ogden that, contrary to the public interest, city water would be owned by speculators. In response to its citizens' concerns, a newly elected city government in 1910 bonded for $100,000 and repurchased the water works, resolving to learn a lesson and not let control of city water pass into speculative hands again.

The Bear River Canal was Utah's first and last major project carried out under the speculative system created by the law of 1880. George Thomas wrote that it was "the first large commercial irrigation undertaking in Utah and one of the first in the West. It is typical of many irrigation schemes in the arid west, and shows how financially disastrous an irrigation scheme may be [when] uncontrolled by the State or the Federal Government."[6]

Some modern historians, including Thomas Alexander, are not so critical of the 1880 law, despite some of its failures. Alexander sees it as "evidence of the demise of the cooperative commonwealth characteristics of early Utah and the subsequent movement to a more capitalistic economy."[7] This change allowed water to be used for purposes other than agriculture, such as mining and manufacturing.

After this experience, Utah water laws enacted from 1897 through 1919 moved toward public control of water by the state and implementation of federal Bureau of Reclamation programs. In 1897 the water laws of 1880 were replaced by two new laws. The first addressed the acquisition of water rights, and the other created the office of the state engineer and defined its duties. These were the first water laws passed under the Utah State Constitution of 1896, which in Article XVII dealt rather succinctly with the water question by stating that "all existing rights to the use of any of the waters of this State for any useful or beneficial purpose are hereby recognized and confirmed."

The main contribution of these laws was the establishment of the office of the state engineer, whose primary duties were to inspect dams and to "keep a full and complete record of all measurements of streams, and all other valuable information in relation to irrigation matters that may come to his knowledge." The law promised greater public control over the state's water by means of its measurement, but lack of appropriations made the position of state engineer almost useless. The first to serve in the office was Robert C. Gemmell.

The 1897 laws abandoned the distinctive features of early Utah irrigation law to conform more closely to practices common to the other arid states. The first law provided that rights to the use of the unappropriated waters of the state could be acquired by appropriation if the appropriator posted and filed a notice of the intended diversion. It further provided that a person who had acquired rights before the passage of the act could record a declaration of those rights, but that failure to file such a declaration would not cause forfeiture of the rights. Because most of the state's water had already been appropriated, these provisions proved to have little practical use. They did, however, codify customs already established in the territory. The new act also repealed the law of 1865 regarding organization of irrigation water districts, and such a law was not renewed until 1909.

In 1901 a proposed comprehensive system for Utah's water failed to pass the legislature, but a portion of the proposal granting increased power to the state engineer went into effect. It gave the engineer general supervision of the state's water and of officers connected with its distribution as well as authorization to measure streams, make surveys, and collect data on canals and reservoirs. Irrigated lands were also to be measured, and maps of the streams and diverting canals were to be made and kept in the office of the state engineer. County commissioners were directed to create one or more water districts in each county, to appoint a water commissioner for each district who would act in accordance with the prior rights of users, and to employ all necessary steps to conserve as far as possible the natural supply of water. The water commissioners in performing their duties were subject to the control and direction of the state engineer.

The functioning of the commissioners and the state engineer was still hampered to a great degree by a lack of funds to carry out their assigned duties. In addition, proposals concerning adjudication and recording of rights failed to pass with the law, which meant that water rights were not clearly defined until they were contested. As a result a new law was considered in 1903 to take care of these problems.[8]

On June 17, 1902, Congress passed the National Reclamation, or Newlands, Act. This act had been under consideration for several years, and western leaders had pushed hard for its passage. One of the main lobbying groups was the National Irrigation Congress, which held its first meeting in Salt Lake City, September 15–17, 1891, and met annually thereafter to prepare policies and memorialize Congress for water programs beneficial to the western states. The 1902 act created the Reclamation Service within the United States Geological Survey. Within seven years the Reclamation Service became an independent bureau of the Department of the Interior, and in 1923 it became the Bureau of Reclamation.

The National Reclamation Act also authorized the secretary of interior to construct irrigation projects in the sixteen western states and territories and to pay for these projects from the proceeds of public land sales in the West. Water users would be required to repay project construction costs within ten years, with the repayments going into a revolving reclamation fund to support other projects. Settlers could obtain 160 acres of unclaimed land within the project areas. One owner could only receive water for 160 acres. This was to prevent development of large tracts by single owners. Several projects were considered immediately for federal reclamation funding. Utah's first was the Strawberry Valley Project, approved for funding in 1906. The Bureau of Reclamation became the financier of all of Utah's subsequent major projects.[9]

The federal act had a great impact on the Utah water laws that followed. The Utah law of 1903 was much influenced by it. In fact, after some of the main provisions of the 1901 Utah law were defeated, officials of the Geological Survey made it clear "that reformation of water-right laws was a prerequisite to federal reclamation." Since Utah was interested in developing several projects with federal support, Governor Heber M. Wells urged the 1903 legislature to revise the state's water laws in line with the recommendations of the state engineer. Governor Wells said, "we cannot too soon place our State in a position to realize the benefits of national reclamation." Encouraged by this message, the legislature passed the Utah Water Code of 1903. In accordance with the Wyoming system of acquiring water rights, this law emphasized the relationship of the appropriator to the welfare of the state. The interest of the state or the community came first, that of the individual irrigator second. The law's prior appropriation concept of water rights also followed the Colorado system by declaring the waters of the state "to be the property of the public."

One of the weaknesses of the old law of 1880 was that water rights were never settled, since anyone could divert water from a stream until challenged by another claimant. The 1903 law remedied this by placing the water of the state under the control of the state engineer and "providing that rights shall be acquired by grant from the engineer." This meant that no one could go to a stream and divert water at will; rather, anyone wishing to acquire a right had to apply for and receive a permit from the state engineer before commencing work or taking water. "In other words the new law returns to the principle of the first law, and makes the State owner of the water instead of possessed of police power only."[10]

The 1903 law specified the precise steps to file and record claims. Water rights still could be transferred separately from land, but transfers had to be by recorded deed or the rights would be void against those of any subsequent purchaser who did record a deed. "The record of water rights will then be complete—the court will define existing rights, all transfers will be of record, and certificates of all new rights will be issued and recorded."

Another section of the law redefined the powers and duties of the state engineer to include general supervision over the public waters of the state, their measurements, apportionments, and appropriation; power to make the necessary rules and regulations pertaining to the public waters; and a mandate to complete a "hydrographic survey of each river system and water source of the State, beginning with the streams and source most in use." After completing the survey, the state engineer was to file a report with the clerk of the state district court in the district where the water source was located, and then the court had "exclusive jurisdiction in the determination of all water rights on the stream or other source."

The state engineer also was to separate the state into water divisions with superintendents appointed and paid by the state to oversee their operations. The divisions would be partitioned further into subdivisions with supervisors paid by the counties and water users to "divide the water among the several canals from the natural streams and to distribute the flow among the users of the canals."

Another important aspect of this law, which was at first peculiar to Utah, was that it defined all water rights in terms of cubic feet per second, rather than in fractional parts of the whole supply, but retained a provision for limiting the amount of time the water could be used. The granting of rights to continuous flow or to a guaranteed amount of volume which may be used as wanted has been the source of more injustice and controversy in the arid states than any other one cause, and Utah avoided most of this controversy by providing for time limits and volume amounts in the water rights.

The 1903 water code was the first comprehensive water law for Utah, but it was not fulfilled because of slow action in completing the process it mandated. The state engineer carried out his hydrographic surveys and reported them to the courts, "but nothing happened." The courts failed to follow up and continue the adjudication.

> In 1914, eleven years after passage of the act, the state engineer complained, "Over $75,000 has been spent in hydrographic survey on the Weber, Logan, Virgin, San Rafael, and Sevier river systems. The hydrographic survey of the Weber River system has been in the court since 1908, the Logan river system since 1912. So far not a single water right has been adjudicated as a result of these surveys."[11]

What had caused the law not to be executed? Robert Dunbar concludes that

> the Utah farmers were relatively satisfied with the definition of rights by the Mormon system, which had been in existence for half a century before the passage of the 1903 act. Under the supervision of the church, rivers and creeks had been apportioned by arbitrators, conferences, agreements, and court decisions. As a consequence, the average farmer did not feel the need for the imported institution.

Moreover, "the legislature failed to appropriate money to fund the adjudication—to pay for the services of the judges, referees, and stenographers—and the average farmer did not wish to pay for something he believed he did not need."[12]

There were few additional changes in Utah water laws until 1919. At that time, after some prompting from the National Irrigation Congress and observation of the success of Oregon's laws, Utah revamped the adjudication procedures of the water law. The law of 1919 had three main components: a new means of determining water rights, a reestablished provision for irrigation districts, which would operate under state supervision, and clauses concerning drainage water.

The new law departed from the former in regards to water rights by following closely the Oregon system of adjudication. It placed authority for both preliminary investigation and initial determination of water rights under the state engineer, transferring power to that office from the courts. The state engineer was again empowered to study and report claims to water rights, but now would also determine water allotments and award new rights to any unappropriated water. Water users had the right to appeal decisions to the state courts. Decisions on rights were still to be based on priority of appropriation, applicants had to go through the same kind of filing and recording procedures as before, and water rights were still recognized as private property. However, the water running in "well known and defined channels is the property of the public," subject to beneficial use, and companies could obtain property by eminent domain to establish canals needed to benefit the public interest.

The law of 1919 reaffirmed the role of irrigation districts and continued the broad participation of the citizenry: "On the petition of fifty or the majority of landowners or upon the request of the Governor of the State the County Commissioners may take the necessary steps to organize an irrigation district." On the receipt of such a petition, county commissioners were to make a survey and allot the water of a district: "The survey is made for the purpose of determining and allotting the maximum amounts of water which could be beneficially used upon such lands; each forty acre or smaller tracts in separate ownership shall be separately surveyed and the allotment made therefore." When lands were listed and plotted, the county commission would call an election at which the landowners would decide if they desired a district and select directors to begin organizing and supervising the functions of the district.

The directors had the power to lease or rent the water not needed by the users of the district. They could with the approval of the landowners build new systems, purchase a system already in existence, or enter into contracts for water with the United States government. Under the law Utah irrigation districts entered into several such contracts with the United States government and repaid costs of the projects. Over the years one of the weaknesses of the irrigation districts was that farmers and other recognized water users alone assumed project costs whereas the entire population received benefits without contributing to the repayment. This situation led to the invention of the water conservancy district, a unit derived from the experiences of the states of New Mexico and Colorado. Utah followed these states by passing a water conservancy act in 1941. Under this law, the property of all citizens is taxed to pay for water projects of the district.[13] The new organizational superstructure of the water conservancy district became the contracting agent for the Weber Basin Project.

Adjudication of water rights went slowly under the law of 1919, as it had under earlier laws, but it was a process that had to be accomplished, and persistent efforts finally completed the task on the Weber River. Adjudication was necessary before any project such as the Weber Basin Project could be contracted. Difficulties in determining rights to the Weber River had been preceded by similar problems on the Ogden River. Ogden River claims began with those of Amos Stone in 1848. Claims by individuals and irrigation companies increased until the Ogden suffered its first water shortage in 1878 "when the Plain city Irrigation Company connected the Ogden River with the Weber river by means of a canal and supplied themselves with more water." As time went on the increasing needs for water due to the expansion of cultivated land and a growing population created more dissension on water rights because no claims had been filed and recorded.

In 1889, William Geddes and others representing the Plain City Irrigation Company sued the Mound Fort Water Association, Slaterville Irrigation Company, Harrisville Irrigation Company, Lynn Irrigation Company, North Bench Canal, Lindsey and Shaw, Eden Irrigation Company, Ambrose Shaw, Joseph A. Taylor, and the North Ogden Irrigation Company. This case became complicated by exaggerated claims for water and spurious dates of filing claims. However, the court issued the Geddes Decree on September 10, 1892, which defined the water rights and fixed the dates of filing claims and the amount of water and acreage irrigated.

This lawsuit was the "forerunner of numerous other cases as a result of this indiscriminate filing of water."[14] One of the main cases was between the Plain City Irrigation Company against the Eden Irrigation Company. Plain City Irrigation Company had filed claims on the Ogden River three years prior to the Eden Irrigation Company; therefore, when the Eden company constructed a canal and a dam across the North Fork of the Ogden River and put the water on Eden farms, the Plain City company demanded release of the water to them, which was refused. Therefore, the Western Irrigation Company and the Plain City Irrigation Company took action to forcibly tear out the dam, which brought on the lawsuit. The decision of the court led to the measurement of the Ogden River at the North, Middle, and South Forks. The Eden Irrigation Company convinced the district court and later the Utah Supreme Court that water from their canal and lands seeped back into the Ogden River and, therefore, their use was not diminishing water for the claims of the Plain City and Western companies. The courts issued a decree in favor of the Eden company, "providing however, that if the water at the inter-section of the forks was not equal to that taken in at the Eden dam, that the Eden Company should provide some channel other than the North Fork to convey the water into Ogden River proper."

After this decision, problems over water claims diminished for a brief period, "as though the matter had been disposed of once and for all," and people went on living as usual. Newcomers arrived in the Ogden River Basin; new lands were cultivated; ditches were lengthened; and the water was spread thinner and thinner. "And so ten more years passed without a violent outbreak of growing pains in the realm of water."[15]

As a result of the litigation, it became evident that claims on the Ogden needed to be filed and recorded, and beginning in 1905, 148 claims for 2,410.17 second feet of water to irrigate 69,674 acres of land were filed. As a result, by 1918 the best distribution of water was made "that has ever been made" in the system. This was short lived, however, because more growth created more demands for water out of the Ogden River.

In 1921 the same kind of process of defining, filing, and recording claims to the Weber River began. It was generated by the case in the Second District Court of *Plain City Irrigation Company, a Corporation, Plaintiff v. Hooper Irrigation Company, a Corporation, North Ogden Irrigation, a Corporation, the State of Utah, and all other Claimants to the right to the use of the water of the Weber River System, Defendants.* This case derived from Chapter 67 of the *Laws of Utah of 1919,* which gave the state engineer authority to determine water rights on the Weber River. The law and the state engineer called for water rights to be "filed with the court in this action" with "statements giving the names and addresses of all the claimants to the use of water from the river system involved in this action." The action had begun with a petition to the state engineer from A. P. Bigelow, D. D. Harris, Levi Pearson, A. V. Johnson, and Lewis Spaulding, who made up a general committee representing all the water users of the Weber River System, "praying that a final judgment and decree be entered herein, as set out in said petition, and upon the stipulation of all the respective claimants that final judgment and decree be entered in accordance with said petition." P. H. Neeley acted as the attorney for the users on the Upper Weber River Division, E. R. Callister acted for the Central Weber River Division, and W. H. Reeder, Jr., and J. A. Howell represented the Lower Weber River Division.

The court action began on January 18, 1921, but continued for years afterwards, indicating the complexity of determining water rights to the Weber. On August 28, 1923, the state engineer filed with the court all the names and statements of claimants, and they became defendants in the action. The engineer made efforts until December 21, 1923, to contact any other water right holders who had not responded. After the filings were recorded, the state engineer surveyed the claims on paper, personally examined the water systems, "formulated a proposed determination of all rights to the use of the water," mailed copies to claimants to which they could respond if they disagreed, and began distributing the water in accordance with the proposed determination. Objections were heard in the county courthouses of Summit, Morgan, and Weber counties on May 28, 1931. Committees were appointed to carry out the determinations as fairly as possible. One committee served for the portion of the Weber River and its tributaries lying above the Echo Reservoir, a second committee for the Weber River and its tributaries below Echo Reservoir and above Gateway, and a third covered the remaining portion of the river system, exclusive of the Ogden River. A fourth, general committee addressed the entire Weber River. The division committees made recommendations, and the general committee approved a schedule of revisions.

The main revisions concerned return flow and needs for more data and a trial period and recommended that a definition be made of the

> relationship between high and low water rights under varying priorities and in order to conform to best and most satisfactory practice, water covered by rights subsequent to March 1903 should be cut off before the supply to rights with earlier priorities than 1903 is diminished; flood water rights with earlier priorities than 1903 should be shut off in order of priority until no flood water rights are being used.

The trial was to end on May 23, 1933, but this was extended first to 1935 and then until the court issued Judge Lester A. Wade's final judgment on June 2, 1937. This decree was significant as it tabulated all the water claims on the Weber River, amounting to 1,013 claims from 1850 to 1937. It specified the amounts of allotted water, measured in "second feet," and required that limitations on use be designated as "flood water rights, high water rights, and low water rights." It also provided for four general classes of water: class 1 included all rights with priorities up to and including the year 1875; class 2, all rights with priorities from 1876 to 1890 inclusive; class 3, all rights with priorities from 1891 to March 11, 1903, inclusive; and class 4, all rights that were initiated by application to the office of the state engineer and whose priorities were subsequent to March 11, 1903. This adjudication brought some litigation in response to its determinations, but overall it was a significant step forward in settling problems on the Weber River—a goal that had been pursued for many years.[16] The decision had a significant impact on later developments. Without it, future projects would not have started.

There were some bitter feelings over the determinations, but such feelings were neither new nor peculiar to the Weber Basin. The words river and rival have common roots; rival derives from the Latin *rivalis*, "one who uses the same brook as another," which suggests the long history of water conflicts. As in other times and places, water troubles along the Weber River continued through the years. They would only subside when and if water became plentiful. As Donald D. McKay concluded, "the only solution is an abundance of water."[17]

The most serious trouble was a 1936 dispute over water in a Kanesville ditch of the Davis and Weber Counties Canal, which led to the killing of one of the disputants. On May 19, 1936, David H. Johnson shot his neighbor John Kap, Sr., in the chest as Kap attempted to take water for his 6:00 P.M. turn from water Johnson was using. Johnson had disputed Kap's water rights for several years, but even though the local watermaster, Charles Thompson, had verified that Kap had shares in the ditch, Johnson shot his neighbor. In the trial Johnson tried to prove that Kap had fired the first shot, but the jury thought otherwise and found him guilty of second degree murder. He was sentenced to serve the rest of his natural life in the state prison. His defense had been complicated by earlier troubles over water. On June 2, 1929, he had been arrested on a charge of assault with a deadly weapon against O. D. Johnson, and in May of 1932, he had been arrested upon complaint of Chris Bouwhuis for carrying a concealed weapon.[18] These instances showed the extremes to which conflicts over water could be carried. The 1937 adjudication of water rights addressed a long-term need to settle such disputes.

THE DAVIS AND WEBER CANAL AND THE EAST CANYON DAM

In the period from 1880 to 1940, several new organizations formed and new canals, dams, and reservoirs were built to meet growing demands on the Weber Basin water supply from increasing population, industry, and agriculture and due to settlement of claims for water. Water companies and ditch companies were organized for almost every portion of the river system. The Weber River Water Users Association and the Ogden River Water Users Association also came into being in the latter part of this period. After 1941, water conservancy districts organized to participate in federal projects.

The major projects undertaken by water organizations included the Davis and Weber Canal, East Canyon Dam, Pioneer Electric Power Company, Bonneville Canal Company, Echo Dam, and Pineview Dam. The initial drive for these projects came from local rather than federal sources. This type of democratic involvement in the Weber Basin differed greatly from the historical model of centrally controlled water development that Karl Wittfogel portrayed in his *Oriental Despotism.*

The Davis and Weber Counties Canal Company undertook one of the earliest and most significant projects in the period. For a long time, as noted in an earlier chapter, interest in bringing water from the Weber River into Davis County had been strong. In 1856, Brigham Young had proposed a canal along the bench line from the Weber River to Bountiful, but early efforts to dig it failed because of the difficulty of tunneling through the sand ridge in north Davis County.

In 1869, the Central Canal Company was established to bring water to a tract of land southeast of Ogden City and into northern Davis County. This company failed to obtain sufficient financial support to complete the canal. On December 29, 1884, local leaders formed the new Davis and Weber Canal Company. The prime organizers were Feramorz Little, William R. Smith, William Jennings, Anson Call, and William H. Hooper. The capital stock originally consisted of three thousand shares at $50 per share.

The company built a canal that started in the Weber River "about four-fifths of a mile from its mouth and one and one-half miles below the Devil's Gate" and extended down the south ridge of the Weber River with a channel three feet deep and twenty feet wide at the bottom. It reached the top of the bench at a distance of nine miles from its head and there descended onto the foothills of the bench and divided into three branches: "One branch, three feet deep and twelve feet wide at the bottom, takes the water to Kaysville; another three feet deep and ten feet wide, conveys water to Ogden; the third, three feet deep and eight feet wide, conveys water to Hooperville."[19]

The canal was dug mostly with "slip scrapers," which were pieces of metal, each with two handles, that were pulled by horse teams and could move about a quarter of a yard of dirt each load. The loads were dumped to build up the lower bank. Water was run in a ditch behind a scraper, and the fall of the ditch was determined by how the water ran in the newly dug portion. If the water did not run, the workers took a "clean shave" off the bottom with the scraper until the water continued along. The Davis and Weber Canal cut through yellow clay that had to be trampled solid to prevent it from washing down the canal and depositing an infertile layer on fields. "This canal was beneficial to the communities of northern Davis County and eventually went all the way to Kaysville, some 23 miles in length, but it had serious problems because it could not deliver the stream of water throughout the entire growing season."[20]

Since the canal company was one of the late appropriators of Weber River water, it was entitled to take water only when there was a surplus in the river. Around July 1 of each year, when flow diminished and primary claims on the river had to be met, the company was compelled to close its head gates and not use any water for the remainder of the season. Grain, hay, and sometimes a second crop of alfalfa could be raised with early irrigation, but not orchard trees or late vegetables with any assurance of success.

Some ten years after the canal was started, the Davis and Weber Counties Canal Company, under the leadership of early officials and directors such as E. P. Ellison, John R. Barnes, J. C. Nye, James G. Wood, John Flint, A. W. Agee, C. A. Rundquist, W. J. Parker, Thomas J. Steed, and Heber Hodson, began to develop a system of storage reservoirs upstream to provide water for the later part of the growing season. In the summer of 1894, an excellent site for a dam was identified twelve miles south of Morgan in the Narrows, or Red Rock Gorge, of East Canyon Creek, the largest tributary of the upper Weber River, and two years later, work on the first structure began. There would be five major construction efforts on this dam through the years. In 1899, the first dam was completed; in 1900, the dam was raised twenty-five feet; in 1902, another seventeen feet were added; and in 1916, an arched, reinforced-concrete dam was completed below the original dam to further increase capacity. Finally, in 1966, a new dam, the present East Canyon Dam, was constructed with federal money.

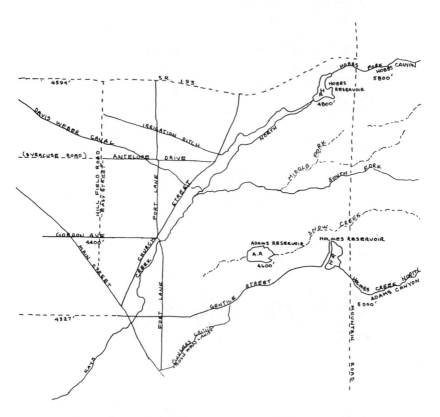

Modern-day map of Layton area showing the major streams and reservoirs—Hobbs, Adams, and Holmes reservoirs. (From Dan Carlsruh and Eve Carlsruh, eds., *Layton, Utah: Historic Viewpoints* [Salt Lake City: Moench Printing, 1985].)

Aerial view looking northwest from Bountiful towards Centerville and
Farmington, Davis County. Photo shows some of the farms watered by
mountain streams prior to the Weber Basin Project. In the distance is the
Farmington Bay of the Great Salt Lake. (1950 photograph by R. J.
Brown. Courtesy United States Bureau of Reclamation.)

1986 photograph of Adams Company Pond, which claims to be the
oldest reservoir in Utah. Built in 1852 by Elias Adams and sons. (Photograph courtesy Dana Love.)

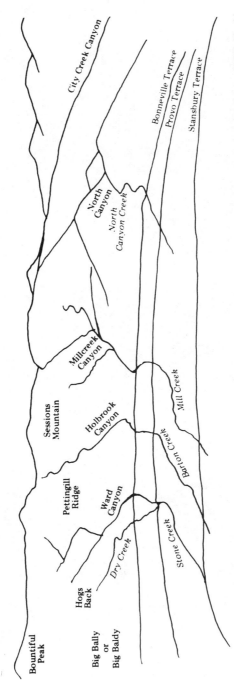

Drawn map showing the canyons and streams of the Bountiful area. (From Leslie T. Foy, *The City Bountiful* [Bountiful: Horizon Publishers, 1975]. Used by permission of Horizon Publishers.)

Hobbs Reservoir in East Layton. (1949 photograph by S. J. Baker. Courtesy United States Bureau of Reclamation.)

Farmington Creek measuring flume, looking downstream from wooden bridge at head of flume. (1949 photograph by S. J. Baker. Courtesy United States Bureau of Reclamation.)

A section of the North Cottonwood Irrigation
Company ditch in Farmington, Davis County, be-
fore the construction of the West Farmington
Trunkline of the Weber Basin Project. (Photo-
graph by D. A. Ellsworth. Courtesy United States
Bureau of Reclamation.)

Diversion structure on Centerville Creek before the construction of
the Weber Basin Project. Here water could be diverted into various
water lines to meet irrigation schedules. (1949 photograph by S. J.
Baker. Courtesy United States Bureau of Reclamation.)

Headgate at the mouth of Barton Creek, which diverted the stream for culinary water of Bountiful City and irrigation for Barton Creek Irrigation Company, as it was prior to the completion of the Davis Aqueduct of the Weber Basin Project. (1960 photograph by D. A. Ellsworth. Courtesy United States Bureau of Reclamation.)

Onion field in Centerville, Davis County, watered by Deuel Creek Irrigation Company water and artesian well water. (Circa 1930 photograph from the Utah State Historical Society.)

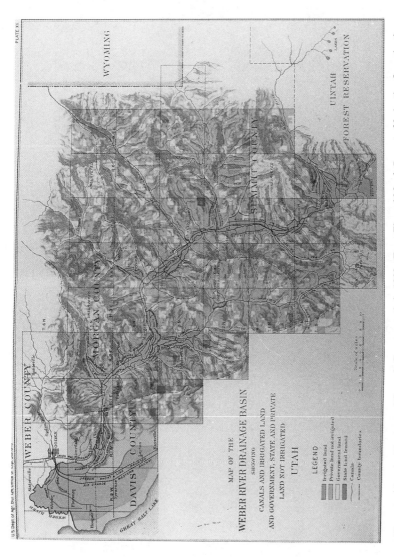

Maps showing the Morgan and Summit County canals in 1903. (From Elwood Mead, *Report of Irrigation Investigations in Utah* [Washington: Government Printing Office, 1903].)

Late-nineteenth-century photo showing Devil's Gate near the mouth of the canyon where the Weber River enters into the valley and makes its way to the Great Salt Lake. This U-bend in the river was a difficult place in the canyon to pass in early days, and was only made efficient for travelling in recent years when Interstate Highway 84 was constructed to by-pass the bend. At this place the Weber River debouches an average of 100,000 acre-feet of water to Weber and Davis counties. (Photograph from the Utah State Historical Society.)

View looking north and northwest across Morgan Valley from the city ceme- tery. The Weber River flowing in the area of the treeline at the foot of the hills in the background provided the resources necessary to water this rather large valley along the river. Mount Ogden is seen in the right background. (1948 photograph by F. M. Warnick. Courtesy United States Bureau of Reclamation.)

Irrigated valley of alfalfa below Echo Dam, Summit County, along Weber River. A part of the area serviced by the Weber River Basin Project. (1945 photograph by P. E. Norine. Courtesy United States Bureau of Reclamation.)

Looking north towards Peoa along Beaver Creek, a tributary of the Weber River. This is a rocky, seeped pasture land in the bottoms, used mainly for pasture or meadow hay crops. (1951 photograph by H. E. Nielsen. Courtesy United States Bureau of Reclamation.)

Weber River drainage above Oakley, Utah, at an elevation of about 7,000 feet. (1950 photograph by W. C. Goines. Courtesy United States Bureau of Reclamation.)

Smith and Morehouse Reservoir as it was in 1950 before the construction expansion of the Weber Basin Project in 1989. This reservoir provided much of the needed water for the farms and ranches in Summit County. (1950 photograph by W. C. Goines. Courtesy United States Bureau of Reclamation.)

View looking southeast toward the main channel of the Weber River flowing from the junction of Gardner and Middle creeks, which are the main headwater tributaries that form the Weber River. (1947 photograph by C. D. Gessel. Courtesy United States Bureau of Reclamation.)

Section of the Ogden-Brigham City Canal of the Ogden River Project, which later became part of the Weber Basin Project. Photo shows wooden check and turnout structure in the canal. (1956 photograph by William Maxey, Jr. Courtesy United States Bureau of Reclamation.)

Board of Directors of the Davis-Weber Canal Company who developed the Davis-Weber Canal and the East Canyon Dam to water the farmlands of Davis and Weber counties. *Left to right:* E. O. Larsen, John Maw, E. P. Ellison, J. R. Alexander, A. P. Bigelow, and J. G. M. Barnes. (Photograph courtesy Davis-Weber Canal Company.)

View of construction of concrete lining of the Davis-Weber Canal in 1911 in south Weber County. (Photograph from the 1911 Engineer's Report, Davis-Weber Canal Company.)

Looking north toward Brigham City from high point on north side of Perry Canyon. The contrast between irrigated and nonirrigated lands is clearly shown. The Ogden-Brigham Canal separates the two areas. Wellsville Peak is in the background. (1948 photograph by F. H. Anderson. Courtesy United States Bureau of Reclamation.)

Diversion point of the Davis-Weber Canal from the Weber River near the
mouth of the canyon. (1965 photograph courtesy Davis-Weber Canal Company.)

Bonneville Canal pump lift, showing the Lower Bonneville Canal in the foreground and the Jordan Canal and pump house in the background. The Bonneville Canal ran from this site in North Salt Lake in two branches—one along
the east bench of Bountiful to Pages Lane in Centerville and a lower branch
running parallel to Highway 89 to West Bountiful (1949 photograph by S. J.
Baker. Courtesy United States Bureau of Reclamation.)

1899 photograph of East Canyon Dam built in 1896, a rock and ground fill dam with wooden spillway. (Photograph courtesy Davis-Weber Canal Company.)

East Canyon concrete dam and reservoir built in 1915 as viewed at point downstream from right abutment. (1948 photograph by F. M. Warnick. Courtesy United States Bureau of Reclamation.)

View of East Canyon Dam taken from upstream right abutment. (1950 photograph by G. K. Wallace. Courtesy United States Bureau of Reclamation.)

Aerial view of East Canyon Dam and Reservoir as it existed in 1958 before the enlargement project of the Weber Basin Project. (1958 photograph by Stan Rasmussen. Courtesy United States Bureau of Reclamation.)

The first dam was constructed from 1896 to 1899. The funds to build the structure, which amounted to $50,000, were subscribed by members of the company at $50 a share. When completed in the spring of 1899, the dam rose sixty-eight feet above the bed of the creek, and its top length was 120 feet. The core of the dam was vertical steel plate riveted, caulked, and set firmly in a concrete foundation. The ends of the dam were set one and one-half feet into the rock walls of the canyon, with which they were united by filling the grooves with asphaltum concrete. The faces of the steel plate were also coated with asphaltum concrete four inches thick. The stability of the steel and concrete structure was secured by dry rubble masonry that was ten feet wide on top, with a slope of one to two on the downstream face, and one and one-half to one on the upstream face. A tunnel in the solid rock at the north end of the dam provided the outlet of the reservoir.

The construction contractor was Perham Brothers and Parker of Butte, Montana. Around one hundred local employees dug and moved dirt and fill by wheelbarrow, dirt scrapers, and teams of horses. Others drilled and blasted with explosives in the rock sides. Octave Ursenbach described some of this work:

> The heavy shooting at the early stage of the dam was to build a tremendous background for a steel core that was to go across the canyon. I recall a huge stone weighing perhaps 100 tons, that slid into the slope, lodging in key-like on either side of the canyon. This alone would be strong enough to withstand the water that would be forced against it. . . .
>
> A tunnel was being bored through a huge cliff on the east side of the dam itself, as an outlet for the water through steel-screw gates, and the work was progressing nicely. Excavation for the steel core had been dug to bed-rock, that is where the cliffs on either side met. Here we placed several thousand bags of cement and sand until the bed-rock was raised to water level. This, however, proved to have been perhaps premature, for the water was rising so rapidly and would soon flood the tunnel. The only solution was to do deep water blasting [so] that an outlet [could] be opened on the rubble masonry. This I mention that for seventy-two hours we worked shoulder deep in water in one shift, the longest shift of my experience. It was twenty degrees below zero, and the water was warmer than the air.[21]

The original reservoir area was 226 acres. The capacity of the reservoir was 167 million cubic feet or 3,834 acre feet. The dam was completed in time to store water for the farming season of 1900. On June 23 of that year, the gates of the dam were opened, and the water flowed a distance of about thirty miles down East Canyon Creek and the Weber River to the head gate of the Davis and Weber

Canal. During that first year, the commingling of the reservoir water and the natural flow of the Weber was a source of much contention because of the difficulty in knowing how much reservoir water was lost through seepage and other means in its transit from the dam to the head gate. This question would not be adequately answered until better measurements and studies were carried out.

This private enterprise, organized and led by local water users, had proven a success. The company was able to deliver water to twelve thousand acres of land in Davis and Weber counties, which otherwise would have had limited use. Total valuation of the company, which was estimated in the early 1890s not to have exceeded $250,000, had by 1910 risen to well over $1 million. In the same period stocks in the company had risen from $10 to $12 to $175 and higher a share. In 1903 it was calculated that the $50,000 originally invested had earned a 120 percent return, or $60,000, the first year and that after two years of use the reservoir had paid for itself nearly two and one-half times.

With that kind of success the Davis and Weber company invested again to increase the storage capability of the original dam. The twenty-five-feet increase in the height of the dam in 1900 cost $35,000 and increased the capacity to 387 million cubic feet, enough to supply farmers with a continuous flow of seventy cubic feet per second for sixty days. After the dam was raised again two years later to a total height of 145 feet above bedrock, the reservoir was 100-feet deep and had a capacity of about 600 million cubic feet or 13,800 acre feet. By 1910 this dam provided water to twelve thousand acres of farm land in Davis County alone.

Engineers Samuel Fortier, William M. Bostaph, and A. F. Parker studied the need for further improvements and the feasibility of greater reservoir capacity. Their studies led in 1913 to the enlargement and lining with cement of the main channel of the Davis and Weber Canal and to the building of an electric power plant on the Weber River that took its water from the canal at a point eight miles from the head gate, from where the water ran to the plant in two sixteen-hundred-foot pipes. This was sometimes called the Bostaph Electric Plant after William M. Bostaph, the engineer who proposed it. The studies also led to replacement of the original East Canyon Dam.

An arched, reinforced-concrete dam located just below the old dam was started in May 1915 and completed in the summer of 1916 by the Ogden-based Utah Construction Company from plans prepared by engineers Samuel Fortier and A. F. Parker. The cost was $175,054.74. Construction materials such as cement and steel were shipped by rail to Morgan City and moved by team and wagon to the building site. Aggregate for making concrete was found close by, about one-half mile east of the project. The new structure rose forty-five feet above the top of the old dam and 140 feet above its outlet. It had a capacity of twenty-eight thousand acre feet of water and served well until 1964 when construction on the present dam downstream from it began with Bureau of Reclamation funding.

Successful construction of the various East Canyon dams brought a reliable source of water to southwestern Weber County and northern Davis County and encouraged more irrigation companies to organize to contract for it. Among the most important examples was the contract the Kaysville Irrigation Company and Haight's Creek Irrigation Company negotiated in 1928 to extend the main Davis and Weber Canal from Clearfield to the southeast part of Kaysville, where it met Haight's Creek. This project was completed in 1931, and added nearly two thousand acres of cultivated land to the Layton-Kaysville area. The 1930s would bring even more water to the area from Echo Dam.

The increased supply of water from East Canyon provided the means to expand agriculture for a period, support World War II government facilities, and respond to demands for culinary water from expanding communities in the area.[22] The success of the cooperative effort of Davis and Weber counties in completing East Canyon Dam had a major effect on how counties would later work together in the Weber Basin Project. The East Canyon experience certainly affected the thinking of such people as Joe Johnson, Ezra Fjelsted, and DeLore Nichols, leaders in the Weber Basin Project who proposed the same kind of cooperation.

THE PIONEER ELECTRIC COMPANY AND THE BONNEVILLE IRRIGATION DISTRICT

Two other water projects built with private investments need to be considered before examining Echo Dam, the first major federal reclamation project in the Weber Basin. The two private enterprises were the Pioneer Electric Company plant on the Ogden River and the Bonneville Canal Company in south Davis County. One successfully brought electrical power to the area, but the other failed to satisfy hopes that it would bring a beneficial supply of water to south Davis County.

The Pioneer Electric Power Company, organized in Ogden on November 27, 1893, resulted from the efforts of a group of Weber Valley men led by a "remarkable young engineer," Charles K. Bannister, who realized the great potential for hydroelectric power produced from the water resources of the Wasatch Range. The company planned to dam the Ogden River about ten miles east of Ogden City and thereby supply not only water for households and for irrigation of twenty thousand acres in the northwestern part of Weber Valley but also electric power for Ogden City, for developing factories, and for electric railways in Ogden and Salt Lake City. The Mormon church invested $520,000 out of the $1,000,000 total raised locally for the company, and George Q. Cannon, first counselor in the Mormon church presidency, was appointed president of the company. Pioneer Electric also sought outside investors, and Joseph Bannigan, a

Rhode Island rubber magnate, invested $1,500,000 to give the company a total capitalization of $2,500,000.

A temporary dam, strong enough to allow the generation of electricity, was completed just below Wheeler Canyon by the end of 1896. Completion of the permanent dam, at a cost of $250,000, followed in the spring of 1898. It was 340 feet long from canyon wall to canyon wall, 93 feet thick, and 60 feet high. The Pioneer Power Plant, located downstream near the mouth of Ogden Canyon at 12th Street and Harrison Boulevard, had a ten-thousand-horsepower capacity and was completed in 1897. To get a reliable supply of water from the dam to the power house, a six-mile-long wooden-stave pipe reinforced with bands of steel was laid down the canyon.

"The construction was a major engineering achievement." Machine shops were built to fashion steel for the pipe, which had a diameter of six feet. Special planing mills were erected to produce pipe staves from two million cubic feet of special knot-free Oregon pine shipped from Portland and Seattle. "The laying of the pipeline required cutting through towering cliffs, and one tunnel 600 feet long was cut through solid rock." An electric line was extended thirty-seven miles from Ogden to Salt Lake City; at the time it was the longest power line in the United States. After the completion of the system, the Pioneer Company merged, in August 1897, with the Union Light and Power Company of Salt Lake City, and later this combination became the Utah Light and Power Company, a predecessor to the Utah Power and Light Corporation.[23]

World War I brought temporary economic prosperity to the farmers of the United States, and Utah farmers were no exception. New areas of agriculture were developed. Davis County farmers supplied much of the truck-garden food supply for a growing Salt Lake City population. The Salt Lake Growers Market, stocked mainly with Davis County produce, was an important industry from World War I until after World War II. Getting water to Davis County farms was crucial. In south Davis County, the waters of local canyon streams had been appropriated, and farming could only expand into new areas if additional water became available. To meet the demand for more water, the Bonneville Irrigation District was organized (sometimes also referred to as the Bonneville Canal Company). It proposed to take water by canal from the Jordan River to the Davis and Salt Lake counties boundary and pump it into two canals that would deliver it as far as Pages Lane in Centerville. The proposal had tremendous support. "Only three property owners in the greater Bountiful area voted against the planned canal in January, 1920." The franchise of the Bonneville Irrigation District, which was organized for the purpose of contracting for the canal, was approved on April 13, 1921. Clarence Eldridge, Charles R. Mabey, and Fred Odell served as the first directors of the company; Charles R. Mabey was elected as president; and Israel E. Willey was selected as company attorney.

The state engineer allotted water to the canal company, but these allotments were poorly assigned: "People with patches of oak brush or creek beds on their land had to pay for their unused property. The directors failed to check the allotments and the Bonneville Irrigation District was soon embroiled in law suits."[24] Hard economic times in the 1920s for Davis County farmers also created difficulties. As agricultural prices fell, the farmers could not sell their produce or pay for their water allotments. The landowners of the irrigation district had two debts to pay: the bond issue for construction of the canals and the operation expenses after the project was completed. The first construction bond was for $600,000. The contract was let out to the Shattuck Construction Company of California on a cost-plus-15-percent basis. Local farmer participation in labor could have cut construction costs but was not used. Waste and overhead soon expended the initial $600,000. The district bonded for $125,000 more to pay for the remaining construction.

The project constructed two canals, known as the Upper and Lower canals. The Upper Canal supplied water to Bountiful and to Pages Lane in Centerville. The Lower Canal furnished water to Woods Cross, South Bountiful, and West Bountiful. Two pumps at the present location of the Standard Oil Refinery on the Salt Lake and Davis county border pumped water from a Jordan River canal into two pipes, one of which connected to the Upper Canal and one to the Lower Canal. The Upper Canal ran north along the Bonneville bench to 350 East Center Street in North Salt Lake, near the Odell Reservoir, and then to 2200 South and Penman Lane and 677 East 4th North in Bountiful and along that level to Pages Lane in Centerville, where it ended at a metal water box. The cemented Upper Canal was three feet wide at the bottom, ten feet wide at the top, and four feet deep. Parts of it were piped across roads. The Lower Canal ran north below the bench from about 952 West 4400 South in North Salt Lake to 425 West 5th South in Bountiful, ending nearby. It was an open ditch with no cement and only culverts to carry the water under roads.

The Anglo California Bank had purchased the Bonneville bonds. The bank's president, who had previous financial dealings with the Mormon church, assumed that if the Bonneville Irrigation District could not pay off its indebtedness, the church would, but the church played no part in the business. "The people of Bountiful began to worry about the nature of the bond. It was not certain if the bond holders owned a blanket mortgage on the real property of the land owner or not." Newly elected directors under president Samuel Howard adopted the policy that people should not pay the improvement tax until the nature of the bond had been settled in court: "They were advised to keep the new tax money in savings accounts until the question could be resolved."[25] In 1929 the Utah State Supreme Court ruled that the bank did not hold a blanket mortgage on all the 250,000 acres of land of south Davis County but only on those tracts of land that had been allotted water. This provided little protection, however, because most of the southern part of the county had been allotted water.

New company directors were elected, and they formed a settlement committee to work out the financial problems with the bondholder. E. O. Muir was appointed chairman of the committee, and its members were Moses L. Holbrook, Burt Parkin, Leland Sessions, James E. Burns, James Howard, George Salter, John Ledingham, Orson Sessions, Lorus Manwaring, Charles Chrisman, John Barlow, and Thomas Burningham. The indebtedness of the district in 1929 consisted of $750,000 in bonds, $275,000 interest on the bonds, and $121,462.52 from other warrants against the district.

In 1931, the settlement committee went to California to negotiate an adjustment on the payment of the debt. Moses Holbrook was spokesman for the group. The Anglo California Bank, recognizing the depressed economy, agreed to accept 20 percent of the Bonneville Irrigation District's indebtedness and appointed an attorney, Wendell B. Hammond, to make individual settlements with each landowner in the district. Because of the extremely hard times, it was six years before the attorney could complete the settlements. When they were done, Hammond filed suit in district court to clear the liens. The court agreed, and the lengthy financial battles of the Bonneville Irrigation District ended.

Some Bountiful residents, discouraged by the project's financial demands, sold their property to avoid losing everything. Others stayed in the district and faced additional problems. The cost of electric power had to be met before Utah Power and Light Company would turn on the pumps. The water users, therefore, usually paid that cost before anything else so that they could receive water. Sometimes there was not enough money to run the pumps for both canals. De-Lore Nichols, Davis County agricultural agent, and Oscar Wood, south Davis farmer, remembered that "they ran both pumps for a while, but then it became harder and harder to collect the money annually, and it took considerable money to run the top pump, so then they run [*sic*] the lower pump, the lower canal several years afterward, after they discontinued the higher canal."[26] This meant fields higher or farther north did not get the water they needed.

The water from the Jordan River was a problem itself. The murky river became polluted with alkali, moss, and silt as it made its way from Utah Lake to the Great Salt Lake. Oscar Wood said the water "brought ground weeds that we had never seen before. We had Russian thistle and these sand burrs and stuff." The bad effect the water had on the land concerned Bonneville Canal water users, especially those with lower lands that received run off from neighboring lots. The canal water was impregnated with a clay that was called a "light salt." Wood's neighbors on lower ground were concerned "where I would water . . . down below us on the Johnson property where the water, when I watered heavy, would come out on their ground, and it would just be white, white like paper plaque, and so the people were getting anxious to do something. That's what helped us get Weber [Basin Project] in. After a few years the alkali build-up was intolerable."[27]

With its users discouraged by these many problems, the Bonneville Canal quit functioning in the 1940s during the presidential term of Lawrence Moss, a Bountiful farmer. When the water from the Weber Basin Water Conservancy District became available in the 1950s, the stockholders voted to dissolve the Bonneville Irrigation District, replacing bad water with good and ending their disheartening struggle.[28]

THE ECHO DAM

The 1930s saw completion of two significant projects, Echo and Pineview dams, that used federal reclamation funds for the first time in the Weber Basin area. The federal government had taken an interest in western water projects since the Carey Act of 1894, which allocated national lands to the state governments to promote irrigation. With the founding of the Reclamation Service in 1902, the government became even more involved, encouraging the building of projects with federal funding. In 1904, the Reclamation Service created District Four, with jurisdiction over the Bear, Weber, Ogden, and Provo rivers. The bureau's commissioners, in 1904 and 1905, conducted surveys in northern Utah to assess water problems and find reservoir sites. Several reservoir sites were located, but the only construction undertaken was of stream gauging stations by the Geological Survey.

Not to be outdone by the federal government, the leaders of a local organization of Davis County and Weber County officials known as the Weber River Irrigation Project began studying in 1904 possible storage dams and canals on their own. On December 4, 1907, Willard Young and Frank C. Kelsey, civil engineers from Salt Lake City who were consulting with the Weber and Davis County interests, reported on their investigations. They recommended that "a private irrigation company construct the necessary storage dam, canals, and laterals and purchase the land which would be served. The land would then be subdivided into tracts and sold with a water right." Recommended dam sites were at Echo, Rockport, Smith and Morehouse, and Larrabee; dams have since been built at all but Larrabee. These preparations were first referred to as the Salt Lake Basin Project but, because of their emphasis on the Weber, became known as the Weber River Project.

Interest in the project subsided for several years because of an economic recession and then World War I but revived in the 1920s. On January 3, 1922, the Reclamation Service and the Utah Water Storage Commission signed a contract, which was renewed as late as 1929, that provided for federal investigations of irrigation projects in the entire Great Salt Lake Basin of northern Utah. These studies, funded by state and federal funds, produced specific construction proposals, among them Echo Dam. E. P. Larson was the director of Region Four of the Bureau of Reclamation under whom the investigations of the Weber River were carried out. Federal reclamation engineer William M. Green made the official report, dated December 1922, that led to the selection of the Echo Dam and Reservoir site in 1924. The recommended site was located forty-two miles southeast of Ogden on the Weber just below Coalville. The engineers also recommended a diversion near Oakley for a nine-mile canal to transfer water from the Weber to the Provo River and thus to Utah County. As approved by Congress in 1924, Echo Dam was to provide supplemental irrigation water to sixty thousand acres of land in the lower Weber and Ogden Valley and twenty thousand acres in the Provo Valley.

Although the site had been selected, the federal government would not issue contracts until water users organized to subscribe for water and to pay back the project's debts. Canal companies and irrigation, culinary, stock water, and industrial users from several counties had been interested in this project as early as the investigations of 1904. A great deal of effort was expended to unite these water users so as to obtain contracts from the Bureau of Reclamation. The leaders most responsible for the organizing were W. M. R. Wallace of the Bureau of Reclamation; W. W. Armstrong of the Old Copper Bank; Joseph R. Murdock of Provo, who was interested in the canal to divert flood water from the Weber to the Provo River; Richard R. Lyman of Salt Lake City, an apostle of the Mormon church; A. P. Bigelow, a banker from Ogden and long-time advocate of Weber River development; John R. Barnes of Kaysville, a banker and board member of the Davis and Weber Canal; E. P. Ellison of Layton; T. R. Jones, a farmer from Kanesville; P. F. Kirkendall, mayor of Ogden; John Maw, a farmer from Plain City; John T. Bybee, a farmer from Riverdale; Thomas Harding, a farmer from Morgan; and Leroy Peterson and Levi Pearson, farmers from Oakley. The incorporation of the Weber River Water Users' Association on January 15, 1926, was their accomplishment.

The association crossed county boundaries to center on river systems and was a predecessor to the similar water conservancy districts that would be formed under the law of 1941. The water users associations of the Weber River and Ogden River were significant steps in the progression towards the Weber Basin Project. Out of these multi-county cooperative efforts came the impetus and legal capability to support that more expansive project. The incorporators of the association were Bigelow, Murdock, Jones, Ellison, Kirkendall, Bybee, Maw, Harding, and Pearson. On December 16, 1926, Secretary of Interior Hubert Work signed a contract with the Weber River Water Users' Association to construct the dam and canal, and President Calvin Coolidge approved the project on January 8, 1927, under the terms of Section 4 of the Congressional Act of June 25, 1910, and Subsection B, Section 4 of the Act of December 5, 1924.

The United States "agreed to create a reservoir with an estimated capacity of 74,000 acre-feet, to construct a canal with a capacity of 210 second-feet, and to furnish the members of the private corporation with water between April 1 and October 31 of each year." As the other party to the contract, the association

agreed to distribute the water in compliance with the federal reclamation law to individual stockholders in the association (individuals
and canal companies) and to operate and maintain the reservoir at its
own expense after construction, and to pay the United States in annual fixed installments the cost of the facilities and supply. That cost,
not to exceed $3 million, was to be paid in twenty equal annual in-
stallments (the first to be made on December 1 of the year of completion).

Before construction would start, the Bureau of Reclamation required the Weber River Water Users' Association to obtain pledges to buy 80 percent of the
proposed reservoir capacity. By the time the contracts were filed on March 19,
1927, thirty-two canal companies, which had eighty thousand acres of land under
their control, had subscribed to 59,411 shares of stock. The major stockholders in
the Echo project were the Provo Reservation Water Users Company, the Davis
and Weber Counties Canal Company, and the Hooper, North Ogden, Plain City,
Warren, Western, and Wilson irrigation companies.

Construction on Echo Dam began on November 26, 1927, and continued until
its completion on October 7, 1930. The contractor was the A. Guthrie Company
of Portland, Oregon, whose contract covered construction of the dam and a nine-
mile diversion canal connecting the Weber and Provo rivers. The canal was built
with a 210-second-feet capacity, with the expectation that it would be enlarged
later. The United States government retained control of the canal until its enlargement to one thousand cubic feet per second in 1947 as part of the Provo
River Project. Construction also involved removal of portions of the Union Pacific Railroad track and the Lincoln Highway from the Echo Reservoir site.

Echo is an earth- and rock-filled dam with a height of 158 feet and a crest
length of 1,887 feet, including the eighty-feet-wide spillway channel. The zoned,
1,540,000-cubic-yard structure was constructed by sprinkling and rolling eight-
inch layers of clay, sand, and gravel. Both upstream and downstream slopes are
heavily covered by conglomerate rock fill and cobbles. A concrete cut-off wall,
bonded to bedrock, extends the full length of the dam. Flood waters are handled
by a spillway with a capacity of fifteen thousand cubic feet per second, which is
fed by a concrete-lined horseshoe tunnel conduit, with the flow regulated by four
eighteen-by-seventeen-feet radial gates.

The reservoir behind the dam is four and one-half miles long and has a maximum width of one mile. Its surface area is 1,470 acres with an active capacity of
73,900 acre feet. The reservoir provided supplemental water for irrigation of
ninety thousand acres of agricultural land in Morgan, Davis, and Weber counties.

Some water also was used by exchange in Summit, Utah, and Salt Lake counties via the Provo Canal. The subscribers consisted of eighty-five individual water users, fifty-nine irrigation companies, and other organized water users.

In a letter dated June 30, 1931, the Bureau of Reclamation informed the Weber River Water Users' Association that the Echo project was essentially completed and that on July 5, 1931, the association should take over operation and maintenance of the Weber-Provo Diversion Canal and the Echo Reservoir. The total cost of the project was $2,875,871.83. The Weber River Water Users' Association annually paid $88,192.90 (although some payments were deferred during 1932 to 1936) on its debt until June 11, 1966, when the final payment was celebrated with ceremonies at the Echo Dam. Appropriate comments and remarks were made on that occasion by some of the leaders in completing the project, including Charles L. Heslop, president of the Weber River Water Users' Association; J. R. Bingham, director, Utah Water Users' Association; E. O. Larsen, former regional director, Bureau of Reclamation; D. Earl Harris, secretary and manager of the Weber River Water Users' Association; and Ralph A. Richards, director of the association.

Except for the early years, 1931–32, when a severe drought cut water resources and the full water supply could not be delivered, the Echo Dam has been very valuable. Its supply of water helped stimulate a great amount of agricultural, industrial, and urban expansion. The estimated value of crops watered by Echo Reservoir in 1961 was $9.5 million, and the project by 1971 provided supplemental irrigation water to 109,000 acres on 3,900 farms. The Weber River Project was basically a single-purpose facility, and only in recent times has its development had any direct relationship to the extensive Weber Basin Project. As Steven Merrill concluded,

> even though the identity and significance of the Weber River Project seemed diminished beside this vast new federal enterprise [Weber Basin Project] it is undoubtedly true that the project is important not simply for its benefits to agriculture, nor merely for its initiation of federal reclamation work in Northern Utah, but also certainly for contributions to the entire development of the Weber Valley.[29]

THE PINEVIEW DAM

There was also a great deal of interest in the early 1900s in developing additional water from the Ogden River. As has been noted, there was, as early as 1878, insufficient water to meet claims on the Ogden. This had brought on the series of legal battles already described and had led to the organization of water associations and to investigations that tried to find solutions to the water shortage. On

July 31, 1915, a meeting of agricultural leaders from Weber County was held in the courthouse in Ogden. Called by the Ogden Valley Commercial Club, which had been promoting agriculture in the area, the meeting's purpose was to "represent all rural interests so that a better working relationship might exist between canners, packers, businessmen, manufacturers and consumers in the County with the farmers." A committee made up of J. L. Robson of Plain City, Thomas Fowles of Hooper, R. T. Rhees of Pleasant View, and Chairman D. D. McKay from Huntsville organized, on August 31, 1915, the Weber County Farm Bureau. One of the main problems dealt with by the organization was the development of the Ogden River.

The interest this organization of widely varied members took in the area's water supply indicates how reliant the entire community was on the resource. Water dictated the growth and direction that the Weber County communities would have, and businessmen as well as farmers shaped its development. Interested parties from south Box Elder County also began to negotiate for Ogden River water, and the concerns of a single county grew to those of several counties and eventually coalesced in the Weber Basin Project. This was a movement of grass roots democracy. It was not a "hydraulic empire" of the kind Karl Wittfogel traced in Asia but was an inland empire representative of the Rocky Mountain West, perhaps the strongest and most extensive in Utah, including as it did Weber, Davis, Morgan, Summit, and Box Elder counties. Looked at in these terms, what was happening in the Weber Basin surpassed anything that developed in Salt Lake County or Utah County during the same period.

The Weber County Farm Bureau carried out investigations and surveys and formulated proposals. Especially active in these functions were Robson, who is credited with doing most of the work; Rhees of Pleasant View; M. P. Brown of Roy; J. R. Beus, Hooper; Norton Bowns, Slaterville; T. R. Jones, Kanesville; and H. B. Stallings, Eden. As investigative work proceeded, O. W. Israelson, professor of irrigation and drainage at Utah State Agricultural College; L. M. Winsor, engineer in the United States Department of Agriculture; and W. P. Thomas, county agriculture agent, acted as consultants. Later, as plans for building a dam were considered, federal and other advisors and engineers were consulted. They included A. F. Parker, engineer and builder of the East Canyon Dam and of irrigation projects in Wyoming; Dr. Samuel Fortier of the United States Department of Agriculture; William S. Green of the Bureau of Reclamation; and C. O. Roskelly, S. G. Margets, W. M. Bostaph, W. W. McLaughlin, Brice McBride, and William Peterson.

In 1919 the plans of the "Weber County Irrigation Committee" were endorsed by "about twenty large canal companies." Donald D. McKay was elected permanent chairman and Angus T. Wright secretary-treasurer. The plans included finding prospective dam sites and reservoirs on the Weber and Ogden rivers, possibly securing water from the Bear River, developing underground water, evaluating the water available and water needed for undeveloped lands, and establishing the types of organization required to support a movement for water storage dams.

Several investigations were carried out to find dam sites on the Ogden. Most notable were reports by engineers Fortier and Green. In a report on February 7, 1917, Fortier pointed out to Mayor A. R. Heywood of Ogden City that an investigation in 1913 had shown that a proposed site at Cobble Creek was unacceptable. He suggested instead that a dam holding five thousand acre feet of storage water could be built on the South Fork of the Ogden River, "a short distance below the junction of Skull Crack Creek," at an economical cost. This would later be the site of Causey Dam. In cooperation with the Utah Water Storage Commission, Green investigated several sites. In 1922, 1923, and 1924, he gave thorough consideration to the Magpie site on the South Fork above Huntsville.

Efforts to develop a dam site, however, died down for a time because of the court fight over upper and lower Ogden River claims, the lack of a strong organization to carry out the project, and a depressed local economy, which reduced the possibility of funding a project. In the early 1930s, though, a series of events finally led to the construction of the Pineview Dam. In 1931 and 1932 the Bureau of Reclamation investigations that had continued intermittently during the 1920s focused on the so-called Huntsville dam and reservoir site, from which a supplemental water supply could be furnished for Weber County lands and lands as far north as Brigham City. The Bureau of Reclamation engineers decided that here, about one thousand feet upstream from the Pioneer Power Dam, a new dam would provide "lower cost water for irrigation" and a greater supply than any other site previously considered. This decision overruled the Magpie site, even though, in 1928, subscribers for water had been solicited for a proposed dam there. The Magpie site would have required a 250-feet-high dam to create a capacity of 45,000 acre feet. As funds became available, they were directed at studies of the Huntsville Dam, the present Pineview Dam.

Although the site had been determined, something more was needed to get the project moving. D. D. McKay wrote that

> in 1934, the County was in the midst of a serious depression, jobs were scarce and labor was cheap. The United States Government was supplying money liberally for make-work projects. The power company's wood pipe line down Ogden Canyon had about reached the end of its usefulness, and in addition, that was an exceptionally dry year. Crop losses were heavy and thin livestock was being bought by the thousands for immediate slaughter.

Ironically, then, the worsening of the poor economic conditions that had previously delayed efforts was among the factors that prompted construction of the Pineview Dam.

In 1933 the organization that would contract for that project, the Ogden River

Water Users' Association, was founded. Important organizers included Frank Francis, editor of the *Standard Examiner*; A. P. Bigelow, president of the Ogden State Bank; and Samuel G. Dye, vice president of Commercial Security Bank— all of whom had advocated development of the Weber River. The main purpose of this nonprofit corporation was to sponsor the construction of the Ogden River Reclamation Project (the Pineview Dam) and to act as the agency to impound and distribute surplus waters of the Ogden for irrigation of lands in Weber and southern Box Elder counties. After its establishment, sixteen Weber County irrigation companies, four municipalities (Ogden City, North Ogden City, Willard City, and Brigham City), and two conservation districts (the Weber-Box Elder Conservation District and the South Ogden Conservation District) subscribed to stock in the corporation. The subscribers held stock in varied amounts.

The two conservation districts were organized to subscribe for Pineview water that could be distributed in areas with few water resources. The Weber-Box Elder Conservation District successfully petitioned the Weber County commissioners for incorporation on November 9, 1933, and a board of directors was elected on March 12, 1934. The first directors were James E. Randall of North Ogden, O. H. Ward of Willard, and Alf N. Olsen of Brigham City. The district hoped to provide water to lands "interspaced between the benchlands and lying below the Ogden-Brigham Canal . . . which heretofore had never been irrigated or cultivated." On September 25, 1934, the district contracted with the Ogden River Water Users' Association for 6,799.48 acre feet of Pineview storage water that had been allotted by the state engineer and approved in an election by district landowners. The Pineview project included construction of a twenty-five-mile canal from Ogden to Brigham City that supplied water to the ditches of many previously organized irrigation and canal companies.

The South Ogden Conservation District was organized in 1934. Its first board of directors—William P. Stephens, John M. Mills, and Joseph E. Wright—were elected under the jurisdiction of the county commissioners as provided by law. The district contracted on July 22, 1935, for 4,871.35 acre feet of Pineview water for 2,169 acres of land, also allotted by the state engineer and approved by the water users. The South Ogden District at the time contained agricultural land, but it was soon evident that much of the area would become residential and that an elaborate distribution system would be needed. Consequently, the Bureau of Reclamation, on May 13, 1940, agreed to loan $345,000 for construction of such a system, which would consist entirely of high-pressure steel pipe, thirty-five miles in length and varying from four to twenty-two inches in diameter, and would also include two large cement-lined equalizing reservoirs. This was considered to be the first time such a pressurized system had been used. It turned out to be extremely successful, much to the credit of Norman Olsen and Carl Vetter, the Bureau of Reclamation engineers who devised it.

On May 31, 1934, the Ogden River Water Users' Association entered into a contract with the United States government for construction and repayment of the cost of Pineview Dam. Although it was not a New Deal allocation, contrived specifically and solely to put the unemployed to work, the money that funded the project was advanced by the Public Works Administration (PWA), which was established by the National Industrial Recovery Act of June 16, 1933, and was administered by Harold L. Ickes; and much of the construction labor was provided by the Civilian Conservation Corps (CCC).

The initial cost of the project was estimated at $2.9 million by the federal engineer, E. O. Larsen, but this proved to be too low, which created difficulties later as more costs were contracted and the assessments of water users were increased accordingly. The Ogden River Water Users' Association signed a supplemental contract for $500,000 and another for $600,000, bringing the total project costs to $4 million.

The construction contract went to Utah Construction Company of Ogden and the Morrison Knudsen Company, Inc. of Boise, Idaho, who submitted a joint low bid of $677,898. The Bureau of Reclamation appointed J. R. Iakisch as the engineer with overall supervisory powers on the project. Groundbreaking ceremonies took place at the dam site on September 29, 1934, with Governor Henry H. Blood turning the first spade of dirt. About one thousand people attended and heard speakers praise those who had promoted the project and the generosity of the federal government in making funds available under conditions favorable to the water users. Speakers also expressed much anticipation that the dam would stop the annual cycle of flood waters crashing "rampantly through the canyons" in the spring, only to "diminish to a trickle in the heat of summer."

When construction began in October 1934, a priority was dealing with the artesian wells in the area of the dam. When filled, the reservoir would bury the functioning Artesian Park beneath thirty to forty feet of water. These wells were the source of Ogden City's water supply. The first well had been drilled in 1889 on the land of James Ririe, and in 1914, thirteen more were drilled to provide additional culinary water for Ogden. The well water was praised for its purity; little treatment was necessary to make it safe for consumption. It was piped from each well to a central collection basin and from there into the city water system. From 1914 to 1933 Ogden drilled fifty-one wells in this portion of the valley. Forty-seven of them were in the landscaped Artesian Park.

To make way for the reservoir, the wells were cut off and capped underground. By this action, the discharge point of the wells was lowered by approximately ten feet, which compensated for the removal of air lift equipment that had been on the wells. "A collection system of steel pipe laterals from four inches to twenty inches in diameter collected the discharge from the various wells into a collecting tank from which a 38 inch outside diameter steel pipe 8,535 feet in length carried the water under the reservoir to a point below the dam." The wells have since provided excellent water over an extended period of time, although the reservoir has been drained a few times to check and repair the system—first in 1944, when some devices to measure the movement of the system were installed and repairs were made; again in 1955–56, when the dam was raised and new connections were installed on the water system; and in 1970 and 1992, for repairs and maintenance.

The original plan for Pineview included the dam; a seventy-five-inch diameter wood-stave pipeline from the dam to the mouth of the canyon (approximately five and one-half miles); the concrete-lined South Ogden Highline Canal from the end of the pipeline south along the hillside east of Ogden for about six and one-half miles to the Ogden Municipal Airport, where the Saint Benedict Hospital is now located at 5475 South 5th East, South Ogden; the concrete-lined Ogden-Brigham Canal from the mouth of the canyon north about twenty-four miles to Brigham City; the reconstruction of the Ogden City culinary-water supply line from the Artesian Park about nine thousand feet upstream from the dam to a point about one thousand feet downstream from the dam; and the replacement of the highway flooded by the dam with about eleven miles of oil-surfaced road skirting both sides of the reservoir.

When completed, Pineview Dam caught the waters of the North, Middle, and South forks of the Ogden River; created a reservoir with a capacity of 44,170 acre feet; and provided supplemental irrigation water to 22,867 acres of land and supplemental municipal water for Ogden City. The height of the dam was sixty-one feet above the bed of the river, and 103 feet above the lowest excavation foundation. Its crest was thirty feet wide and about five hundred feet long; its width from the upstream to the downstream toe was about 440 feet. The dam's capacity was increased to 110,200 feet in 1955.

Pineview Dam was completed in June 1937, and the Ogden River Water Users' Association took over operation and maintenance of the project on August 1, 1937. Construction continued from 1938 to 1941 on the South Ogden Highline Canal and on a distribution system consisting of seven miles of concrete-lined canal and thirty-five miles of pressurized steel pipe. The first water was delivered to farmers during the 1937 irrigation season.

At dedicatory services held at the dam on August 14, 1937, W. R. Wallace, chairman of the Utah Storage Commission, expressed confidence in the project: "I have no fear for the economic development of the northern Utah area. If the return from new crops to be made possible with Pineview irrigation waters were diverted to one cause, they would be sufficient to pay for the entire project in a period of five years." At a later visit of Harold L. Ickes, secretary of the interior, to the dam, Ora Bundy, president of the Ogden River Users' Association and mayor of Ogden, said,

> The project offered, perhaps, more unique engineer problems than
> any one of its size ever built in the United States. We hope it will

mean security of the Weber and Box Elder districts over the 27-mile strip which it serves for years to come. In fact, reports from farmers who have been using the water during the 1937 season are most gratifying, and they are all high in their praise of the asset which this additional water will mean to every water user in years to come.[30]

Although there was general acceptance of the project because it alleviated water shortages, some objected. Donald McKay complained that, at the expense of the upper Ogden River valley, which he represented, "the entire capacity of the reservoir has been taken by Ogden City and users in the lower valley and Box Elder County." In addition, he felt that the Ogden City springs under the dam were in a precarious situation, that the power company's claim for water was too large to run through their electric generating plant, and that south Box Elder County should not draw its water from Pineview.

Arlie Campbell, who served as secretary of the Weber-Box Elder Conservation District in its early years, recalled other problems. In 1936, some of the farmers staged a tax revolt caused by announced increases in taxes to cover project costs that had increased from $2.9 million to $4 million. Campbell said,

when the farmers, particularly the farmers in Box Elder County, found out that it was going to cost more because of this error in the original estimate they organized a tax strike, and they refused to pay their taxes one year. And they stormed into our office, and some of them were quite irate about it. I was just the secretary; I wasn't responsible for it. But some of them refused to pay their taxes.[31]

This was before any water had been delivered: "as time went on and as soon as they got the water onto the land and they found out the beneficial effects of this water that tax strike just solved itself, and one by one unknown to . . . [their] neighbor, the farmers would sneak up there and pick up their delinquent water taxes." An eight-to-five differential between the charges for water from the high-line canals, which cost more to maintain, and those that Ogden City paid for water to replace supplies previously obtained from the old irrigation companies also caused controversy.

In contrast to such objections, the Ogden River Project's benefits included more water for further agricultural, industrial, and urban development in Weber and south Box Elder counties; a solution to most of the earlier conflicts over water claims to the Ogden; a stable water supply for Ogden City; and expanded recreational opportunities on the river. Pineview Dam's success motivated its expansion in 1955 under the Weber Basin Project.[32]

FLOOD CONTROL

During the 1880 to 1940 period, flooding in the Weber Basin and surrounding region frequently caused heavy losses to property. Floods that raged out of the short, steep canyons along the Wasatch Range brought major damage to communities located along the Weber and Ogden rivers and elsewhere. The most destructive floods were in 1850, 1861, 1862, 1866, 1873, 1875, 1896, 1912, 1923, 1930, 1936, 1952, 1983, and 1984. The floods of 1923 and 1930 had the most telling effect because they eventually led to preventive governmental action.

In 1923 the deluge resulted from heavy rainstorms along the Wasatch. Overgrazing and destruction of the soil cover allowed storm waters to run off and erode mountainsides. On August 13, 1923, mud and rocks cascaded off the west face of the Wasatch damaging everything in their path and taking some lives. The flood deposited boulders near the mouth of Willard Canyon that were estimated to weigh fifty tons and more. One large boulder landed on and destroyed the Willard Municipal Power Plant. Smaller rocks and debris washed through the town and orchards as far as the Oregon Shortline Railroad tracks and left a heavy deposit of sediment. Two people were killed in the Willard flood, and thousands of dollars of property was damaged.

Gravel and mud from Perry Canyon covered the highway and orchards north of Willard, but damage was not as extensive as elsewhere. A torrent from Cutler Creek, a tributary of the North Fork of the Ogden River pushed debris, rocks, and gravel into the North Fork and narrowly missed campers and a girls camp. Above Ogden, a deluge of boulders, rocks, and gravel rolled down Waterfall Canyon to inundate about a two-hundred-acre triangle that extended from the mouth of the canyon into Mount Ogden Park and as far as Polk Avenue. All the debris had gone over the 160-foot waterfall, which would have been quite a spectacular sight if anyone had been there to see it. Investigation showed that sheep and cattle had eaten essentially all of the vegetation in Malan's Basin and had been a major cause of the flood.

Farmington, Steed, Davis, and Ford canyons in Davis County had severe floods in 1923. The flood in Farmington Canyon killed four Boy Scouts, Walter Wright, and his pregnant wife, who were all camped at Cottonwood Grove. Debris was carried as far as the Lagoon Resort well below the mouth of the canyon. Mud and rock slides cut the electrical plant at the mouth of the canyon in two and completely destroyed its generating machinery. The overflow in Davis, Steed, and Ford canyons was not as devastating, but the largest boulder dislodged anywhere by the floods came down Davis Creek and was deposited at the mouth of the canyon. The weight of the boulder has been estimated at two hundred to three hundred tons.

Although the 1923 disaster inspired several investigations that pointed out the relationship between vegetation destruction and flooding, the general public was apathetic and nothing was done to correct the problems. Thus, flooding occurred again on July 10, August 11, and August 13, 1930, with most of the damage concentrated at the mouths of Parrish, Ford, Davis, and Steed canyons. "The flood from Parrish Creek was a real 'blockbuster.'"[33] It destroyed several homes, damaged the Centerville School, and covered several hundred acres of the most valuable suburban and farm land in Davis County with boulders, gravel, and mud. Early estimates of damage amounted to several hundred thousand dollars.

After appeals for aid, Governor George H. Dern met with a group of Davis County commissioners, Utah Road Commissioner Henry H. Blood, and other local leaders to devise means of assistance. An appeal to the Red Cross brought immediate aid to flood victims, and $35,000 was raised for them. Other responses came from the state and county in the form of equipment and personnel, including Caterpillar tractors to clean debris off farms and inmates from the state prison to work at restoring them.

Long-term results came from the flood. Governor Dern appointed a seventeen-member committee headed by Sylvester Q. Cannon, city engineer of Salt Lake City, to study all aspects of the flood problem in Davis County and throughout the state. Committee members included Reed W. Bailey, assistant professor of geology at Utah State Agricultural College, and Clarence L. Forsling, director of the United States Intermountain Forest and Range Experiment Station, who helped create significant studies and programs that changed the entire nature of handling grazing, erosion, and floods.

In addition, Davis County organized the Davis County Flood Control Committee. DeLore Nichols, Davis County agricultural agent; William G. Barton, county commissioner; Henry A. Cleveland, mayor of Centerville; and chairman John Ford of Centerville were important members of the committee. This group paid special attention to causes of the floods in the upper mountains instead of concentrating on downstream engineering, mainly in the form of debris basins, to solve the problems. As a result of the committee's work, Section 2, a mile-square section of land just north of the head of Parrish Creek, was designated for experimental study as the Davis County Experimental Watershed Project. Instrumental in developing that project were Forsling, Bailey, Wallace M. Johnson of the Intermountain Forest and Range Experiment Station, and A. R. Croft, the project leader.

With the help of the CCC, which established a "spike camp" near the head of Parrish Canyon, a great number of conservation projects were carried out. They included an inventory of vegetation in Section 2, a contour map of the area, a flood damage survey, new mountain roads and trails, debris basins and water channels, flumes and weirs for water measurements, and contour trenches to alter runoff from the high mountain slopes. These trenches turned out to be a major development that would be studied and used in other parts of the United States and the world. They were described, as first used, as

> constructed large enough to store 80 per cent runoff from a 2-inch rainstorm in 30 minutes. Distance between trenches was determined

largely by this size specification. As a "rule of thumb" this would be about 25 feet, more or less, depending on slope, soil, plant cover, if any, and other factors. Low check dams were to be constructed at intervals of about 25 feet along the trench so that in the event of trench breakage, only the water from one compartment would be lost. Maximum slope of ground was about 25 per cent.[34]

Larger trenches that were more susceptible to excavation on 40 to 50 percent slopes later were dug. These became successful on national forest watersheds throughout the West.

A. R. Croft recalled that Reed M. Bailey developed the concept of the contour trench and that Clarence L. Forsling, Arnold Standing, and others helped in its development. Croft eventually took over much of the trench work on Parrish Creek and in Farmington Canyon and was principally responsible for the development of the larger trenching projects and for the spread of the program to other forest and mountain areas in the West.[35]

Flooding problems along the Wasatch Front had a signal impact on the water resources of the region. As the area became more compacted with population and improved properties, flooding became one of the issues that the Weber Basin Project had to confront.[36]

The period of 1880 to 1940 served as a transition period between Utah's settlement and modern eras. It was a time when water technology changed from primitive diversions and ditches to more advanced dams and canals that could transport water considerable distances from its sources to locations throughout the five-county area that would participate in the Weber Basin Project. More extensive projects were necessary to make additional water available to meet the growing demands from agriculture and from the increasing shift to industrial and municipal uses of water. This period also brought significant changes in water law. Secular control replaced church control, and the water projects of individuals and small irrigation companies gave way to those of water user associations and water conservancy districts, which drew on taxes and other funding from entire communities. Another significant change was in the sponsorship of water projects. Small private developments and early cooperative efforts sponsored by the Mormon church were supplanted by large regional projects with the federal government, through the Bureau of Reclamation, as the major sponsor. Early cooperative experiences and intermediate projects developed by combinations of counties laid the foundation for the larger federal projects. The period 1880 to 1940 in many ways presaged the largest development of all, the Weber Basin Project.

Chapter 6
The 1940s: An Era of Foresight

O ver the years, local water development on the Weber River and its tributaries moved from private and church-sponsored efforts to involvement with the federal Bureau of Reclamation in the larger projects of constructing the Echo Reservoir (1927–1930) and, during the 1930s, the original Pineview Dam and Reservoir. This reflected the recognition, as early as the turn of the century, by both private and public officials of the need for water conservation in the Weber Basin.

During the 1920s, William Green and E. O. Larson, engineers for the Bureau of Reclamation, had studied the Weber drainage and recommended the construction of Echo Reservoir on the Weber and Pineview on the Ogden River. The Green and Larson studies were financed in large part by the state of Utah.[1] From 1938 to 1942, H. E. Wilbert, I. F. Richards, and N. T. Olsen—all engineers for the Bureau of Reclamation—and others investigated various plans for development of Weber Basin water. In 1942 these engineers issued the only report they completed during this period. It addressed assorted schemes for supplying Ogden city with water. The onset of World War II terminated most water development studies.

Although the war stopped the investigations, it increased greatly the need for water along the Wasatch Front, particularly in Davis and Weber counties. The War Department had, in 1920, purchased twelve hundred acres at Sunset in Davis County for a U.S. Army arsenal. When the arsenal was activated in 1936, the army sought a water supply for it; one source was Military Springs near the mouth of Weber Canyon.[2] In 1939, as construction of the $8 million air base to be known as Hill Field began, the army needed water for it. Captain Kester L. Hastings, the constructing quartermaster, noted, "No construction is being held up by lack of water. We are faced with the problem of acquiring additional water sources for the air base, but we have a half dozen places to obtain it and it is simply a matter of deciding which is the most convenient and economical."[3]

The army's establishment of a supply depot two miles northwest of Ogden on sixteen hundred acres acquired in August 1940 placed further demands on existing water resources. The depot, estimated to cost $18 million, was an important new source of income for Weber County. Salt Lake City mayor Ab Jenkins complained, "I hope that Ogden's chamber of commerce and business men do not, in their zeal to develop Ogden, attempt to 'lift' the state capitol building and the L.D.S. temple from Salt Lake."[4]

The population of Ogden had grown from 16,313 in 1900 to 32,804 in 1920 and 43,688 in 1940. It would reach 57,112 by 1950, showing the impact of the war years, and 70,197 by 1960. Davis County and other areas of Weber County kept pace with Ogden's increase. The needs of an expanding population as well as those of agriculture caused many civic and business leaders to seek long-term water solutions for Weber and Davis counties.

At a meeting in the Ben Lomond Hotel in December 1940, Dr. J. Morris Godfrey, president of the Associated Civic Clubs of Northern Utah, demanded more water development, especially for agricultural use. He noted that northern Utah had a population of 125,766, nearly one-fourth of the state's total, and contained between eighty thousand and ninety thousand acres of irrigable land without available water. Godfrey urged the civic club members and state legislators who were present to develop a plan of water development.[5]

Davis was the most active of the northern counties in pursuing water during the mid-1940s. Many Weber County residents felt that they had sufficient water for their existing needs. On February 19, 1945, a message carried on the back of a penny post card was mailed to many of the water users of Davis County:

> To ALL USERS of WATER: An important WATER MEETING will be held at the COURT HOUSE, Farmington, Friday evening FEBRUARY 23RD at 8 O'CLOCK.
>
> The purpose is to organize the county water users and affiliate with a state organization to look after water usage and development in Utah.
>
> A representation of every ditch company, city, town, and community using water should attend—water is important. Please be present.
>
> Yours very truly,
> John H. Burningham, Joseph W. Johnson
> Davis Water Committee appointed to perfect organization[6]

Forty-two men attended the meeting, where Burningham and D. D. Harris explained the need for Davis County water users to organize and for Utahans to work in unison for water projects. After some discussion the assembly chose D. D. Harris, William Dawson, Joseph Johnson, Roy Smith, John H. Burningham, H. B. Parkin, R. O. Kirland, Homer Holmgreen, and DeLore Nichols as a committee to recommend the organizational scheme for the Davis County Water Users Association. This committee's recommendation was presented and accepted on March 16, 1945.

Eleven directors were selected, including DeLore Nichols as secretary of the association. Nichols resided in Farmington and had been the county agricultural agent since moving there in February 1927, prior to which he had been the county agent in Morgan County since April 12, 1920. He was a strong advocate of developing water for agricultural and culinary use.[7] Robert Harding of Bountiful was chosen president of the association, and he directed the meetings of the board of directors.

The directors of the Davis County association focused particularly on local water needs, but they also worked to protect water interests throughout the state. Harding, for example, wrote to State Senator Warwick C. Lamoreaux on April 11, 1945, noting that the association opposed, in its then current form, a proposal to settle rights to the Colorado River and hoped that "Utah does not get 'sold down' the Colorado River."[8]

At their April 18, 1945, meeting the directors discussed numerous county water problems and solutions, including "a high line canal out of the Weber River" that could meet the rapidly growing needs of Davis County cities. A proposal aimed primarily at meeting Bountiful's needs involved diverting Weber River water south via a wide canal near the shore of the Great Salt Lake and then pumping the needed water eastward. Seeking alternatives, the directors asked Ed H. Watson, state engineer, in August 1945 to ascertain the status of ground water in the Bountiful area. Bountiful and Davis County would each pay $1,000, dividing the cost of the survey.[9]

Joe Johnson and DeLore Nichols continued to provide the driving force on water issues in Davis County. They put together membership lists, set agendas for meetings, and drummed up support. But, as Harding acknowledged in a March 6, 1946, letter, support for the association from Davis County water users had been insufficient. He recommended that individuals who were not "bonafide water users" earning "their livelihood from the farmlands in Davis County" should not serve as association officers. He was a consulting engineer with an office in Salt Lake City himself and therefore announced he would resign at the next meeting to allow "bonafide water users to be placed in positions of authority."[10]

Although Harding regretted the lack of backing from some quarters, he noted that several potential projects had been outlined by the association, including the following:

1. Extension southward of the Davis and Weber Canal
2. Routing of a new high-line canal from Weber River to Haights Bench
3. A shoreline canal near the Great Salt Lake to intercept spring water and waste water
4. Extensions of the Salt Lake Aqueduct to Davis County lands

5. Use in the Bonneville Canal of early runoff water from streams
6. Further spreading of spring runoff water from canyons on to the high benches
7. Conservation of water in artesian basins and underground survey of artesian basins in the county
8. Study of proposed Colorado River Authority and its effects on individual water rights[11]

Harding's letter was filled with his frustration concerning lack of support and action, and he resigned his position on March 18, 1946.

Calling, as association secretary, for attendance at the March meeting of the Davis County Water Users in Farmington—where George D. Clyde, who was in charge of water conservation and development for the Department of Agriculture in the eleven western states, would speak—DeLore Nichols urged attention to the county's water concerns: "There is no greater problem confronting Davis County than that of water development. We have reached the limits of expansion in most districts without more water . . . . We need your full support in lending influence and numbers to this program . . . . We will get no where short of a full united county front."12 As county agricultural agent, Nichols kept abreast of water problems related to agriculture, including the salinity of water reaching farms.[13] The higher-than-advisable saline content of the Jordan River water in the Bonneville Canal was a particular problem.

The frustration of many of the water leaders of Davis County had become evident by the spring of 1946. Their population was growing along with their water needs, yet they couldn't seem to gain the interest of many in the county. They also were frustrated with those who controlled water development and usage on a state and regional basis. Nichols, Johnson, and other leaders had given up hope of putting together a comprehensive plan for the county with their limited resources. They had decided that private means either were not available or were too difficult to arrange. They turned their attention to the national government and specifically to the Bureau of Reclamation. In order to apply pressure to state and federal leaders, they arranged a public meeting for May 3. Letters of invitation noted that the meeting would develop a single water program for the county and that input would be sought from all in attendance.

Nichols, wanting to guarantee a well-attended meeting, sent both invitations and reminders to William R. Wallace, president of the Utah Water Users Association; Ed H. Watson, state engineer; M. T. Wilson of the United States Geological Survey (USGS); Ralf Wooley, senior hydraulic engineer for the USGS; C. P. Starr, district soil conservationist for the United States Department of Agriculture; George D. Clyde of the United States Soil Conservation Service and Utah State Agricultural College; F. C. Koziol, forest supervisor of the Wasatch National Forest; J. Howard Maughan and Orson W. Israelson of the Soil Conservation Service; D. D. Harris, secretary and manager of the Davis and Weber Counties Canal Company; Ray A. Hales, irrigation specialist at Utah State Agricultural College, and E. O. Larson, regional director of the Bureau of Reclamation. Larson promised to attend and bring Reid Jerman, regional planning engineer, with him. Nichols then proceeded to rally the county water users to attend the meeting, reiterating in letters and reminders that the meeting would be the "one big opportunity to pack the court room with county people interested in the water program" and to demonstrate to the visitors the concern of county water users. It also was going to provide a platform for Nichols and Johnson to carry forth their water crusade. They were concerned about Bureau of Reclamation commitments along the Colorado River and wanted to redirect attention to assistance in northern Utah.

Johnson took charge of the meeting, noting they hoped to elicit a positive response from each invited visitor. Nichols then asked each guest to tell the packed Farmington courtroom "What can you do for Davis County?"[14] Many of the state and regional water officials had expected to speak, but instead they were asked only the one question. Johnson and Nichols orchestrated the meeting with the primary goal of obtaining a positive response from E. O. (Ollie) Larson of the Bureau of Reclamation to the county's needs. Larson appeared angry and suggested he was short of a budget and very involved with the Colorado River project.[15] Nichols recalled that although Larson was ruffled, he talked with him the next day:

> Now Ollie, I said, I believe you get in with us and you've got all of these people in back of you. You'll make a name for yourself. I said you've got four counties with you here. . . . Someday you'll come back and thank us for it. Now these men have intentions, but we were sitting here for more than ten years, Ollie; nobody would do a thing for us. We started out in '34, and you know how that was.
>
> All these dams were built—you even built the Hyrum Dam over there. . . . Here's four counties involved, some of the most important land in the state, and the population will be right here—the center of the population. As Utah grows we can't help but grow in these three or four counties here, especially Weber County and Davis County.[16]

Regardless of how ill-treated he felt, Larson suggested at the meeting that if he could get a pledge of support from the people and $5,000, he would move ahead with a feasibility study. Within a week, Davis County commissioners Eugene Ford and Eugene Tolman came up with $5,000 and Morgan County donated $500. Weber County, however, was not yet feeling a need for new water and refused to contribute money. It took more than a year to get Weber County involved in the project.

The Bureau of Reclamation and E. O. Larson were not strangers to the Weber

River. During the decade of the 1920s Larson and William Green had undertaken the state-supported studies of the Weber noted previously. Then had come the 1938–1942, Bureau of Reclamation studies by Wilbert, Nielsen, Richards, and Olsen. Now during the summer of 1946 a new bureau study of the river and its possibilities was begun.

The study inventoried agricultural lands that could be served with water and classified every acre of irrigated and irrigable land from the Great Salt Lake to the headwaters of the Ogden and Weber rivers. The inventory was compiled during the years 1946-1948; meanwhile, the bureau also studied the water rights situation in the basin, including existing water uses; inventoried surplus basin water; and undertook numerous reservoir operation studies using different combinations of systems to see how much water could be developed. A major effort was made to bring water users up and down the river together with Bureau of Reclamation personnel and to develop mutual trust so that bureau predictions would be accurate. Literally hundreds of meetings were held with irrigation groups so that the bureau could obtain the background it needed to put together a comprehensive water resource development plan.[17]

Existing storage facilities as well as potential reservoir sites were visited and investigated, from glacial lakes enlarged by dams in the Uinta Mountains to a proposed storage site adjacent to the Great Salt Lake and to the town of Willard. N. T. Olsen, who had managed the bureau's office in Ogden in the early 1940s, apparently first suggested the idea of a large diked fresh-water reservoir adjacent to the Great Salt Lake. After some modifications this idea developed into the Arthur V. Watkins Dam and Reservoir, commonly called Willard Bay.[18]

The legwork with local water users and the feasibility studies for the bureau were spearheaded during 1946-1948 by Francis Warnick and other engineers.[19] Warnick worked closely with E. J. Fjeldsted, secretary of the Ogden Chamber of Commerce; DeLore Nichols and Joe Johnson in Davis County; David Scott, manager of the Ogden River Water Users Association; D. D. Harris of the Weber River Water Users Association; Harold Clark of Morgan County; Ed Sorenson of Summit County; and Phil Sorenson, who was the water commissioner on the Weber River. All were intensely interested in the success of the Weber Basin project.

As the Bureau of Reclamation engineers studied the river system, the project was being sold at the local level. In Davis County, Johnson noted:

> Mr. Nichols and myself spent most of our time in the evenings going out and visiting the different communities and settlements and promoting this thing. The first meeting we called was down in Bountiful. Some of the landowners down there had some little private projects and a little spring they owned; why, they was the ones that opposed us. But, after we found out what our opposition was, why we

had another meeting with the opposition . . . . Ward Holbrook became one of our better promoters in the area because we showed him
what we was going to do, we wasn't going to rob him of anything.[20]

Contacts were made with local, state, and national political figures to inform
them of the proposed project and obtain their support. After a tour of the proposed project with Johnson and Nichols, Senator Elbert D. Thomas told them to
work through Senator Arthur Watkins on this project, and "when he gets to the
point where you need my help, it will be there."[21]

Local water problems continued as the new Weber Basin plan was investigated. During the summer of 1946 the underground water situation was widely
discussed in Davis County. Harold E. Thomas, geologist of the USGS, proposed
to DeLore Nichols, on June 13, 1946, a comprehensive groundwater resource
survey of the east shore of the Great Salt Lake in Davis and Weber Counties.[22]
Ed Watson, the engineer for the state of Utah, agreed with the Geological Survey
to conduct the needed study.

Watson also became involved with Davis County water users concerning the
related issue of wells during the summer of 1946, which reflected the mounting
pressure concerning water. Several letters criticized Watson for allowing "a very
considerable number of temporary permits to drill wells and tap the underground
water basin at the southern part of Davis County,"[23] and strongly urged him to
stop issuing new permits because of the effect they seemed to have on older
wells. Watson had earlier suggested to Davis County well owners that if they
could not control the water flow from their wells—in particular, if the wells
leaked—the wells should be plugged.[24] After two months of discussing Watson's
well-plugging order, Nichols wrote and urged restraint in an unusually dry year.[25]
Considered together, the water issues of the summer of 1946 indicated an urgent
need for new water supplies, particularly for agricultural purposes, in Davis
County. The need for more supplies for urban development was also underlined
in a report of the Davis County Water Users Association to the State of Utah
Water Users Association during October of 1946.[26]

During 1946 there were twenty-four irrigation companies in Davis County
that served areas in excess of three hundred acres and ten that served less than
that. Of the twenty-four large companies, five had pronounced needs for supplemental water during an average year and fifteen suggested their water supplies
were usually adequate. Of the ten smaller companies, seven suggested that they
had pronounced needs for more water, and one suggested its supply was adequate. The remaining companies suggested they had moderate need of water during an average year.[27] Most irrigation companies noted that, on the average,
seven days passed between water turns for irrigation.

One of the reasons Nichols, Johnson, and others pushed for more and better
supplies of water was the agricultural base of Davis County. In 1945 the county

was the leading producer of apricots in the state, producing 53,242 bushels valued at $154,402 from 42,123 trees. In 1946, the county led the state in tomato production with 2,934 acres in tomatoes, compared to 8,238 acres for the entire state. About 36 percent of the state's tomato acreage was located in Davis County. During 1947, Davis County produced 7,500 tons of onions to lead the state. Farmers were interested in expanding crop acreage and in extending the crop year with more water, and the Weber Basin Project seemed a certain source for this additional supply.

Johnson, Nichols, and their colleagues continued to push for rapid development of the project and to pressure the Bureau of Reclamation directly and through elected officials. Johnson wrote Senator Watkins to gain leverage with the bureau. Watkins responded that he would personally question E. O. Larson and stress the importance of the project: "Mr. Larson will be in Washington within a few days, and I am going to hold your letter on my desk and ask him about the various questions you have raised. I will not involve you in any way but will put the questions squarely up to him. I am very interested in reclamation. That is one of the reasons I am in Washington."[28] In February, Johnson indicated to William R. Wallace, president of the Utah Water Users Association, that "everything indicates we are on the road to a new water supply in the county."[29]

Nichols used his positions of county agricultural agent and secretary of the Davis County Water Users Association to nudge county citizens toward his views on water. He and Johnson arranged monthly meetings to rally support, reaching even into the neighboring county. Referring to a March 7, 1947, meeting of the water users of Weber County at the Utah Power and Light Company office in Ogden, he said that at this gathering "We are hoping to sell a cooperative effort toward the development of water on the Weber River system. We are of the opinion a cooperative effort with Weber County might lend greater influence to terminate the decision on these projects."[30] Weber County was slowly convinced that a joint effort concerning water development would be beneficial over the long haul and, on future occasions, would even claim to have initiated the project.[31]

Nichols was also beginning to look ahead to how local water entities would repay the federal government for eventual water developments as well as how they would exclude from a project individuals who did not wish to participate. During March of 1947, E. O. Larson informed DeLore Nichols that there were two types of organizations which were commonly used as contracting agencies between the federal government and local water users for new developments. The first type was a water users association, a corporation formed under state law. This type was exemplified by the Weber River Water Users Association, which was paying for Echo Reservoir. "In this form of organization the various canal and irrigation companies subscribe for shares of stock in the water users association. The indebtedness to the Government is guaranteed by mortgages on the existing water rights and physical properties of the canal company. There is no lien on the titles to the individually owned lands."[32] Larson went on to explain the second type of organization—an irrigation district. He noted that

> This method is more commonly used for areas in which extensive tracts of new lands are located which have no water rights or constructed works to give as security. The district is empowered to levy taxes on the lands within its boundaries. Likewise, the land within the district can be sold for nonpayment of taxes. Small irrigation districts may be and frequently are members of a water users association.[33]

Larson reminded Nichols that the investigations of the bureau had not yet proceeded to a point where it could make definite recommendations as to the desirable form of repayment organization. Larson suggested that all parties maintain "open minds" as to the eventual repayment method.

Johnson continued his campaign of correspondence, writing regularly to Utah Congressman William Dawson to urge assistance in the form of funds and of influence with the Bureau of Reclamation. Johnson reminded Dawson that there were two projects in which the Davis County water users were interested: a sufficient supply of storage water, which involved the bureau, and the underground water survey being conducted by the USGS. He urged Dawson to see that sufficient funds were made available to complete both studies during the next year. Johnson noted that H. E. Wilbert, engineer in charge of the Weber River Survey, had indicated that a congressional cut in funds would stall the survey and the project, and pleaded with Dawson to do as much as possible to fund the bureau's investigation. To underline the water situation in the county, he informed Dawson that every available share of Davis and Weber Canal water had been rented for 1947.

Dawson responded that he had been in "the thick of the reclamation fight" and hoped that his efforts would be fruitful for the work in Davis County.[34] He also informed Johnson that although serious cuts had been made in the survey budget by the House of Representatives, he was hopeful that some restoration would be made by the Senate. The Department of Interior had requested $5 million for general survey work, but this had been cut by the House to $125,000.

Johnson and Nichols also urged Utah's congressional delegation to visit the project area during the late summer or early fall of 1947. During the latter part of September, Senator Elbert D. Thomas visited portions of Davis County, noting increased growth and water needs. While inspecting the watershed area of the Wasatch Mountains, Thomas urged flood control as well as water development and stated, "It has been said water is the life's blood of our Utah communities and this applies to Davis County as much as anywhere." Johnson, Nichols, and

Ward Holbrook accompanied Senator Thomas on the tour along with Dr. H. L. Marshall of the University of Utah and Russell Croft of the Forest Service. Johnson, an expert local politician, sent Senator Thomas newspaper clippings which recounted the inspection tour of the Davis County watershed.

Johnson's frequent and lengthy letters to Congressman Dawson continued. In a letter dated November 6, 1947, he asked for a special $25,000 interim appropriation to allow the Bureau of Reclamation to finish its survey work by July 1948. Part of Johnson's argument for the needed money is revealing in terms of the growth of the project:

> When we first started on this water program for Davis County, we did not vision that it would develop into a five-county program and one which would develop the entire water resources of the Weber and Ogden Rivers. Summit, Morgan, Weber, Box Elder and Davis Counties will all greatly benefit with the overall development program. The plan is to develop a full water supply for the cities and towns and also for the other counties, and have their support because of the benefits they will receive. While, apparently, we are leading out we intend to act as a unit. The unit overall plan is such that any part of the program can be developed at a time and it will still be a step toward the complete development program.[35]

Johnson further urged Dawson that without the special appropriation, "We are either faced with a complete stoppage or a long delay, neither of which is desirable." Dawson responded to Johnson's urgent request by suggesting that a special appropriation was "impossible" but that he would push hard to have it approved as part of the regular appropriation of the Bureau of Reclamation.

The correspondence with the Utah delegation was reported to the November 25, 1947, meeting of the Davis County Water Users Association, where it was decided to circulate petitions in Davis County to show unified support for the project. Johnson urged strong support from the group and "that we must not wait to get this additional money either one way or another. As a last resort we must think in terms of local funds if other methods fail to raise the necessary money."[36] This meeting concluded on the note that money spent to develop more water should be looked upon as an investment to produce additional thousands of dollars to Davis County.

Within four days a petition was sent to Michael Strauss, the commissioner of the Bureau of Reclamation. The petition, dated November 29, 1947, stated,

> We the undersigned, citizens of Davis County, State of Utah, do hereby petition you to take the necessary steps to make available suf-

ficient funds to complete all planned investigational work on the Weber River, by July 1, 1948.

With half of our agricultural lands needing supplemental water, many homes being built without a dependable water supply, and all cities urgently needing water, it has created an emergency which requires an immediate solution. Any delay will be disastrous to the welfare of our county.[37]

The petition was signed by 1235 Davis County residents (Bountiful, 496; Clearfield, 216; Layton, 191; Syracuse, 76; Kaysville, 114; Centerville, 49; and Farmington, 93) who noted their specific place of residence and 223 more who listed their place of residence only as Davis County. Letters urging financial help for the Weber River water survey were also sent to Commissioner Strauss by the Layton Town Board, the Clearfield and Layton Kiwanis clubs, and the Mayor of Kaysville (Emil M. Whitesides). Copies of the petitions were sent to all the members of Utah's congressional delegation, and Congressman William Dawson was asked to deliver the petitions to Commissioner Strauss.

Dawson later reported that he met with the engineers of the Bureau of Reclamation on December 19 and with Commissioner Michael Strauss on December 20, and that no new funds would be available until new appropriations from Congress, but that Strauss would order a special report on the project from E. O. Larson. Dawson promised to stay on top of the project in Washington and said that he told Strauss "I was going to follow this very closely because it was in my home county and backed by some of the best people I knew of." Finally the congressman reported that "The view of the engineers was that it would be very difficult to proceed with the Davis County or lower end of the Weber River project without going ahead on the whole project at once. In view of this fact they felt it would be almost impossible to have the complete investigation made by July 1, 1948; their opinion was that it would be nearer July 1, 1949 before the full survey could be made."[38]

It appeared that the project would be a year later than had been hoped, but in a letter to Congressman Dawson, DeLore Nichols wrote that he appreciated the "splendid" work Dawson had performed and asked him to continue. Nichols also noted that "You likely have noticed that all correspondence to you is signed by Joe Johnson. I am still acting as secretary of this water committee, and Joe and I go into a huddle on every move made. Our instructions from headquarters are that no U.S. employee is allowed to write direct to Congressmen to influence their action on local project. This you likely already know, so I shall have to continue to hide behind Joe in the future."[39]

Johnson was appointed in 1947 to serve on the newly created Utah State Water and Power Board. He was also Davis County assessor and helped develop the slogan of the county water users, "Water Development—Develops Davis

County." During August a meeting was held at Morgan with the water users of Morgan County. Johnson, Nichols, and reclamation officials were in attendance to put together a unified front on the Weber River Project.[40]

The Associated Civic Clubs of Northern Utah, who represented civic clubs in Utah's six northern counties, had as its motto "The Pulse of a Progressive Era." Orson Christensen, the manager of the Utah-Idaho Sugar Company, served as chairman for the Reclamation and Irrigation Committee of the association. Six members, one from each county, served on the committee with Christensen. Johnson wrote Christensen to urge widespread support of Weber River development, particularly for culinary water for Davis County.[41] Johnson sent a copy of his letter to Christensen to E. J. Fjeldsted, secretary of the Ogden Chamber of Commerce, requesting Fjeldsted's support and that of Ogden for developing the Weber.[42] During the fall of 1947, Senator Arthur Watkins also met with some of the Weber County leaders to hear their feelings concerning water development.

By December of 1947, Fjeldsted had been converted to working with Davis County and began to push water development in Weber County. A meeting was arranged for December 16 at the Utah Power and Light Company auditorium to present a plan for water development for the two counties. Judge J. A. Howell, chairman of the Ogden Chamber of Commerce Water and Conservation Committee, and Win Templeton, Ogden City engineer, presented the plan to the group, which included all of the mayors in Weber County and George Smeath of the Weber County and Ogden City Planning Commission. A special invitation was sent to Joe Johnson to attend the meeting and to "bring DeLore Nichols and any other members of your committee" to the meeting. This meeting formally began a unified approach to pushing Weber River water development, although prior to the December meeting, Davis and Weber county officials had been making some preparations.[43]

Fjeldsted informed Johnson that Ogden was particularly interested in culinary water development and that the chairman of the committee to develop further culinary water was the Ogden city engineer, Win Templeton. Fjeldsted further noted that Weber and Davis counties had a lot in common with regard to municipal water problems.

The Johnson-Nichols team had gained a crucial victory for Davis County by gaining the involvement of Weber County interests. The chairman of the December 16 meeting was Judge Howell, who, together with Win Templeton and Johnson, prodded the Weber County mayors to raise funds and work for Weber River water development in harmony with Davis County. The mayors present at the meeting agreed that the controlling factor in their communities was the limitation of water. Fjeldsted followed up the meeting with a personal letter to each of the mayors and town board presidents in Weber County requesting a strong and unified approach with the Bureau of Reclamation and Davis County to the water question.

On February 20, 1948, at the offices of the Ogden Chamber of Commerce, representatives from the incorporated towns in Weber County met to organize the Weber County Water Development Executive Committee. Elected to the committee were the following: Thomas East, Ogden City commissioner, as chairman; Doren B. Boyce of South Ogden; Dean Parker of Roy; M. W. Maycock of Pleasant View; and Alma Ellis of Riverdale. E. J. Fjeldsted was appointed secretary of the group.

On March 9, 1948, Fjeldsted wrote to Joe Johnson and suggested the formation of a Davis and Weber counties metropolitan water group. A meeting was arranged for Wednesday, March 24, at the Davis County Court House to get the representatives of the two counties together. Mayor Harold Ellison of Layton headed the Davis County municipal water users group, and members of the Davis County Water Users Association were also present. Reid Jerman of the Bureau of Reclamation also attended. Jerman, the assistant regional director, gave a detailed report to the group concerning the studies being made by the bureau of the Weber River. He also requested specific information on community water needs and desires that could be worked into the bureau's plans. The second half of the evening included a meeting of the representatives of the municipal water users. A six-member executive committee was appointed to represent both counties, which included Harold Ellison from Layton as chairman and two other representatives from Davis County (Alton P. Rose of Farmington and Mayor Vee Waddoups of Bountiful) and Thomas East, Doren Boyce, and E. J. Fjeldsted from Weber County.

Jerman indicated to the entire group that the whole Weber River Basin was being surveyed and that several reservoir sites had been identified in a tentative fashion. These included the Jeremy site on East Canyon Creek with a possibility of sixty-five thousand acre feet of water storage, Perdue on the upper Weber with sixty-five thousand acre feet, Lost Creek with twenty thousand acre feet, Dry Creek at Mountain Green with nine thousand acre feet, and the Magpie site on the south fork of the Ogden River. Jerman reported that tentative plans included an idea to divide the water at the mouth of Weber Canyon into three streams: one to flow north into Weber County for irrigation, one to flow south into Davis County for irrigation, and the third to be treated for culinary water for the communities of the two counties.

Thus the Davis-Weber Counties Municipal Water Development Association was born. With Harold Ellison as its first chairman, its purposes included stimulating water development along the Weber River and unifying Davis and Weber counties in the process. The association bridged the gap between ditch companies, water-user associations, and counties and was the precursor to the Weber Basin Water Conservancy District. It also became the spokesman for all of the water users, who on previous occasions had been fragmented. It was another in a series of pragmatic approaches to Weber River water, which had begun with community efforts in pioneer times. Development had moved into private hands in the early twentieth century, but by 1948 the pendulum had swung back to the communities. Water projects were now being pushed by local organizations that had banded together in the Davis-Weber counties association. The federal government would be brought in as a partner to help finance and build, but the local people were in the driver's seat.

Bylaws for the Davis-Weber Counties Municipal Water Development Association were drawn up during April 1948. The stated goals of the organization included "to conserve, protect, develop, and promote by all legitimate means all domestic and culinary water now used or to be used by the present or future members of the Association, including cooperation to that end with any governmental, state, or other agency."[44] All incorporated cities and towns of the counties were eligible for membership by payment of a $25 fee. Each municipality would be allowed one member. Officers included a board of directors and an executive committee. The five-page set of bylaws were accepted by a meeting of the group on April 8, 1948.

E. J. Fjeldsted, the secretary of the association, operated out of his office at the Ogden Chamber of Commerce, and his chamber stationery included the notation "Ogden—Utah's Industrial and Railroad Center." At the April 8 meeting, he was reelected secretary-treasurer. Harold Ellison was elected president, Thomas East first vice president, and Mayor Alton P. Rose of Farmington second vice president. Other elected members of the executive committee included Alma Ellis of the Riverdale Town Board; Elmer Carver, president of the Plain City Town Board, and Vee Waddoups, mayor of Bountiful.

Joe Johnson continued to pressure Utah's congressmen to get funds for the Bureau of Reclamation to complete the survey work along the Weber River. Johnson did not want to take no for an answer nor did he want to be told to wait for the next fiscal year. He wrote to Senator Elbert Thomas, Congressman W. K. Granger, and Senator Arthur Watkins: "There seems to be a stoppage somewhere along the line, and we would like to see it broken loose, if possible. Is Utah getting its just share of these funds? Is there a stalling for time and funds, causing unnecessary slow-downs and delays? What is the status of District #4 in relation to other districts? Have our projects been allowed to fall into second place?" Johnson was even more forceful in his letters to Representative Dawson stressing the need for immediate funds. "Has something miss-fired and put us on the small end? If additional funds are made available now, the investigational work can be sped up, and a report can be completed by the deadline for which we are working. What can be done to induce the Bureau of Reclamation to make a re-check of the whole program, allotment of funds, personnel and all phases, to clear the way from bottlenecks?"[45] Johnson wanted immediate action on funding and wanted test drilling on the dam sites to begin during the winter rather than wait for summer. He asked Dawson if some reclamation regions were treated in a more favorable fashion than Utah's. The congressional representatives all wrote encouraging replies, stating that they would continue to expedite reclamation on the Weber River from Washington D.C.[46]

On January 22, 1948, Michael Strauss, commissioner of the Bureau of Reclamation, wrote to Johnson, noting the petitions from Davis County as well as the ongoing interest of Utah's congressional delegation. Strauss underlined that the bureau was progressing as fast as funds were available but that no money was available to do any test drilling at the proposed dam sites during the current fiscal year. Without the drilling tests, preliminary plans and estimates for the proposed dams could not be prepared. Strauss assured Johnson that the regional director, E. O. Larson, had definite plans to file a complete report by July 1, 1949.[47]

Dawson wrote Johnson that it was possible that their region had not been treated fairly in the distribution of appropriated funds and that he would tell Commissioner Strauss that if he failed to give the region the share they had requested for investigation work, he could not expect the support of the Utah delegation on the appropriation.[48] Johnson responded in a favorable manner and again urged Dawson to keep the pressure on so that funds and equipment would not be diverted from the Weber project to other areas.

Obviously, Johnson and Nichols were not content with just waiting for action from the national government, or with applying pressure in Washington. Continuing their campaign they called a special meeting of the Davis County Water Users Association at the courthouse in Farmington on February 3, 1948. As well as the directors of the association, the three county commissioners attended (Eugene Tolman, Eugene Fort, and W. Alvin Nalder) as did representatives from Syracuse, Farmington, Kaysville, Layton, Bountiful, and Centerville. Johnson reported to the group that there was a "serious problem," an "emergency." He stated that the Bureau of Reclamation in District Four did not have enough money to test drill at even one dam site and that the money should be raised locally; this would allow the Bureau of Reclamation sufficient backing to start drilling and avoid the proposed plan of moving the drilling equipment to the Navajo Reservation. All present agreed that Weber County, Ogden City, Morgan County, and Summit County should be asked to pay their proportional shares, but that Davis County would immediately guarantee to District Four of the Bureau of Reclamation, if necessary, that the county would produce $5,000. The motions at the meeting all passed unanimously.[49]

To authorize and fund the Weber project, Congressman William Dawson introduced a bill in the House of Representatives on February 9, 1948, to allow for the construction, operation, and maintenance of the Weber Basin reclamation project. A supplement to the Federal Reclamation Act of 1902, Dawson's bill not only provided for construction of reservoirs, irrigation and drainage works, power plants, and transmission lines in the four counties of the Weber Basin but also was "condition precedent" to any construction. The Weber Basin area was to establish an organization that would repay the United States government for the cost of construction. This entity would have power to tax and the power to contract with the national government for the needed reclamation projects. Dawson also proposed that federal monies be appropriated for the project which would later be repaid by the conservancy district.[50]

Dawson's proposed bill provided a beginning. In Utah, on February 16, Joseph Johnson deposited a check of $2,500 with the state engineer, Ed Watson. Johnson instructed Watson that this was an emergency fund from Davis County to be used exclusively for drilling at contemplated dam sites in Weber River basin. He also noted that monies, which would increase the fund to $5,000, would arrive shortly from the other three counties of the basin. By the middle of March, Morgan County, to meet its obligation, had deposited $500 with the state engineer.

Since 1946, the Bureau of Reclamation had been conducting studies along the river to determine the best places to build reservoirs and control the river for irrigation and culinary water. Initially the engineer in charge of the project was Harry Wilbert, who was replaced by R. C. Johnson, who was, in 1948, replaced by Francis Warnick. By that summer, most of the inventory work was completed on the river, including possible reservoir sites, present water uses, present land uses, and possible land developments in the basin. The next step was to investigate all of the possible alternatives on the system for water use and storage. This led to a determination that there was a need for a reservoir to be located near the bottom of the drainage. The plan for a reservoir at the west end of Morgan Valley was eventually abandoned because of the high cost of relocating the Union Pacific railroad rails in that area.

As planning proceeded, a proposal by N. T. Olsen, in charge of construction on the Ogden River from 1940 to 1943, was re-thought. Olsen had suggested building a dike or dam along the east shore of the Great Salt Lake where both the Bear River, flowing south, and the Jordan River empty into the lake. This dam would capture all the fresh water flowing into the lake along the Wasatch front. This idea—or rather concept—became the kernel that would grow into the Willard Reservoir.

As the Bureau of Reclamation continued its surveys, opposition to development on the river came from the Utah State Fish and Game Department and the Federal Fish and Wildlife Service. Dawson's bill had proposed developing the river for, among other purposes, "the preservation and propagation of fish and wildlife, and the provision and improvement of recreational facilities." Thus the project was to include wildlife and recreation as well as reservoirs and flood control facilities.[51]

During 1948, engineers from the Bureau of Reclamation continued to inventory agricultural lands from the Great Salt Lake to the headwaters of the basin to determine which were currently being irrigated and which might potentially be irrigated. They also identified who held water rights, who currently used the water, and the amount of surplus water available. Numerous reservoir operation studies were made to see how much water could be developed and hundreds of meetings were held with small irrigation groups to explain the investigations and to attempt to gather background information. All of this was done to prepare a comprehensive water development plan that would sell to the people. Inspection of all current water facilities took place, including those high Uinta Mountain lakes that had been dammed to impound water over the past century. Most of the dams on the Uinta lakes on the Weber drainage had been built in the late summer and early fall and had taken from one to three seasons to complete.[52]

The task of getting public approval for the project was a team effort that included engineers from the bureau like Warnick, as well as local water users such as Nichols and Johnson in Davis County, Fjeldsted in Weber County, Harold Clark in Morgan County, and Edward Sorenson from Summit County. During a conversation with one businessman, who Joe Johnson thought opposed the project, Johnson began to retreat, "Well if that's the case, why, then we have delivered our message and we'll go." The businessman said, "No, don't go, don't go. The thing I am opposed to is using the federal government for these types of developments." Johnson said, "You are entitled to that, but will you answer one question? Where else can we get it?" Johnson recalled that the man looked at them for a minute or two and then reconsidered, "Well if it's too big for a private enterprise, why maybe the federal government is the right place. I don't want to discourage you guys at all. We will not oppose it. It will bring in industry. It will bring in more homes. It will make the conditions better here."[53]

Johnson and his colleagues not only advocated water development but they also made periodic trips up the Weber River to check on Bureau of Reclamation progress, particularly the drilling tests at reservoir sites. To make the bureau's report as complete and accurate as possible, regional director E. O. Larson asked local water users to make their own independent surveys of their present water uses, supplies, resources, and future needs. Municipalities were also asked to make their independent studies. All of the reports were to be merged into the overall report on the Weber Basin.[54]

The survey of water use in Davis and Weber counties was pursued by the Davis-Weber Counties Municipal Water Development Association. This group asked Win Templeton, Ogden City engineer and an associate of the Salt Lake Engineering firm of Caldwell, Richards and Sorenson, to estimate the cost of a "comprehensive and complete engineering report on the Culinary, Industrial, and Irrigation water requirements for the area in Davis and Weber Counties from Bountiful on the south to Plain City on the north." Templeton suggested that a fee of between $15,000 and $20,000 should be provided to complete the survey within six months. The Davis-Weber Counties Municipal Water Development

Association agreed on May 11, 1948, to contract with Caldwell, Richards, and Sorensen for $17,500. The association further agreed that each town and city would pay its share of the fee and that county commissions would be asked to make certain that all fees were paid and to make up any differences in payment.

The engineering study was to be as comprehensive as possible and include the following:

1. Population trend of each town, past and future
2. Valuation trend of each town, past and future
3. Present incorporated limits and possible future limits
4. Present usage of water (culinary, industrial, and irrigation)
5. Future usage of water (culinary, industrial, and irrigation)
6. Present water rights of each town (springs, wells, and streams)
7. Present water right of irrigation companies (springs, wells, and side streams)
8. Measure flow from all side streams, springs, and large wells
9. Future requirements of each incorporated area
10. Possibility of exchange of water between towns and irrigation companies
11. Estimate of cost of project from mouth of Weber Canyon to town limits, to include filtration and chlorination plant
12. Feasibility of project from standpoint of necessity and ability of the district to finance[55]

Letters were sent to all towns and cities in Davis and Weber counties urging each to become a member of the Davis-Weber Counties Municipal Water Development Association. The letter asked prospective members to pay a $25 initiation fee and encouraged them to participate in the engineering survey to help collect information that would dovetail with the surveys being performed by the Bureau of Reclamation. At the same time, the surveys being performed by the bureau were being stalled by lack of funds.

In a May 21, 1948, letter, bureau regional director E. O. Larson informed Johnson that because of the "accelerated" surveying program advanced by the state, Utah would be "overexpended" by June 1. Larson said he regretted having to use local money on the survey work, but conceded that "it appears that this is our only recourse to continue the work at its present rate." Johnson was to advise Larson whether it was agreeable to use part or all of the monies deposited with the state engineer to continue the bureau's work.[56] Johnson then met with Reid Jerman of the Bureau of Reclamation on June 2 and discussed the use of the locally raised monies. By return mail on the same day, Larson sent Johnson a budget for June that proposed expenditures of $2,350 for office personnel, $1,000 for a survey party, and $1,010 for a test pit crew.[57] Johnson notified Larson by letter that his organization would authorize spending only $1,010 for the test drilling crew (test pit crew) plus a reasonable amount for traveling expenses; the other expenditures suggested by Larson would not be covered. Johnson further authorized Larson to use the monies already deposited with the state engineer to cover the drilling expenses.[58]

In order to raise the nearly $20,000 needed for the comprehensive water survey it had authorized, the Davis-Weber Counties Municipal Water Development Association on June 18, 1948, again contacted the towns and cities of the two counties and offered a detailed letter of explanation of the survey. The letter included an assessment to be paid by each member to the association to fund the survey. The assessed amounts were based on a quarter mill tax that, if paid in full by all areas, would raise $20,364.63: Davis County would raise $5,354.48 and Weber County would provide $15,010.11. Ogden City was assessed $10,432.13—more than half of the total amount—while Fruit Heights was to pay $23.33, Woods Cross, $28.23, and Bountiful received the highest assessment in Davis County, $599.50.

The Davis-Weber Counties Water Association was also asked to determine what portion of the $1100 advanced to the Bureau of Reclamation during June 1948 should be paid by Weber County. Morgan County had deposited $500 with the state engineer's office while Davis County had deposited $2500. In a letter to E. J. Fjeldsted, Joe Johnson applied pressure to Weber County to pay its proportional share of the drilling monies used during June. Fjeldsted, as secretary of both the Ogden Chamber of Commerce and the Davis-Weber Counties Municipal Water Development Association, was in a position to apply pressure to Weber County and bring about a payment of monies to the other two counties. Johnson and Nichols had gained a reputation of being relentless in applying pressure, whether it was on the Bureau of Reclamation, Weber County, water users in Davis County, or elected representatives, especially when it best served water development. Johnson and Nichols, sometimes recognized as those "irritating" grains of sand, in this case produced a pearl in the form of the water developments along the Weber River during the next two decades.

Fjeldsted, responding to Johnson with noticeable pique in his letter, said he was certain the Weber County Commission would follow through on their share of the June drilling expense, but that it was more important to first approach them about their $3,329.34 assessment by the Davis-Weber Counties Municipal Water Development Association.

Fjeldsted then scheduled a meeting with the Weber County Commission at 10:00 A.M., July 29. Harold Ellison, president of the municipal water development group, assisted by Fjeldsted, was to make the presentation. Alma Ellis of Riverdale, Elmer Carver of Plain City, and John Reese and Doren Boyce of South Ogden also attended the meeting to urge the Weber County Commissioners to fund the proposed assessment. Though the members of the water development group did not receive a firm commitment from the county commission they did leave with a secure feeling that their presentation would be rewarded.

During the summer and fall of 1948, Fjeldsted contacted many of the communities who had not yet paid their assessments to the Davis-Weber Counties Municipal Water Development Association. Blanket delinquent statements were sent to all non-paying towns on July 30 and October 7. Harold R. Howard, president of the town of North Salt Lake in Davis County, responded not with immediate payment but with an explanation of how difficult it was to raise the money, especially with other concerns such as railroad taxes, and a promise that their assessment of $81.50 would be forthcoming.[59] By August 5, only $2,960.81 of the $20,000 assessed had been collected. Ogden City, for one, seemed to be hesitant about being involved in the water development project. At a meeting of the Davis-Weber Counties Municipal Water Development Association held on August 5, President Harold Ellison spoke to the group of twenty-four representatives and told them that he was pessimistic about Ogden City's participation in the water development program. Thomas East, a member of the water association's executive committee and also a commissioner in Ogden City, assured the group that he would make every effort to get Ogden to become involved with the program, particularly in light of the funding needed for the water survey being conducted under the direction of Win Templeton. The association agreed that the water development and the survey should proceed with or without the assistance of Ogden. Nevertheless, those in attendance expressed their hopes that Ogden would participate fully in the project.

Ogden was specifically concerned about a particular aspect of the water development program: flood control on an ongoing basis. Fjeldsted had written to Senator Arthur Watkins during June 1948 asking for assistance in flood protection along the Weber and Ogden rivers. During the latter part of May, much of the Weber below the confluence of the Ogden River had overflowed its banks into rich agricultural and pasture land. Watkins was informed in a letter from R. A. Wheeler, lieutenant general and chief of army engineers, that the rivers had recently been surveyed under the direction of Colonel Joseph S. Gorlinski, the district engineer at Sacramento, California. Gorlinski had suggested that the Weber River could be cleared of brush, trees, and sandbars from the 24th Street bridge westward for eleven miles for $125,000, but added that such an extensive improvement would do little in terms of long-term flood control. The improved river channel would still not be able to handle the extensive flood flows of the stream. The Weber Basin plan being put together by the Bureau of Reclamation, which involved new upstream reservoirs, then became an attractive option as a flood control program;[60] and this part of the bureau plan helped to gain Ogden's support.

The Weber County Planning Commission urged Weber County Commissioners to participate fully in the water development program. Howard Widdison, chairman of the commission, and Ernest R. Ekins, vice chairman, indicated that, among other reasons, "Properties in the southeast section of the county could be developed and choice residential and farming areas utilized to greater advantage if water were available. Many property owners in this area are only waiting for water to develop property in this section."[61]

The association received a boost when, on August 12, Win Templeton met with the executive committee of the Davis-Weber Municipal Water Development Association and informed them that after beginning the two county water survey and noting what had been done, he believed the job would cost no more than $9,000. This was a most welcome piece of news for the group who had been pushing hard to raise twice that amount. Templeton further suggested that the city of Ogden might compile much of the data, further reducing the cost of the project. The committee encouraged Templeton to proceed with his survey and promised they would raise the needed monies.[62] The optimism encouraged Fruit Heights to contribute their assessment of $23.33, but did not convince them to join the bi-county organization.

By the first of October, three tentative dam sites had been drilled on the upper Weber River and the drilling machinery was being moved to drill at the Lost Creek site. The project had received ongoing local and national attention. On September 8, the Salt Lake Chamber of Commerce hosted a dinner of reclamation officials and national politicians. The next day twenty-five of them—including Harold Ellison, Joseph Johnson, DeLore Nichols, Congressman Ben Jensen of Iowa, chairman of the Appropriations Committee for the Reclamation Bureau of the seventeen western states, Congressman William Dawson of Utah, Congressman John Sanborn of Idaho, William R. Wallace, chairman of the Utah Water and Power Board, E. O. Larson, chief engineer of District Four of the Bureau of Reclamation, and Ed H. Watson, Utah state engineer—traveled over the Weber River system to inspect parts of the proposed project area.

Although the efforts of Johnson, Nichols, Fjeldsted, Ellison, and their fellow water users seemed to be a bit disjointed and perhaps unorganized, their common goal was Weber basin water development in all of its facets in as short a time as possible. Initially, Johnson and Nichols were not certain as to the best method to bring development about. They looked at the possibilities of private and public cooperation, and as time went on they realized that one county could not carry the program. They decided then to use a cooperative approach between counties in the Weber River Basin while they continued to exert as much pressure as possible on politicians, water users, Bureau of Reclamation officials, and the public.

Johnson and Nichols and later Fjeldsted and Ellison were all like jugglers who wanted to keep all of their balls in the air. They planned, schemed, visited, and pressured any and all who might assist in water development. They seemed to be tireless in their efforts to promote, at every level, water development on the Weber River. Johnson and Nichols, at first, worked to increase resources for agricultural water, but soon they were working to include culinary water, water for defense industries, water projects to assist in flood control, and finally recreational and fish and game-oriented water projects. Those involved both at the beginning and at the finish were like Daniel Boorstin's "Boosters" and "Go-Getter's." They were grass roots pragmatic politicians who would have fit in well half a century earlier in the ranks of the Populist party.

On September 28, E. J. Fjeldsted appeared before the Ogden City Board of Commissioners to again plead the case of water development needs. Fjeldsted, assisted by Fred Abbott, explained that in their proposal they had requested a large sum of money from Ogden, but that the original sum was only proposed and tentative, and that any monetary assistance for the survey would be appreciated. Commissioner Saunders suggested that such water information as requested had been compiled numerous times and this fact should be considered in any monies granted. After some discussion the Ogden City Board of Commissioners agreed to subscribe $1,000 to the Davis-Weber Counties Municipal Water Development Association for the proposed water survey. They further agreed that all water information available in the city's files would be offered for use in the survey. Fjeldsted also sought remittance from communities in the two counties who had not paid their assessments by telling them that it was an "emergency" and their help was needed.

During October 1948, Joseph Johnson corresponded with Reva Beck Bosone, then one of the four judges of the city court of Salt Lake City and a candidate for congress in the second district. Bosone, who defeated the incumbent William Dawson in the November election, responded to Johnson's inquiry about her interest in water matters, particularly the Weber River Project, by noting her long-standing interest in water matters and suggested she "should like to see Davis County get all the water it can."[63] Bosone informed Johnson that she had sponsored the Metropolitan Water Project in the special session of the 1935 Utah Legislature. Johnson also received words of encouragement on the Weber project from Amasa R. Howard, president of the Bountiful Irrigation Company and participant in Colorado River Pact affairs, and numerous other Davis County residents during the fall of 1948.

As the study by the Bureau of Reclamation neared a close, arguments for acceptance of the proposal and its benefits multiplied. One case compiled in 1950 presented several compelling points: Davis County's population had doubled during the 1940s, up from 15,774 in 1940 to 30,771 in 1950, with one town's population multiplying by 400 percent; early construction of reservoirs would save water but it would also begin a flood relief program; expansion of housing for defense workers would require more water; and the area (particularly Davis County) was losing money in individual wealth, taxable property, and industrial production because of the lack of water. Proponents contended that the project would provide work for the unemployed and would be a permanent source of income for the state and national government. Finally, it argued that the area was growing in both agriculture and industry, and that the repayment sources to the national government for construction of the project were sound.[64]

Water rates throughout Utah were analyzed in preparing justification for the project. Rates in Weber Basin towns needed to be somewhat consistent with their neighbors, but also needed to cover water production costs as well as construction costs. The 1948 survey noted that most towns were moving from charging a flat rate to installing water meters. Variety prevailed. Minimum water charges for cities and towns included an $18 a year charge ($1.50 a month) for Centerville, Clearfield, Huntsville, Layton, Kaysville, North Ogden, South Ogden, and Syracuse. Morgan charged $1 a month per family, while Bountiful, Roy, and South Weber charged $14, $15, and $13.20, respectively, a year. These charges compared favorably with yearly rates in most towns across the state: Cedar City $16, Brigham City $9, Logan $13.20, Salt Lake City $12, and Nephi $7.20. The residents of the towns of Hiawatha and Castle Gate paid no water charges on a monthly or an annual basis. The town of Aurora charged $3.75 a quarter for homes without bathrooms and $4.50 for homes with bathrooms. Widows in Aurora were charged one-half price. Other towns such as Goshen and Springville had complicated water schedules, which included different rates for bakeries, parks, barbershops, billiard halls, butcher shops, drug stores, dance halls, churches, schools, urinals, water closets, soda fountains, and bath houses. Farmington's rates included an annual charge of $5.25 for a home with a bath and a toilet and a .0125 cent charge for each square yard of lawn. Provo charged 2 cents per square yard for lawn. Many towns stipulated that if the users paid their bills promptly (during the first half of the next month) they would receive a 10 percent discount.[65]

Another part of the overall plan to justify the development of the Weber Basin program was the in-depth study of current water resources. Such studies were performed throughout the basin; Win Templeton's analysis of Ogden's water supply was typical of these evaluations. Completed in May of 1948, his survey showed that the larger portion of Ogden's supply came from forty-seven artesian wells located eleven miles east of Ogden in Ogden Valley. These wells, which were fully recharged even after drought years, produced twenty-eight to thirty-three cubic feet of water per second, which was carried down the canyon by two water mains. Water was also gathered into the Ogden system from canyons adjacent to the city. A well at 23rd Street and Van Buren Avenue (583 feet in depth) and two others at the municipal airport added water directly to the supply system. Templeton noted that additional water supplies would be most helpful to Ogden's continued growth.[66]

Forty-one representatives from throughout the four county Weber Basin area (Summit, Morgan, Weber, and Davis counties) attended a meeting at the Layton town hall on Wednesday, December 15, 1948, to discuss the types of organizations that would properly effect the development and utilization of the Weber River System. Harold Ellison chaired the meeting and introduced Judge J. A. Howell of Ogden to the group. Howell stated that after reviewing in detail the provisions of a conservation district, a metropolitan district, a water users association, and an irrigation conservancy district, his study found that the organization of an irrigation conservancy district would best serve the needs of the area. DeLore Nichols made a motion that a special committee be formed to study the proposed organization. The representatives passed the motion and appointed the necessary committee: E. J. Fjeldsted from Ogden; Joseph B. Andrews of Kamas; Horald G. Clark of Morgan; D. D. Harris and Joseph W. Johnson from Layton; Alton P. Rose from Farmington; Harold Welch and Thomas East from Ogden; and J. A. Howell as an ex-officio member of the committee.

At the same meeting Templeton discussed the status of the project and noted that a change in plans had just been negotiated with the Bureau of Reclamation. Instead of having two means of transporting water north and south from the mouth of Weber Canyon (open ditch and pipeline), all water would be carried by pipeline for both irrigation and culinary use. Templeton noted that both initial costs and operating expenses would be less. F. M. Warnick from the Bureau of Reclamation reviewed for the group the bureau's progress, and said that the report would be compiled and ready for printing by February 1, 1949. All groups involved were pushing for the end of the investigative phase of the project and looking forward to construction and to forming organizations for operation. Such organizations, they knew, would need to take into account the need to dovetail with already existent water users such as the Ogden River Water Users Association.

By February 1949, Templeton had completed his study of water in Davis and Weber counties. His forty-four page report, illustrated with pictures, maps, and tables, evaluated past water use in the two county area, indicated present water use and availability, and then forecast the future needs of the area based on population projections for next two decades. He wove the suggested plan for water development along the river into his report and indicated the feasibility of the proposals in terms of finances and water use. After extensive study Templeton stated, "The Davis-Weber area must have irrigation, municipal and industrial water for continued growth. The only practical means of developing the required water is through participation in the potential Weber Basin Project now under investigation by the Bureau of Reclamation."[67] Templeton's report analyzed agricultural and mineral development as well as proposing a method and schedule for repayment for the proposed water developments.

Just before Templeton's report came out, during January 1949, Representative Walter Granger and Senator Arthur Watkins both introduced bills to the U.S. Congress that called for funding of the Weber Basin Project. This funding, however, was contingent upon the introduction and acceptance of the plan for the project prepared by the Bureau of Reclamation. Francis Warnick gave a report on the completed bureau plan at a meeting of the Davis-Weber Counties Municipal Water Development Association held in the Farmington court house on March 7.

Warnick explained that the report, which had taken three years to complete, was ready to present to Congress. He warned that the congressional process could be a long one, but "short-cuts can be effected and a lot of time saved" if the Utah congressional delegation would get completely behind the project. Warnick urged local citizens to contact their congressmen and encourage them to support the project fully.

Hollis Hunt of the Bureau of Reclamation staff spoke briefly to the group and explained. He indicated the cost would be $65 million and that at present only 60 percent of the water was being stored, but that the new plan would provide a method to reclaim most of the remaining 40 percent. Hunt indicated that as well as the enlargement of Pineview, new dams would be built at Perdue, Jeremy, Lost Creek, Magpie, and Willard Bay.[68] Morgan Kreek, also of the Bureau of Reclamation, discussed the costs of the project and suggested that it would take about sixty-five years to pay off the first phase of the project. He noted that the bureau recommended agricultural land be divided into two categories—dairy and truck garden—and that dairy farmers be assessed $1.80 per acre foot of water and truck gardeners $3.30 per acre foot. Kreek also recommended that some of the costs be borne by the national government because of the flood controls built into the project. This two-hour meeting gave water users in Weber and Davis counties an extensive, up-to-date view of the proposed water developments. Joe Johnson remained active in the process and directed Davis County's water efforts and continued to write anyone with influence to expedite the water process.[69] Meetings concerning the proposed projects continued to be held and received support from the residents of the four-county Weber Basin.

On March 22, 1949, at the Layton town hall, the directors of the Davis-Weber Counties Municipal Water Development Association met with representatives from most of the communities in the two counties. Harold Ellison, as president, conducted the meeting in which he gave a history of the association and a review of the finances. Of the $9,643.69 collected by the group, $8,800 had been paid to Win Templeton, largely for his engineering services related to the preparation of a report concerning water resources and use in the two counties. Templeton then spoke to the group and introduced the question of forming the area into a conservancy district that would be able to contract with the Bureau of Reclamation. Owen Hughes of Bountiful moved to accept Templeton's report and his suggestion to form a conservancy district; Lawrence Jenkins of Plain City seconded the motion. The motion passed, and it was the intent of the motion that the directors of the Davis-Weber Counties Municipal Water Development Association work closely with the officials of the Davis and Weber County water users groups to formulate the conservancy district.

E. J. Fjeldsted, as secretary of the Davis-Weber Counties Municipal Water Development Association, wrote to the Utah congressional delegation following the March 22 meeting. Senators Arthur Watkins and Elbert Thomas and Representatives Reva Beck Bosone and Walter K. Granger were urged by Fjeldsted to "use your best efforts" to have the materials concerning the Weber Basin Project properly presented to Congress and accepted and funded by that body. Other groups of water users continued to meet and sanction the program proposed by the Bureau of Reclamation. Each group saw the need to advance from the exploration and planning stages to an all out effort to move the proposal through Congress. Many of the groups sent letters of commendation and support to the Bureau of Reclamation offices in Salt Lake City. Groups sending such letters and comments included the Ogden Chamber of Commerce, the Davis County Water Users Association, the directors of various ditch companies, the Davis County Bankers Association, the Davis County Commissioners, the Davis County municipal officers, and Davis County service clubs. Not all the letters, however, supported the project—many suggested alternatives other than those proposed by the Bureau of Reclamation.

Letters from congressional representatives to Davis and Weber County citizens reflected the pressure the congressional delegates were beginning to feel. Representative Granger wrote to E. J. Fjeldsted that he would "do everything humanly possible to expedite the legislation."[70] Arthur Watkins, in a letter to Joseph Johnson, noted that he and Senator Thomas were both following the project closely in Congress, and Reva Beck Bosone wrote Johnson that she was deeply interested in the project.[71]

The Davis-Weber Counties Municipal Water Development Association with Harold Welch, president of the Weber County Water Users Association, and Joseph W. Johnson, who held a similar position in Davis County, proceeded to organize a conservancy district. At the association's April 26, 1949, meeting, they decided to assess the communities in the two counties one-tenth of one mill of the 1948 assessed valuations to cover the cost of organizing into a conservancy district and to promote the Weber Basin water program in Congress. Judge J. A. Howell and his associates were to be engaged as the "counsel" for the conservancy district. All legal costs for organizing this district were not to exceed $1,000. E. J. Fjeldsted contacted Judge Howell, then on business in Washington, D.C., and urged him to work with Congress as counsel for the hoped for district.[72]

By the end of April and into the first part of May, it was apparent that both the citizens of the Weber Basin counties and Utah's congressional delegation were anxious for the passage of the Weber Basin legislation. Letters from one group to another intensified in number and by early May both groups were sending numerous telegrams to inform each of the other's activities.

The problem that emerged by early May, however, came from the Bureau of Reclamation. By the spring of 1949, the Salt Lake office of the bureau had finished the preliminary report on the project and planned to complete the final report by the end of year. Yet pressures from the local county groups and the congressional delegation from Utah would change those plans. Both groups began to apply more pressure on the bureau through E. O. Larson, director of the Bureau of Reclamation offices in Salt Lake City. Congressman Granger's telegrams to E. J. Fjeldsted and Joseph Johnson sent on May 9 summarized the current problems and recommended solutions. Granger suggested that since only the preliminary report from the Bureau of Reclamation had been received in Washington, E. O. Larson would need encouragement to expedite a final report. Granger suggested that constant pushing from the congressional delegation in Washington and from interested Utahans back home would ensure a timely final report. Granger, who signed his telegrams "MC" (Member of Congress), noted that he was "anxious for action" on the project. And in response to Granger's telegrams, Fjeldsted and Johnson met with Larson on May 10, and received assurances from him that the report in its final form would be expedited.[73]

Senator Arthur Watkins, a close friend of E. O. Larson, also pushed to hurry the presentation of the final report. Watkins knew water problems well, having served as attorney for the Provo River Water Users Association. During the early spring of 1949, Watkins asked Larson for a meeting with the Bureau of Reclamation people in Salt Lake City to discuss the status of the final report. Members of the Salt Lake bureau staff told Watkins they planned to complete the report by the end of the year; Watkins responded noting that would not be satisfactory because legislation authorizing the project was already before Congress, which meant the final report needed to be in his hands in Washington by early summer. Larson told his staff in no uncertain terms that Watkins's timetable needed to be followed. The report Watkins received by mid-June was generally complete although flood control problems had not been thoroughly worked out with the Army Corps of Engineers. It was finalized by July 1.

On July 5, two bureau officials—Paul Sant and F. M. Warnick—hand carried the report to Washington from Salt Lake City. July in Washington was not only hot and humid but most eventful for the Weber Basin Bill. The Department of the Interior and the Bureau of the Budget had to approve the report before it could be submitted to Congress. Once these two agencies signed off, Warnick, who was the area planning engineer, presented the case for the Bureau of Reclamation before the Joint House and Senate Interior Committees. This was unusual since the regional director for the bureau usually acted as chief witness, but in this case Larson opted for Warnick to be the bureau's expert. The presentation, though intense and detailed, did have moments of levity. When Senator Long from Louisiana asked Warnick what kinds of crops were grown in Utah, Warnick told the senator alfalfa, sugar beets, grain, and potatoes. Long then asked if they were Idaho potatoes and Warnick responded, "No they are potatoes grown in Utah." The whole subcommittee burst into laughter.

The document printed by the Government Printing Office to describe and illustrate the Weber Basin Project was 172 pages long. Labeled Senate Document

No. 147, it provided an in-depth documentation of the area, including geological, economic, historical, and engineering data. Flood control, water supplies, and potential power sources were noted. Maps along with numerous graphs and tables helped to clarify the document. Congressman Granger in a July 21 letter to Joseph Johnson reported the initial success of the program: "The speed with which we obtained the approval of the sub-committee and the full committee on this project is almost unprecedented for such an outlay of funds and I assure you that it was only accomplished by the very willing and helpful cooperation of my colleague, Representative Reva Beck Bosone."[74] The Weber Basin Bill passed the Senate on August 9. Johnson and Fjeldsted were informed of this action by telegrams from Senator Watkins. Johnson and Ellison in turn sent telegrams to Granger and Bosone urging passage of the measure by the House of Representatives prior to adjournment.

The developers though had overlooked an important and influential voice, one that would throw a monkey wrench into the process. R. L. Turpin, director of the Utah Fish and Game Commission, sent a two-page airmail special delivery letter to Bosone and Granger on August 10, 1949. Turpin objected to the project because his agency had not been allowed to submit a complete report assessing the impact of the project, particularly to the Ogden Bay and Farmington Bay refuges. Turpin contended that the project could proceed as long as it in no way damaged the present wildlife habitat. At this point the Department of Agriculture joined in opposition to the bill's passage, noting that they had not had time to evaluate the project or submit their comments to Congress. After a good amount of back room negotiating, the enabling bill passed the House on August 15.

To promote the bill, Granger and Bosone both made important speeches outlining the bill and its intended impact. Granger argued in part that Weber and Davis counties "compromise the new defense area" and the water provided for municipal use there would be most beneficial to the national defense.[75] In writing to Joseph Johnson about the passage of the bill, Bosone penned a postscript, "of course—there is yet the appropriation; but the bad part is over and Granger and I are so happy. This bill has some interesting history—believe me. R.B."[76]

With the passage of the bill, rumors began that the Agriculture Department was pressuring President Truman to veto the bill. Another letter writing and telegram sending campaign began in Utah, this time focusing specifically on the president. At the same time Utah's congressional delegation began an intense lobbying effort with the president. Eventually, party politics influenced Truman's actions and decision. Bosone, Granger, and Thomas, all Democrats, had Truman's ear much more than Watkins did. Drew Pearson suggested that Truman was miffed because of some of Watkins's activities in the matter.[77] On August 29, Elbert Thomas sent a wire to E. J. Fjeldsted, "White House just called me informing me that the President has just signed the Weber Basin Project Bill."

This was particularly good news as Truman had just vetoed a similar bill regarding New Mexico. In signing the bill, Truman appended a letter with suggestions from the Department of Agriculture, which needed to be studied prior to the funding of the project. The feelings of many in the Weber Basin were verbalized by Fjeldsted as he wired Congressman Granger on August 29. "One hundred percent of your fellow citizens this area rejoice with you in realization of the accomplishment of many years research and effort culminating in President's signing Weber Basin Bill. This one of great events in history this area. We appreciate your faithful efforts."

Senator Elbert Thomas carefully explained the method used to convince the president to sign the bill even though he objected to some procedures. Thomas suggested to President Truman that presidents usually record their opinions about bills only when they veto them, but that a statement about a bill, "especially parts which the Executive does not like, would be in order in a positive way even when he signs a bill." The presidential statement then would have lasting impact and serve as advice to Congress in the way a bill might be implemented in the future. Truman accepted Thomas's advice and issued a strong statement about the bill as well as signing it. Thomas went on to say that even though Truman suggested that the bill had been rushed through the review process, that discussions with members of Congress from Utah suggested to him that the project was basically sound and should be approved.

Thomas concluded his summary of the bill's approval by noting that it had been a team effort in the nation's capital as it had been in Utah in preparing the project for congressional discussion. In pointing to those who had a need to claim credit for the project, Thomas was reminded of a story from his early years.

> I donated 25 cents, when I was a very small child, to the building of the Brigham Young Monument on Main Street. That has always made me feel that I'd had an interest in the Monument and I have, but I have never gone so far as to tell anyone I built it. I may do so, though, sometimes if I find myself hard up for any excuse for living.[78]

With the approval of the authorization legislation, work was ready to proceed on several fronts. The first order of business was to organize a conservancy district to contract with the Bureau of Reclamation for the construction of the facilities on the river. This organization would have the power to tax in order to repay the U.S. government reimbursable costs allocated to irrigation and municipal water supplies. The construction of facilities designed by the Bureau of Reclamation would have a high priority. As population increased and water demands grew, water allocations and use became a major challenge to be coordinated among ditch companies, water users associations, and the Weber Basin Water

Conservancy District. The efforts of local officials, concerned citizens, state and national congressional representatives and even the president of the United States ensured that the challenge would be met successfully.

Chapter 7

 Construction of the Weber Basin Project, 1951–1969

In December 1951 the Bureau of Reclamation issued a map detailing the Weber Basin Project as it was then conceived. The plan included five new reservoirs: Larrabee on the upper Weber River near Holiday Park, Wanship, Lost Creek, Magpie on the South Fork of the Ogden River, and Willard; two expanded reservoirs, Pineview and East Canyon; and an extensive system of pipelines and canals. This map was part of the plan to secure appropriations from the United States Congress for the project. Although the plan originated with Bureau of Reclamation workers in the field, revisions and compromises were written into it by the regional office, by the commissioners office, and at the departmental level, after review by other Department of Interior agencies and representatives from the Department of Agriculture.[1] The bureau finalized the report on April 30, 1952, and nearly six months later President Harry Truman authorized funding of the project. The negotiations concerning the plan between the Bureau of Reclamation in the Department of Interior and the Department of Agriculture had not only smoothed the way for acceptance of this plan but also for the initial funding of the project at $1.35 million in July of 1952 for the first phase of construction. The bureau was to supervise construction and the recently formed Weber Basin Water Conservancy District would contract to repay the reimbursable project cost and operate the project. The repayment obligation, signed on December 12, amounted to $57,694,000 and represented the largest single repayment contract entered into by the Bureau of Reclamation at that time.

Just one month before, on November 3, J. S. Lee and Sons were awarded the first contract on the project to drill an observation well near the mouth of Weber Canyon so that ground conditions in the area could be studied. The drilling proceeded during November and December to a depth of 217 feet, although the water table had been reached at 172 feet. This well was part of a cooperative ground water investigation, which was to be conducted jointly by the Bureau of Reclamation and the United States Geological Survey (USGS). The bureau, also during 1952, worked to determine the final location for a dam on the upper Weber. Many property owners in Summit County had opposed the construction of a dam at Rockport so other sites considered, which included Perdue, Chalk Creek, and Larrabee. Extensive field surveys were conducted and numerous diamond drill samples were made at each site. By the end of the year, bureau workers decided that the site at Rockport, which would become the Wanship Dam site, would work best because of foundation conditions, availability of construction materials, and estimated cost of construction. Field work at the Wanship site was at times complicated by landowners in the area who objected to the dam.

As the investigations for dam sites continued, construction for the Gateway Tunnel began. The most support for this project had come from Weber and Davis counties because both areas anticipated a large growth in water needs following World War II. In order to provide these counties with water, numerous schemes had been considered including moving water from Morgan Valley to the Ogden Valley and Davis County via tunnels and building a dam at the west end of the Morgan Valley near Mountain Green. For a variety of reasons, including expense, planners chose to build a diversion dam on the Weber River west of Morgan at Stoddard. Water taken from the river at Stoddard would be conveyed by a new canal (The Gateway Canal) built on the south side of Morgan Valley to the site of the Gateway Tunnel.

The tunnel was to be drilled through 3.3 miles of the Wasatch Mountains, with the east portal or inlet beginning at an elevation of 4941 feet above sea level and exiting the mountains at the west portal at 4923.84 feet. The tunnel, designed to be nine feet and four inches in diameter, was lined with reinforced concrete. It would connect at the outlet with reinforced concrete pipe aqueducts that would carry the water to service areas in Davis and Weber counties. The tunnel was to be drained of water seepage and leakage by an eight-inch sewer pipe laid in a gravel filled trench beneath the floor of the tunnel. Estimates for the tunnel lining showed it would take 17,500 cubic yards of concrete to complete. Ten companies bid on the Gateway Tunnel Project in November, with the low bid of $2,486,613 made by Utah Construction Company. This bid came in about 22 percent below the bureau's estimated cost of the project, which was $3,202,207. Construction then began the following month with excavation at the west portal.

By January 9, 1953, the common material had been removed from the west portal and a ground-breaking ceremony was held with representatives of the contractor, the Weber Basin Conservancy District, the Bureau of Reclamation, as well as state and local representatives. William R. Wallace, chairman of the Utah Power and Water Board, pushed the plunger to officially begin construction of the Weber Basin Project. Wallace, who was also associated with the Utah Oil Refining Company and Walker Bank, had been involved with water development in Utah for nearly half a century. Known as "Mr. Water," Wallace was recognized for his wide ranging activities promoting water development, including Strawberry Reservoir and the Colorado River Compact.

Crews began their intense work schedule of three shifts a day, six days a week on January 26, 1953. By year's end, 13,074 feet of the tunnel had been excavated and by March 25, 1954, fourteen months later, the final section was excavated, bringing the total length of the tunnel to 17,203.5 feet, with an average daily advance on the tunnel of 51.2 feet. About 78 percent of the tunnel required structural steel supports. One problem encountered during construction was that the tunnel produced considerably more water than was anticipated. Although the west portal allowed an average flow of about three hundred gallons per minute, flows of five hundred gallons per minute were encountered.

Construction of the concrete lining for the tunnel took place in three stages. The first stage included placing a curb along the sides of the invert (floor of the tunnel) to facilitate the placement of the forms for the concrete arch. Following completion of the curb, the arch was formed and placed through the entire length of the tunnel and finally the invert or floor was placed. The concrete lining required 31,855 cubic yards of concrete containing 44,700 barrels of cement—almost twice as much as the project engineer estimated. The major reason for the additional concrete was that there was more overbreak of rock in the tunnel because there were more fractures in the rock than the geologists had predicted. Despite the complications, all construction work in the Gateway Tunnel reached completion by the end of 1954, almost a full year ahead of the projected and contracted deadline.

Construction on the Weber Basin Project, beginning with the Gateway Tunnel, differed from that undertaken by pioneers a century earlier. Not only were modern methods and equipment used to build the Gateway Tunnel, but workers were drawn from both local and distant areas. Whereas pioneer construction had been slow and tedious and performed only by those who would benefit from the projects, Weber Basin construction proceeded rapidly during all but the winter season. Also, the project was initially financed by the national government, but over time would be repaid by water users from throughout the conservancy district who, like their pioneer forefathers, would benefit greatly from the water.

As the Gateway project continued during 1952, a definite signal, in the form of a spring flood, indicated that other elements of the Weber Basin Project were needed. Flooding started about April 15, causing the rivers to reach their maximum flood stage by May 5. The extent of the flood can be shown by discharge measurements taken at the Gateway gauging station on the Weber River. The all-time maximum flow of 7,980 second feet of water had been recorded at this spot on May 31, 1896, which was before the Echo, East Canyon, or any other major dam had been built on the river. The average recorded flow for May 5, 1952, reached 7,540 second feet, with a maximum flow for that date being 7,710 second feet; by then, however, the flood water was cushioned by both Echo and East Canyon reservoirs. Had these reservoirs not been in existence, surely a new record would have been set.

Total flow past the Gateway station during 1953 was 198 percent of the average annual flow for that station, for a total 828,000 acre feet. Flooded land stood at 21,400 acres, resulting in at an estimated loss of $1,605,700. Most of this was in Morgan Valley, on the lower Weber River, and on the lower Ogden River. Flooding was particularly devastating at the mouth of Weber Canyon where highways, bridges, and railroad tracks and embankments were destroyed. Estimates at the time indicated that flood damage, had all the facilities planned under the Weber Basin Project been completed, would have been reduced by 90 percent.

With the completion of the Gateway Tunnel, four other major projects received the go-ahead in 1954: the Wanship Reservoir, the Davis Aqueduct, the Weber Aqueduct, and the Gateway Canal. One of the difficulties in beginning and completing these and other projects was land acquisition. The right-of-way for projects was acquired through the purchase of easements, outright purchases of land, donations of land and/or easements, and by crossing or relocation agreements. When landowners refused to negotiate, the Weber Basin Water Conservancy District filed suit against them and often called on Bureau of Reclamation personnel to act as expert witnesses in such proceedings. In all, only about two dozen court cases were executed to secure the needed land for water development purposes. Many involved land acquisition proceedings in Summit County where the village of Rockport, including the cemetery, had to be moved to make way for the Wanship Dam and subsequently Rockport Lake.

On June 10, 1954, the Utah Construction Company, who submitted the lowest bid at $2,423,004.10, won the contract to build the Wanship Dam. Construction on the dam, initiated during late July, included excavation for the dam embankment foundation, spillway, and detour road, and the excavation and placement of concrete for the outlet works, tunnel, and stilling basin to provide facilities for diversion of the river as soon as possible.

Another project associated with the Wanship Dam was the relocation of the highway above the proposed dam and reservoir. The Clyde Construction Company won the highway contract with a bid of $683,742.80, and by the end of 1954 the new highway was 68 percent complete. The Clyde Construction Company received payment from the Utah State Road Commission, who was to be reimbursed by the Bureau of Reclamation, who in turn would be paid by the Weber Basin Water Conservancy District.

Winter weather usually stopped construction on such projects for four or five months a year, which proved to be the case in the building of the Wanship Dam. Despite the slow down, the outlet works was completed and the river diverted through it by May 5, 1955. During the same year, the dam embankment was raised to a height of eighty feet. Upon completion of the dam in 1957, the total height of the earth and rock structure reached 156 feet. It spanned 2,100 feet and could store sixty thousand acre feet of water.

Work on the embankment included building it up with materials from the borrow areas within the reservoir basin and compacting the materials as well as grouting the abutments and finishing the spillway. Borrowed materials were usually taken from the area to be covered by the new reservoir. Impervious materials (zone 1 materials such as clays and/or silty clays) were used to form the core of the dam, while semipervious materials (zone 2 materials such as gravel, silt, and clay) were used for stability, and finally rip rap (zone 3 materials such as large rocks) were placed on the outside slopes of the dam. By the end of 1955, the dam was about 69 percent complete. Work on the embankment was finished by November 1956, and during that same year the concrete stilling basin of the spillway was also finished.

Other aspects of the project were also undertaken during 1956. The H and M Construction Company began clearing the reservoir site of buildings, fences, trees, and brush, while the Davis and Butler Construction Company began a 1425 KW hydroelectric plant at the Wanship Dam site, which was completed in 1958.

This dam was the first major dam of the Weber Basin Project to be placed in operation. To commemorate the occasion, dedication services, attended by more than one thousand people, were held at the site on May 9, 1957. The ceremony included a dedicatory address by Senator Arthur V. Watkins, a prayer by Horace A. Sorensen, national president of the Sons of Utah Pioneers, an address by George D. Clyde, governor of Utah, and remarks by Harold Ellison, president of the Weber Basin Water Conservancy District. Following the speeches and prayer, Wilbur A. Dexheimer, U.S. commissioner of Reclamation, closed the gates of the dam to begin water storage in the reservoir. Following the dedication of the dam, Summit County hosted a picnic dinner of barbecued beef and potato salad below the dam along the Weber River for all of those in attendance.

Use of the dam and reservoir began immediately. To allow access, a graveled access road to the future east shore recreation area was constructed during 1957, which encouraged many residents of the area to fish, boat, and water ski during the summer vacation season. Sixty days after the dedication of the reservoir, 260 boats had registered at the facility. Recreation use became an important part of the project.

While construction work had been proceeding on the Wanship Dam, other projects were undertaken. These included the Stoddard Diversion Dam, the Gateway Canal, and the Weber and Davis aqueducts, all of which were completed during 1957.

The contract for the Stoddard Diversion Dam and the first three thousand feet of the Gateway Canal was awarded to the Horner Construction Company of Denver, Colorado, on July 22, 1955. Horner Construction subcontracted the Gateway Canal portion of the contract to Morrison Knudsen Construction Company, who had previously agreed to build the major western portion of the canal.

The Stoddard Diversion Dam is located three miles west of Morgan, and during the remainder of 1955 Horner Construction diverted the river from its regular channel to begin construction on the dam. An unusually mild winter allowed construction on the dam to continue—a fortunate turn of events because all the concrete work on the diversion facility had to be completed before the spring run-off. The favorable conditions meant that by April 23, 1956, the river could be rerouted through the completed structure and the diversion channel refilled before the spring run-off. The remaining work on the structure, including construction of the fish deflector structure, Parshall flume, and numerous irrigation crossings was completed by the end of July 1956. The only aspect of the dam not finished by that time was the installation of ice preventative equipment, which was not available.

The Stoddard Diversion Dam diverted seven hundred cubic feet per second (CFS) of Weber River water into the Gateway Canal, which has the same capacity and carries the water to the Gateway Tunnel. The Gateway Tunnel's capacity of 435 CFS empties into the Weber and Davis aqueducts. The Weber Aqueduct is five miles in length and the Davis Aqueduct is twenty-two miles in length. The Gateway Tunnel and the Davis and Weber Aqueducts form a "T" with the tunnel forming the column of the "T" and the two aqueducts forming its arms. The excess water flowing through the Gateway Canal but not into the Gateway Tunnel flows through the Gateway Power Plant, an outdoor plant with a capacity of 4275 KW.

By September 1954 the Morrison Knudsen Company began construction on the Gateway Canal, which was to be partially concrete lined and cover the 8.5 miles from the Stoddard Division Dam to the Gateway Tunnel. Initial work included ground clearing, fence building, and excavation for the canal. Although crews made a concentrated effort to complete most of the work during 1955, land slides along the steep hillsides adjacent to the canal excavations hindered and slowed their work. Original plans had called for a partially concrete-lined canal, but cost estimates made during 1955 showed that it would be more economical to line the canal completely with concrete during original construction. After the approval of the change order, the process of fully lining the canal with concrete was begun on August 20 at the downstream end of the canal and completed on October 16 at Stoddard.

Despite the land slides and the problem of the concrete walls of the canal buckling during the winter and spring of 1956, clean-up and repair continued as the weather improved. The canal was finished by August 13, 1956, almost ten months ahead of the contracted date.

Since initial construction on the Gateway Canal, land slides and weather have caused serious problems. Portions of the canal were moved with ground water, with some being relocated. Eventually this problem was solved by drilling wells adjacent to the canal and the hill and then pumping the ground water away before it had the opportunity to blow under the canal. Cracked and damaged concrete canal walls have had to be replaced on an ongoing basis—initially by the Bureau of Reclamation but most recently by the Weber Basin Water Conservancy District. The buildup of moss on canal screens both at the beginning and end of the canal have caused an ongoing maintenance problem on this section of the project.

Another of the Weber Basin Projects, the twenty-two-mile-long Davis Aqueduct, though begun in 1954, advanced very little during its first year. Concrete pipe for the project was cast at the United Concrete Pipe Corporation at Pleasant Grove, Utah. The pipe, made in diameters of 42, 48, 60, 66, 78, and 84 inches, was to supply both the Davis and Weber aqueduct projects. But because the pipe producers encountered difficulties making acceptable joints and reinforcement cages, production moved slowly. The pipe had been designed by the Bureau of Reclamation and making the rubber gaskets and fittings for the larger size pipes proved challenging.

Construction for the first three schedules of the Davis Aqueduct was awarded to the United Concrete Pipe Corporation, while schedule four, for the southernmost section, was offered to W. W. Clyde and Company. The work on the aqueduct included trenching, filling the trench with sand as a bed for the pipe, laying the pipe, back filling the trench with selected sandy materials, then consolidating the covering materials with water and vibration, and finally, filling the upper portion of the trench with dirt.

The Davis Aqueduct, after leaving the west portal of the Gateway Tunnel, was designed to remain high on the hill above U. S. Highway 89 east of Layton and above the towns of Kaysville, Farmington, Centerville, and Bountiful. The water was to flow by gravity to the south end of Davis County. The Davis pipeline would cross Highway 89 to the west and then again back to the east near the present Weber Basin office building on Hill Field Road; it would cross the same highway again near the south boundary of Davis County.

To accomplish this, it was necessary to acquire and then clear and bench the land prior to trenching. Here too construction was slowed due to several factors including the steepness of the slopes, the caving or sloughing of the pipe trench in areas of unstable soil, as well as heavy traffic, buried water lines, and high ground water where the trench crossed Highway 89 and the railroad tracks near the southern terminus. By the end of 1956, work on schedules one, two, and three was 90 percent completed.

During the same year, a contract had been issued to Nelson Brothers Construction Company for construction of the East Bountiful Pumping plant and the South Davis Pumping plant. These and other pumping facilities on the project were to use electricity generated on the project and carried through Utah Power and Light Company lines. These two pumping facilities were to provide water for lands located at higher elevations than the Davis Aqueduct. A record 8,700 linear feet of pipeline sized 66, 60, and 48 inches was laid during March of 1957. The first irrigation water flowed through the Davis Aqueduct to the Haight Bench Irrigation Company at Farmington on July 18, 1957.

By the end of the summer of 1957, the Davis and Weber aqueducts delivered water to fifteen irrigation companies, three domestic water treatment plants, and one major industrial water user, the Salt Lake Refining Company. The treatment plants were designed by consulting engineers employed by the conservancy district and were constructed with funds secured by a $7 million bond approved by the voters of the district. The treatment plants and the pipeline from the Davis Aqueduct to the industrial plant had been constructed under the direction of the Weber Basin Water Conservancy District. During 1957, the Weber and Davis aqueducts delivered 6335 acre feet of water for municipal use, 4935.5 acre feet for irrigation use, and 374 acre feet for industrial use. Work would be ongoing on laterals—both pipe and ditch—branching off from the canal and storage reservoirs.

An important aspect of the Davis Aqueduct project completed at this same time was the Farmington Bay Wasteway, which was an enclosed pipe wasteway that carried overflow from the aqueduct into Farmington Bay, adjacent to the Great Salt Lake. Table 1 illustrates the amount of water delivered for municipal, industrial, and agricultural use by the Weber Basin Water Conservancy District in 1957.

The contract for the construction of the five-mile-long Weber Aqueduct, which would carry eighty CFS, was awarded to the Wheelwright Construction Company who began work on the project during 1955. The Weber Aqueduct, similar in structure and purpose to its sister water carrier, the Davis Aqueduct, was to run north and west from the west portal of the Gateway Tunnel. It would be siphoned under the Weber River, the highway at the mouth of Weber Canyon, and under the Union Pacific Railroad tracks, then up onto the Uintah town bench. From this position water could be distributed for both culinary and irrigation purposes into Uintah, South Ogden, and other adjacent areas. Work on the aqueduct concluded by the end of 1956 and included an equalizing and wasteway reservoir located on the Uintah bench. Work on the lateral system extending from this aqueduct, however, continued.

With the completion of the Davis and Weber aqueducts, as well as small storage and wasteway reservoirs, attention turned to completing another feature of the system—the several pumping plants (see Table 2) along the aqueducts that would be used to raise irrigation water to reservoirs so that land above the aqueducts could be irrigated. Power for these pumping plants was conveyed over Utah Power and Light Company lines after being produced at the Wanship Dam Power Plant and the Gateway Power Plant. The initial agreement stated that Utah Power and Light Company would charge the Weber Basin District for wheeling the electricity through their lines.

TABLE 1
Weber Basin Project—Water Delivered During 1957

Use	Amount in Acre Feet
Municipal	
Delivered through Weber Basin Water Conservancy Districts' treatment plants to Ogden, South Ogden, Riverdale, Roy, Washington Terrace, Sunset, Clinton, Clearfield, Layton, Laytona, East Layton, Fruit Heights, Kaysville, Farmington, Centerville, and Bountiful	
Total	6335
Irrigation	
Haight Creek Irrigation Company	500
Shepherd Creek Irrigation Company	100
Haight Bench Irrigation Company	925
Barnard Creek Irrigation Company	83
Centerville-Duell Creek Irrigation Company	125
Parrish Creek Irrigation Company	89
Stone Creek Irrigation Company	400
Barton Creek Irrigation Company	200
Davis Creek Irrigation Company	21
North Canyon Irrigation Company	120
Mill Creek Irrigation Company	1103
South Davis Improvement District	1050
Individual Users on Gateway Canal and Weber River	162
Individual Users on Davis and Weber Aqueducts	57.5
Total	4935.5
Industrial	
Salt Lake Refining Company	
Total	374

TABLE 2
Pumping Plants

Name	Location	Horsepower	Total Capacity	Height of Lift
South Davis	Davis Aqueduct	1,025	14 CFS	580 ft.
East Bountiful	Davis Aqueduct	775	13 CFS	475 ft.
Val Verda	Davis Aqueduct	225	5 CFS	240 ft.
East Layton	Davis Aqueduct	150	4 CFS	165 ft.
Sand Ridge, No. 1	Davis Aqueduct	160	8 CFS	92 ft.
Sand Ridge, No. 2	Davis Aqueduct	325	13 CFS	138 ft.
Uintah Bench	Weber Aqueduct	750	17 CFS	356 ft.

Another major project planned for construction during the first phase of the Weber Basin Project was the enlargement of Pineview Dam located on the Ogden River. As well as raising the dam, the plans called for relocation of numerous facilities, including roadways that would be inundated by the enlarged reservoir and new recreation facilities. Several problems became apparent as the Pineview project proceeded. The original dam was constructed by the Bureau of Reclamation with ownership in the name of the United States; the Ogden River Water Users Association operated the facility and delivered and disposed of the water collected behind the dam. Since both the Ogden River Water Users Association and the Weber Basin Water Conservancy District would use the expanded facilities, it was necessary to write a contract that would provide for this joint use.

The contract needed to spell out the operational procedures to be used with the new dam by both groups as well as provisions for using the enlarged reservoir for flood control. Another problem associated with the enlargement of the dam related to the artesian wells located beneath the reservoir, which served as a prime source of culinary water for Ogden City. Until 1953, the Ogden River Water Users Association had been required, if so demanded by Ogden City, to empty the reservoir for inspection of the well system. Since the enlarged reservoir would have a large carry-over of water, the existing agreement had to be modified. A final problem arose between the Ogden River Water Users Association and the Weber Basin Water Conservancy District when the association presented a proposal to the district in August 1953, asking that the association be given a credit of $745,000 by the district toward the construction of the enlarged dam to compensate for the district's use of the existing dam. Negotiations concerning the matter of credit for the existing investment in the dam continued over the next several years.

By 1955 the Ogden River Water Users Association decided to petition the Bureau of Reclamation to enlarge Pineview Reservoir as part of the existing Ogden River Project with full operating authority to remain with the association. On March 22, the commissioner of reclamation rejected the request of the Ogden River Water Users Association. Subsequent appeals by the association to Senators Arthur Watkins and Wallace Bennett were also rejected. When the Bureau of Reclamation awarded the contract for the Pineview project on June 30, the Ogden River Water Users Association, in a telegram to the secretary of interior, protested the action, noting that all parties concerned had not reached a settlement on the project and threatening legal action.

This legal action became a reality when the association filed suit in the Federal District Court of Utah under the jurisdiction of Judge Willis R. Ritter on August 31, 1955. The association asked for an injunction that construction on the enlarged dam be stopped and that the association be considered the equitable owner of the dam. Judge Ritter dismissed the declaratory judgement suits brought by the association on January 3, 1956, and ruled that there was no cause for action. The Ogden River Water Users Association appealed the ruling in the Circuit Court of Appeals in Denver, Colorado, but the circuit court upheld the decision of the federal court in their decision reached on October 10, 1956.

While contract negotiations continued between Ogden River Water Users Association and the Bureau of Reclamation and the Weber Basin Water Conservancy District, the bureau pushed forward on construction of the project. On June 30, 1955, the Utah Construction Company was awarded the contract for enlarging the dam and during the next two and one-half years they completed the project. The earth and rock dam was being expanded from a capacity of 44,200 acre feet to 110,000 acre feet, which was done by raising the dam to a height of ninety-one feet. Construction plans called for a revision of the spillway, excavation for the revised outlet works, and relocation of the Eden and Huntsville highways. While work proceeded on the dam, an agreement was reached with Ogden City to modify their artesian well system. The modification of the wells was completed on December 22, 1956, with most of the work being accomplished in the late fall when the reservoir was nearly empty.

Smaller yet necessary projects were also accomplished during 1956–57. In order to protect the Huntsville cemetery from inundation and erosion, the Lee Moulding Construction Company worked placing gravel and rip rap on the shores of the reservoir adjacent to the cemetery. A second contract for clearing the reservoir floor and fencing the reservoir was awarded to the Kraling Construction Company of Mahnomen, Minnesota, in November 1957; most of that project was completed during the next year. Essential work on the dam and roads was wrapped up by the end of 1957, though clean-up work on the project continued through the next year. This essential part of the Weber Basin Project more than doubled the size of the existing reservoir.

A unique feature of the project completed during the first phase of construction was the Slaterville diversion dam and the Layton pump intake channel. This operation diverted water from the lower Weber River into the Hooper, Slaterville, Willard, and Layton canals. The Hooper and Slaterville canals already existed and would benefit from an increased supply of water. The Willard Canal would carry water northward into Willard Reservoir and, with pump assistance, southward from Willard Reservoir into the Weber River to be diverted again at the Slaterville Dam. The Layton Canal would carry water diverted into it at Slaterville south and west into Davis County. This canal would provide irrigation water to land located west of Clearfield, Layton, and Kaysville.

The contract for building the Slaterville Dam and Layton pump intake channel was awarded to the Mountain States Construction Company on June 15, 1956. During construction, water from the Weber River had to be diverted from its channel. The Slaterville Dam was designed and constructed to accommodate a flow of 6,600 CFS and the outlet works into the Willard Canal would handle a flow of 1625 CFS. In building it, the contractor chose to provide a cement bound pile curtain in lieu of the conventional steel sheet piling for the upstream cutoff wall. This cement-bound curtain consisted of a row of seventeen-inch diameter concrete piles mixed in place at approximately twelve-inch centers. These piles were to be cemented with special auguring and mixing equipment and would overlap each other. The six-gate dam, including the radial gates and hoists, was completed by the fall of 1957, which paved the way for the Willard Canal and Reservoir.

Another aspect of the first phase of construction included building drains in the western portion of the project area to drain high water from lands to reclaim them for agriculture. Drains were completed at Hooper (1954-56, including a rehabilitation of the Hooper slough), Syracuse (1957) and in south Davis County. This aspect of the project showed that the project not only provided water for new land, but that it also reclaimed what appeared to be old and worn out farm land.

The building of drains was in part a study of ground water in the project area. Another aspect of ground water study included the drilling of observation wells. Observation of the wells yielded information not only about ground water as it existed before the project but also showed the impact of Weber Basin water development on ground water levels. In addition to wells that had already been drilled during 1953, 118 wells 2 inches in diameter and 10 to 20 feet deep were drilled; 21 wells 3/8 inches in diameter from 4 to 30 feet deep were drilled; and 3 wells 3/8 inches in diameter over 30 feet deep were drilled. During 1953, 827 wells were read at least semiannually and 220 wells were read on a monthly basis.

As the Hooper Pilot Drain became operative in 1954, test well levels in the vicinity indicated an immediate lowering of the water table near the drain location. As water measurement continued during the entire project, observation results showed that water rose in the majority of wells from January to July and then declined until the end of the year. Just as ground water pressure and depth studies were being made, other studies including a chemical analysis of the water, the total volume of water, and the transmissivity and storativity of the water generated valuable data. These studies demonstrated both the original characteristics of the ground water and changes brought about by water development. An initial study on ground water was completed by the Bureau of Reclamation during 1957.

During the first phase of construction, Bureau of Reclamation estimates on total costs for the project fluctuated from $69,534,000 in 1949 to $73,513,000 in 1959. The estimate of total costs involved all of the various parts of the project from dam construction to ditch construction, from recreational construction to digging test wells. The yearly estimate changes are illustrated in the following list:

1949 Planning Report on the Authorized Project	$69,534,000
Definite Plan Report 1951-1952	70,385,000
F.Y. 1953 and 1954 Official Estimate	70,385,000
F.Y. 1955 Official Estimate	65,362,000
Revised Definite Plan Report, April 8, 1955	70,110,000
F.Y. 1956 Official Estimate	68,602,000
Revised Definite Plan Report, December 30, 1955	70,340,000
F.Y. 1957 Official Estimate	70,370,000
F.Y. 1957 Revised Estimate	70,670,000
F.Y. 1958 Official Estimate	70,523,000
F.Y. 1959 Official Estimate	73,513,000

Estimated increases resulted from rising prices, the redesign of many lateral systems using more pipe and fewer open ditches, increased cost in right-of-way acquisition, the cost of slide stabilization on the Gateway Canal, and the payment of a claim to the contractor for excessive rock in the Wanship Dam borrow area.

The appropriations received from the national government for project construction are detailed in Table 3.

By the end of 1958, the project was 48 percent completed on the basis of total expenditures.

The initial operation of the project was begun in July of 1957. Costs for operation and maintenance for the first year totaled $17,920; $15,670 came from the sale of water through the conservancy district while $2,250 came from construction funds. During 1958 the total operation and maintenance costs increased to $60,980; $42,330 was advanced by the Weber Basin Conservancy District and $18,650 came from construction funds charged to the heading of Future Year Capacity Provisions.

TABLE 3
Federal Appropriations

Fiscal Year	Withheld or Appropriations	Withdrawn	Net Allotments
1953	$ 1,350,000	$ 47,462	$1,302,538
1954	5,776,000	1,274,268	4,528,732
1955	7,900,000	1,397,000	6,503,000
1956	10,895,000	1,293,314	9,601,686
1957	10,066,000	670,553	9,395,447
1958	6,500,000	4,246,250	2,253,750
1959	5,273,000	———	5,273,000
Totals	$47,760,000	$8,901,847	$38,858,153

The commercial operation of the Wanship Power plant began on August 5, 1958, and the Gateway Power Plant began to produce electricity on December 1, 1958. The expenses for operating these two plants during 1958 totaled $8,641; with $7,814 being generated from the sale of surplus power. The electrical power portion of the project was 90.4 percent self-supporting during its first calendar year of operation, while the water supply portion was 69.4 percent self-supporting during the same year.

The second phase of construction on the Weber Basin Project included the construction of three reservoirs—Willard, Lost Creek, and Causey—and the enlargement of East Canyon Reservoir, which had originally been built during the 1890s by the Davis and Weber Counties Canal Company. Although the Weber and Davis aqueducts had been completed during the first phase of construction, the building of storage reservoirs, drains, and laterals related to those aqueducts continued throughout the second phase of construction. This component of the project made both culinary and irrigation water available to all the residents of Weber and Davis counties.

Laterals to carry both irrigation and culinary water were built of three types of construction: pipe, pipe and ditch, and ditch. By 1970 those laterals built, particularly during the second phase of Weber Basin construction, served 34,605 acres. More than 90 percent of these new laterals were constructed in Davis County.

Another ongoing feature of this project that took more definite form during the second phase of construction was the building of recreational facilities at all of the reservoir sites. The most original idea involved in this development of the

Construction of "new" 1964 East Canyon Dam of the Weber Basin Project. (Photograph courtesy United States Bureau of Reclamation.)

Last bucket of concrete poured on the East Canyon Dam on January 14, 1966. *Back row, left to right:* William White, assistant field engineer; Ray Hardy, field engineer, Bureau of Reclamation; William Groseclose, construction engineer; Rex Greenhalgh, acting project manager, Bureau of Reclamation; Wayne Winegar, manager, Weber Basin; James Ayers, superintendent of construction. *Front row:* unidentified woman; D. Earl Harris, manager, Davis-Weber Canal and Weber Water Users Association; Herbert Barnes, president, Davis-Weber Canal Company; David Scott, Utah Water and Power Board and Pineview Systems; Keith Jensen, Weber Basin Board of Directors. (Photograph courtesy United States Bureau of Reclamation.)

Upstream view of 1915 and 1964 East Canyon Dam. The new dam was 265 feet in height and increased the capacity of the reservoir from 28,000 feet to 52,000 acre-feet of water. (1966 photograph by Mel Davis. Courtesy United States Bureau of Reclamation.)

Flooding was a frequent problem during the spring runoff from the streams in the Weber Basin area. This photo shows some of the debris left by high water near the mouth of Willard Canyon in Box Elder County in the flood of 1930. (Photograph by the United States Forest Service. From the Utah State Historical Society.)

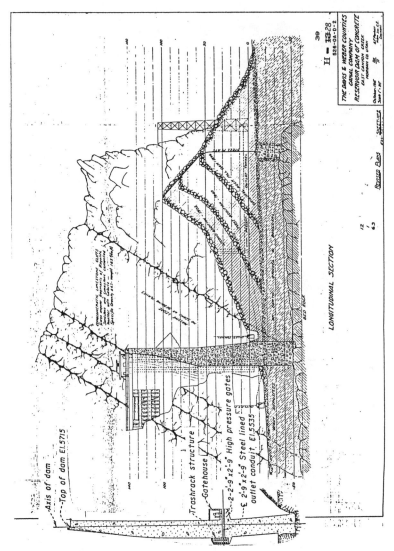

Engineer's schematic drawings showing the five dam sites of 1896, 1900, 1902, 1915, and 1964 in East Canyon. (1915 drawing by A. F. Parker with an addition by the Bureau of Reclamation showing the addition of the 1964 dam on the left of the schematic. Courtesy United States Bureau of Reclamation.)

Large boulder weighing two hundred tons washed out of Parrish Canyon in Davis County by floods of 1930. (Photograph by the United States Forest Service. From the Utah State Historical Society.)

Aerial photo showing the effects of a flood from Parrish Canyon in Centerville in 1930. Several houses were swept out of the path of the waters that left only the school house remaining in the flood plain. (From A. Russell Croft, *Rainstorm Debris Floods: A Problem in Public Welfare* [Tucson: University of Arizona, 1967].)

View near Warren, Weber County, on the lower Weber River indicating how the floods of 1952 inundated the farm lands on the lower areas. (May 5, 1952, photograph. Courtesy United States Bureau of Reclamation.)

Flooding of the Weber River at Kendall's Junction of U. S. Highway 89 and U. S. 30 South in South Weber, Weber County. This photo indicates the high spring runoff of the river on May 5, 1952. (Photograph courtesy United States Bureau of Reclamation.)

1935 photo of irrigated row crop farming taken from approximately 700 North Main Street in Bountiful, Davis County. This area was later watered by the Weber Basin Project. (Photograph from Utah Writers Project, Utah State Historical Society.)

1988 photograph taken from the same area of 700 North Main Street in Bountiful. The comparison of these two photographs indicates the extreme change of water use from irrigation to culinary municipal use. (Photograph by John D. Moyes.)

NOTICE TO THE STOCKHOLDERS OF THE CENTERVILLE DEUEL CREEK IRRIGATION COMPANY.
JULY 17, 1962

DELINQUENT ACCOUNTS:

Paragraph 4 of the Irrigation Agreement between the Company and the Stockholder reads, "The User agrees to pay all costs and charges computed in accordance with Paragraph 1 hereinabove, in advance at the beginning of the irrigation season for which the water is used as determined by the Company. In the event any amounts herein required to be paid are not paid strictly within the time herein specified and if no other time is specified then within thirty days after receipt of invoice or bill, the User consents that his irrigation water supply may be stopped and service to him may be completely suspended until all money due from the User to the Company for water used or subscribed has been paid in full."

The 1962 statements were mailed to each stockholder on March 20, 1962 and the due date for payment was printed on the statement as April 20, 1962. Therefore, any stockholder who has not paid the 1962 statement is now delinquent and subject to having the water turned off. All delinquent accounts are being duly processed and if necessary legal action will be taken. Your cooperation will be appreciated. Payments are to be mailed or taken to L. Peter Nielsen, 94 East 400 South, Centerville, Utah.

APPLICATION FOR NON-USE CREDIT:

At a recent board of directors meeting the following motion was duly made and passed, "That a development period as hereafter determined by the Board, but not extending beyond December 31, 1964, be allowed and that during the first year of such development period a reduction of $2.50 per acre foot be allowed against the annual water charge on all land on which no water from the Company's distribution system has ever been used, provided that such reduction will apply only to those lands where all charges for water service are current and upon written application of the owner therefor or his duly authorized agent presented annually to the company prior to due date of such water charges, and that such development period will in any event terminate when any water from the Company's system is used upon the land described in said application."

Application forms may be had at L. Peter Nielsen's residence.

LEGAL OWNERS OF LAND:

Paragraph 2 of the Irrigation Agreement reads, "This agreement for the right to use water and for the right to have water distributed and made available to the User shall run with the land and shall be binding upon the successors and assigns of the User. If the User sells his entire tract of property as described below, he will be released from this agreement upon the purchaser signing an acceptance of this agreement, and upon the Company approving such assignment. If only part of the land is sold, and the land is thus divided into separate ownerships, then each owner shall be required to have a separate delivery point and any owner thus required to obtain a new delivery point will be required to pay a connection fee for each such new delivery point in accordance with the then established rates of the Company."

The seller and buyer MUST see that this transfer is made with the Centerville Deuel Creek Irrigation Co. In any event the irrigation company will hold the legal holder of property liable for any unfinished business or for any unpaid obligations.

BY THE BOARD OF DIRECTORS

The irrigation company has hired Mr. L. Peter Nielsen of 94 East 400 South, Centerville, Utah to handle the bookkeeping details. He will be available between 5:30 and 6:30 P.M. Monday thru Friday.

Stockholder's notice describing the relationship of land ownership to water rights under the Weber Basin system. (Photograph from the Deuel Creek Irrigation Company notice.)

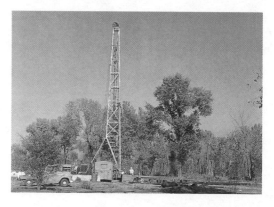

Wells became a major source of water for the Weber Basin area. Underground sources as well as surface water became a source for the overall water supply of the region. This is a view of the drilling machinery driving a well in West Weber County. (1960 photograph by Harold E. Dean. Courtesy United States Bureau of Reclamation.)

First section of pipe laid on Davis Aqueduct in Fruit Heights area. *Left to right*: Clinton D. Woods, project manager, Bureau of Reclamation; Delore Nichols, agent, Davis County; Joseph W. Johnson, Layton; Clyde Adams, Eugene Tolman and Amby Briggs, Davis County Commission; D. D. MacKay, president, Weber Basin Water Conservancy District. (Photograph courtesy United States Bureau of Reclamation.)

Wayne M. Winegar, then secretary-manager of the Weber Basin District, points to the location of the Davis Aqueduct in the Weber Basin Project. (1966 photograph. Courtesy United States Bureau of Reclamation.)

First concrete 48-inch by 16-foot length of pipe laid of Wilson Aqueduct, Weber Basin Project. This line of pipe would carry water under Highway 89 to South Weber County. (1955 photograph by D. A. Ellsworth. Courtesy United States Bureau of Reclamation.)

Driving sections of concrete pipe into position prior to placement
in trench on east bench of south Weber County. (1955 photograph
by D. A. Ellsworth. Courtesy United States Bureau of Reclamation.)

Laying 48-inch concrete cylinder pipe of Weber Aqueduct, Weber Basin Project, in deep
trench through Weber River bed. Water was diverted from Weber River while this pipe was
put in place. (1955 photograph by D. A.
Ellsworth. Courtesy United States Bureau of
Reclamation.)

Digging the Hooper Drain of the Weber Basin Project. Digging was done by a backhoe machine shown in foreground and boom and drag line shown in the distance. (1954 photograph by F. H. Coulter. Courtesy United States Bureau of Reclamation.)

Photo showing Hooper Drain Canal, Weber Basin Project, dug by boom and dragline. Construction was hindered by caving of sand banks and the high water level which filled ditch and required pumping. (1954 photograph by F. H. Coulter. Courtesy United States Bureau of Reclamation.)

Completed Slaterville Diversion Dam in Weber County looking upstream to the six gates which divide the Weber River into Slaterville Canal, Hooper Canal, and Layton Aqueduct. (1957 photograph by Stan Rasmussen. Courtesy United States Bureau of Reclamation.)

View of laying the siphon pipe through Parrish Creek of the Davis Aqueduct. (1956 photograph by A. M. Jensen. Courtesy United States Bureau of Reclamation.)

Completing the concrete floor slabs of Lateral No. 17 reservoir of Ogden River Project in southeast Weber County. (1950 photograph by J. R. Hinchcliff. Courtesy United States Bureau of Reclamation.)

Preparing the trench for the Woods Cross Trunkline. The deep trench and ground water created problems in excavating for the pipe. (1957 photograph by D. A. Ellsworth. Courtesy United States Bureau of Reclamation.)

First water being delivered to Reservoir No. 17.2 through an automatic float valve of the Bountiful Subdistrict Lateral. *Left to right:* Ezra J. Fjeldsted, C. D. Woods, and Golden Stewart. (1960 photograph by Harold E. Dean. Courtesy United States Bureau of Reclamation.)

Looking northwest across completed Bountiful Subdistrict Lateral Reservoir No. 17.2. Dean Ellsworth in foreground. (1960 photograph by Harold E. Dean. Courtesy United States Bureau of Reclamation.)

View of Reservoir No. 18.0 of the Bountiful Subdistrict Lateral of the Weber Basin Project. Reservoir is ready to receive its concrete lining. (1959 photograph by D. A. Ellsworth. Courtesy United States Bureau of Reclamation.)

View of Davis Aqueduct showing a section of 84-inch concrete pipe with a welded tee vent and turn out. The welded tee would later be encased in concrete. (1955 photograph by A. M. Jensen. Courtesy United States Bureau of Reclamation.)

View looking northwest of Stoddard Diversion Dam on the Weber River in Morgan County. On the left is the Gateway Canal that delivers Weber Basin water to the Wasatch Front. (1967 photograph by David Crandall. Courtesy United States Bureau of Reclamation.)

Weber Basin was the creation of Willard Reservoir (officially called Arthur V. Watkins Dam and Reservoir). The original plan called for this off-stream reservoir to be formed in a shallow bay adjacent to the Great Salt Lake. The dam was to be thirty-six feet high and 14.5 miles long in the shape of a rough rectangle. The pad of earth constructed for the dam to sit on was 455 feet wide at its widest point, and when filled to capacity, the reservoir was expected to hold 215,000 acre feet of water.

To supply the reservoir, water would be diverted from the Weber River at the Slaterville Diversion Dam into the 10.7-mile-long Willard Canal. The water would flow by gravity the total length of the canal north into the Willard Reservoir. When the western portions of Weber and Davis counties needed water, it would be pumped back into the Willard Canal, which was built to be nearly level so the water could be pumped through the canal and then pumped back into the Weber River at the Slaterville Dam. From this point it could be pumped into the Layton Canal and carried southward into western Davis County or used in western Weber County via the Weber River.

Construction on the Willard Bay Reservoir began during the summer of 1957 with the building of two drains and the earthen pad for the dam. The drains constructed on the south and the east were to keep water from inflowing streams out of the reservoir; the pad on the west and north was to exclude water from the Great Salt Lake from the reservoir area and provide a more stable base for the dam. Excavating equipment, which operated near the shoreline of the Great Salt Lake, required the support of heavy timber mats. This initial seven-mile-long pad or dike required 534,435 cubic yards of earth material and was built by the Parson Construction Company and completed by the end of 1957.

The completion of the initial pad allowed the reservoir area to dry during the winter of 1957-58. The Hasler and Smith Construction Companies of Los Angeles, California, received the contract for the first stage of the dam in early 1958. This contract provided for the placement of approximately two million cubic yards of embankment. The contractor began his work in June and completed about 75 percent of the work by the end of the year. Winter weather then caused a four to five month slow down—and sometimes a complete shut down—on this and most other projects on the Weber River. Although full-time work resumed in April, wet material in the borrow area for the dam delayed loading operations. To reduce costs and increase the capacity of the reservoir, the borrow materials were largely taken from what would be the bottom of the new reservoir.

Hasler and Smith signed off on the first phase of construction during July 1959, allowing the second phase to begin under the direction of George M. Brewster and Son, Inc. of Bogota, New Jersey. This phase called for the enlargement and extension of drains as well as the hauling of earth to enlarge the dam. During the month of August, crews hauled 810,000 cubic yards of borrow material to the dam site; the next month they brought another 958,000 cubic yards.

During September, certain sections of the dam had reached the desired height of thirty-six feet. The hauling of borrow material continued with 581,000 cubic yards placed in the embankment during October. The contractor also completed work on five piezometer gauging stations, which were spaced along the dam.

In order to excavate material from the bottom of the reservoir to be used for dam construction, Sauerman draglines were set up to drag the bottom and stock-pile the borrow material near the sides of the reservoir. These draglines, though set up in early 1960, because of the wetness of winter and spring weather did not lead to meaningful construction work until June when 265,000 cubic yards of material was placed in the embankment and an additional 130,000 cubic yards stock-piled near the draglines.

Throughout the year, workers placed 2,913,000 cubic yards of dam materials, with the following amounts being placed monthly: April, 56,500 cubic yards; May, 164,500; June, 265,000; July, 405,000; August, 660,000; September, 542,000; October, 440,000; November, 299,000; and December, 81,600. By the end of the year the work was 64 percent complete while only 56 percent of the contract time had elapsed.

During the next year, 26 percent of the work on the second phase of the contract was completed, including the hauling and placing of 3,682,800 cubic yards of borrow materials. As the end of the year approached, construction crews had finished 90 percent of the contract work. To do this they placed the following amounts during 1961: March, 20,000 cubic yards; April, 322,200; May, 583,800; June, 692,800; July, 388,000; August, 663,000; September, 434,000; October, 457,000; November, 123,000. Although more materials were hauled in 1961, earlier aspects of the second phase of the Willard project included more than just hauling materials, which meant that overall 33.5 percent of the project reached completion in 1959, 30.5 percent in 1960, and 26 percent in 1961. This left 10 percent of the second phase to be completed in 1962.

Although work could not begin until April, crews worked rapidly to complete the project by the middle of August. They completed that second phase by August 15, 1962, moving more than one million cubic yards of material during the final year. The drains at the toe of the dam were also completed during this year. This work schedule allowed the third stage of the dam construction to begin during the summer of 1962 under the direction of the W. W. Clyde Company who had issued the low bid of $4,712,470.

The third stage of work took three years to complete and included finalizing the level of the dam, which required hauling and spreading more borrow materials; laying bedding for rip rap and then placing rip rap over the dam surface; installation of 169 settlement measurement devices; excavation of the outlet and inlet channels and concrete work for outlet and inlet features; and construction of the marina facilities including the boat ramp. Although bad weather, difficulty in acquiring material, and labor problems sometimes hindered progress, the project was finished ahead of schedule by the fall of 1964.

The concrete inlet and outlet works were very important to Willard Reservoir because of the unique idea behind its placement and design as both a storage reservoir during the spring and as a supply reservoir in the late summer and fall. Work on the Willard Pumping Plant, which was to pump water back through the Willard Canal, began under the direction of the Gibbons and Reed Construction Company during July 1961. Since the Pumping Plant Number 1 was to take water from the Willard Reservoir and pump it up to the Willard Canal, it was necessary to remove several hundred thousand cubic yards of material to prepare the site for the plant and then to drill wells and drains to keep the site free of water. Work for both the Willard Canal Pumping Plant Number 1 and Willard Canal Pumping Plant Number 2 (located approximately at the midpoint of the canal) was completed during the period 1961-1963. Plant Number 1, with a capacity of five hundred CFS, had to lift the water a total of forty-five feet back into the canal near where it exits into Willard Bay. Pumping Plant Number 2 could hold 250 CFS; its water had to be lifted twenty feet and pushed southward toward the Slaterville Dam where it was diverted either to run down the Weber River or through the Layton Canal. These two pumping plants along with the pumping plant for the Layton Canal are by far the largest pumping facilities in the system. A twenty-ton traveling crane was installed in Willard Canal Pumping Plant Number 1 so that repairs on the pumps could be made. Since the Willard pumping plants were completed nearly two years before the canal, testing of the pumps had to wait until the canal was ready.

The next phase of the Weber Basin Project was the construction of the eleven-mile-long earth-lined Willard Canal, which can hold 1,150 CFS. Work for this canal, built by the Syblon-Reid Construction Company and the Strong Construction Company, began in August of 1961. The project proved challenging for a variety of reasons. Since a high water table existed along much of the canal route, efforts to permanently drain ground water from the canal bed were made during the early part of construction. Also, the caving and sloughing of banks required second and third excavations along some reaches of the canal. And, to accommodate both private and public road crossings, concrete for the many irrigation and drainage crossings had to be placed over the canal. The outlet works for irrigation ditches and canals were also cast in concrete. The excavation for the canal required the removal of several hundred thousand cubic yards of earth material, part of which was compacted and used for the embankments of the canal. Finally, though it was designed as an open canal for most of its length, it was necessary to siphon under where it would pass beneath the Southern Pacific Railroad, state and county roads, and the future site of interstate highway. A major turnout on the canal was constructed for the Warren Canal where radial gates were installed and completed early in 1964. Also during 1964, construction workers lined the canal, placed rip rap, positioned gravel protection at completed turnouts; and by the beginning of June water began to flow into the canal.

The Syblon-Reid Construction Company continued working, but shifted their focus to the Layton Pumping Plant. This plant, designed to pump water into the Layton Canal from the Weber River, with a capacity of 260 feet per second, and lift it some twenty feet, was under construction by 1962 and completed by the end of 1963, within 72 percent of the contracted time. Work on the Layton Canal and the relocation of the Wilson Canal then began during January 1963 by the Wheelwright Construction Company, who had been awarded the construction contract on January 15.

Work on this canal proceeded slowly during the early part of the year because of bad weather and the need to place siphons along the canal route for irrigation ditches and provide for road crossings of the canal. The size of pipe used for siphons was usually twenty-four-inch precast concrete, although some were as large as seventy-two inches. An additional feature of the canal included about 4,215 feet of siphons from twenty-four inches to seventy two inches in diameter, with 5,500 feet of the canal under drains. Much of the canal excavation took place from March of 1962 through the summer of 1964, with about 551,000 cubic yards of material removed from the canal bed. During this same period, 1,986 cubic yards of concrete was placed along the canal in siphons, bridges, turnouts, diversion boxes, and other structures. Although the canal was to be an earth-lined structure, a large number of structures both over and under the canal were cast in concrete, while the farm bridges were constructed of timber.

The project also involved the construction of about 84,500 linear feet (sixteen miles) of fence along the canal. Most of the fence work was done during early 1963. The last half of 1964 and early 1965 saw the final touches of the canal finished, including the compacting of the sides and bottom and the placement of rip rap, with work being completed on the canal on June 14, 1965.

After winning the contract on October 5, 1962, the Fife Construction Company began the initial phase of construction on the Ogden Valley Diversion Dam and Canal immediately. The construction of the dam, located on the South Fork of the Ogden River about three miles above Huntsville, took place during the late summer of 1963. Water would be diverted by this dam into the Ogden Valley Canal, which was to be about six miles in length and built in a northwesterly direction from the diversion dam.

Excavated and constructed during 1963 and 1964, this canal's main purpose was to provide supplemental irrigation water for farmers in the Ogden Valley. The canal was basically an earthen structure with compacted sides and concrete turn out stations. As with other canals, siphons had to be placed along the canal route as well as at farm bridges and road crossings. And to reduce the loss of water from the canal because of the pervious nature of the ground, some portions of the canal had to be overexcavated and then refilled with compacted impervious materials. Work on the canal including fencing was completed during September of 1964.

Another project on the South Fork of the Ogden River was the building of Causey Reservoir. As the early plans evolved for the Weber Basin Project, developers proposed that a dam be built on the South Fork of the Ogden River where Magpie Creek enters the river. At first this reservoir was to be called Magpie, and the Ogden Valley diversion canal associated with the project was identified as the Magpie Canal. After evaluating the geologic explorations and cost considerations, however, planners decided to move the reservoir site four miles upstream and name it after Causey Creek. The dam for Causey Reservoir was to be built of earth and rock like the Pineview and Wanship dams, have a capacity of 6,900 acre feet (the smallest in the system), and be two hundred feet high. The R. A. Heintz Construction Company, who had entered the low bid of $3,836,419.10, secured the contract for the Causey Dam on June 15, 1963. Construction of a new access road to Camp Kiesel, the boy scout camp that would be adjacent to the new reservoir, was performed by C. M. C. Construction Company of Great Falls, Montana, during 1962 and 1963. In building the road and excavating for the dam, construction crews had to blast the solid rock they encountered; this slowed their progress. Much of the material removed in constructing the Camp Kiesel road was stockpiled and eventually used in construction of the dam. Carved out of the steep mountainside adjacent to the reservoir, the Camp Kiesel road provided both access to the camp as well as a magnificent view of the reservoir and surrounding country including high red rock cliffs.

Two other projects initiated during the summer of 1962 that related to the Causey project were the clearing of timber and brush from the proposed reservoir floor and excavating the tunnel for diversion during construction and for the outlet facilities of the dam. Because the area was heavily timbered, brush and trees had to be removed by bulldozers and chainsaws and stacked in piles that would be burned in the fall. The excavation of the tunnel was to be undertaken by the Bud King Construction Company. The tunnel was carved through a formation of dolomitic limestone, which fractured easily but could be held in place by putting rock bolts in the roof of the tunnel and by placing wire mesh throughout the tunnel and steel rib supports; these supports were erected in the inlet portal. Other surprises awaited discovery by the constructions workers. While excavating for the spillway, crews exposed several large caverns (six feet deep, thirty feet long, and fifteen feet wide). These caverns were encountered adjacent to and at about the same level as the Camp Kiesel Road. To secure the area, the caverns were filled with a mixture of lean concrete placed by grout pipes. The concrete lining for much of the diversion tunnel was placed during the summer of 1963 as was the grouting and lining of the tunnel and pressure grouting the rock surrounding the intake and outlet works.

Concrete work on the diversion tunnel was completed by mid-October 1963, so that on October 17 the water from Causey Creek and the South Fork of the Ogden River could be diverted through the tunnel and work on the dam could move forward in earnest. High run-off water during the spring of 1964 (and unusually heavy rainstorms from June 17 to June 23) caused work to move ahead slowly on the dam. Large amounts of materials were placed for the dam during the summer of 1964, with much of it coming from areas adjacent to the dam, though zone 1 materials were obtained from the Weber County borrow area. Due to its slow start, by the end of 1964, the project was only about 70 percent complete although about 90 percent of the contract time had elapsed. Six months later, still only 73.6 percent of the work had been accomplished while 101.5 percent of the contract time had elapsed. In an effort to reestablish a new deadline, project officials considered the extenuating circumstances that affected the construction timetable, including the severity of weather, and extended the due date 145 days. During the same summer, a great deal of dam material was placed so that by the end of the summer more than 90 percent of the dam was completed. The zone 1 section of the dam was completed on September 20 and the zone 2 section on September 22. Work finished during the fall of 1965 included the placement of the remaining zone 3 material, of rip rap, and of the last of the concrete for the spillway. Only a small amount of work remained to be done on the dam during 1966, including electrical installations, gate controls, 4,672 feet of guardrail on the Camp Kiesel road, and the final placing of rip rap. Final work was completed on the dam during August 1966.

The Lost Creek Dam and Reservoir were part of the original plans for the Weber Basin Water Project. This dam was to be located on Lost Creek, a tributary of the Weber River, which enters the river at Devil's Slide. Located about twelve miles above Devil's Slide, the dam was to be constructed of earth and rock to a height of 220 feet, which would allow for a water capacity behind the dam of twenty thousand acre feet. The contract to build the dam was awarded to the Steenberg Construction Company on June 24, 1963; one month later the contractor began working. Early operations on this project included clearing and stripping the Lost Creek Canyon bottom and relocating the county road so that the new road could be located adjacent to the reservoir. By August preparations for the dam site construction included stripping materials from the site and placing zone 2 materials.

The inlet and outlet tunnels were also started at this time. The tunnel was excavated in sound limestone and the strength of the tunnel structure, just as in the Causey project, was insured by the placement of structural steel supports as well as roof bolts and wire fabric. Excavation for the cutoff trench began during the summer of 1963 and efforts to reach bedrock to firmly secure the dam through the trench continued throughout the summer. By September bedrock was exposed on the right abutment but excavation continued through the middle portion of the dam to the left abutment. The cutoff trench was excavated somewhat upstream from what would eventually become the middle of the dam. The dam embankment finally started to take shape during the fall of 1963, with materials that came mostly from excavating other features of the site including the cutoff trench. Drilling and grouting operations were begun during the fall to attach the structure of the dam firmly to bedrock.

During the winter of 1963-64, inclimate weather restricted activity at the dam site. In order to handle the spring run-off through the partially constructed dam, three thirty-six-inch culverts were placed side by side to divert the water to a stream bed on the left abutment. The placing of concrete as lining for the outlet tunnel began in April of 1964 with 180 cubic yards of concrete placed then and 940 cubic yards placed during May. Three grout pipes of varying lengths were then placed in each cavern to allow grout to be put above the concrete lining to fill the voids as necessary. This concrete work was completed by the middle of June and required 1,545 cubic yards for forty-six placements.

By the middle of May, the three bypass culverts in the Lost Creek Stream bed were plugged with debris, allowing flood water to wash into the cutoff trench. Sections of the road that paralleled the stream were washed out as the stream continued to flow out of its bed until the middle of June. Nevertheless, excavation for the boat ramp and its access road was begun and virtually completed during the month of May, while excavation of the spillway continued during June with drilling and rock blasting advancing rapidly during two ten-hour shifts a day.

At the same time excavation took place for the outlet works stilling basin. Concrete was then placed in the gate chamber and as footings for the outlet works stilling basin. As the outlet works tunnel was being grouted, it was necessary to fill caverns adjacent to the tunnel. During July a total of 1,040 sacks of cement and 3,108 cubic feet of grout sand were used to fill the caverns. The backfill grouting of the tunnel lining was completed during August with a total of 477 sacks of cement being used—or approximately a half a sack per linear foot of tunnel.

Work on the spillway and stilling basin of the dam took place during the late summer and fall of 1964. Grouting of the dam foundation was accomplished during the same period, and was sufficiently completed by September 5 when the first zone 1 materials were placed in the bottom of the cutoff trench. The placement of zone 1 and zone 2 materials and the compacting of those materials continued through the fall as did the grouting of the right and left dam abutments.

During October 1,095 lineal feet of holes consumed 8,465 sacks of cement, with one hole alone requiring 3,254 sacks. By the end of October more than 24,000 cubic feet of grout had been placed (along the bottom and sides of the dam foundation), which exceeded the quantity specified for the dam. An additional 4,158 sacks of cement were used in grouting operations along dam abutments in November. The stilling basin walls and the boat ramp were completed with concrete placements by the end of October just before the rain and winter weather halted work by the middle of November.

Optimistically, the Steenberg Construction Company again began operations during the early spring, but they were still subjected to messy weather; six inches of snow fell at the dam site during March, twenty-one inches (containing one and one-half inches of water) came during April, and 1.71 inches of precipitation fell in May in the form of snow and rain. Work resumed fully during the summer and fall when the contractor placed the embankment material for the dam and the concrete for the dam spillway. A total of 2,816 cubic yards of concrete was placed during 1965 on the project with almost all of it going in the spillway.

As the work progressed rapidly during the summer and the dam neared completion, the dam site received an unusual variety of visitors, including local and national water officials as well as five visitors from Ankara, Turkey, during August and two from Afghanistan and two from India during September.

By the end of August 70 percent of the work on the dam had been completed within 69 percent of the time, but during September problems concerning the performance of the Steenberg Construction Company on the Lost Creek Dam started to appear. The construction company determined that it had been losing money on the job and decided it would end its work on the dam during September. The contractor began removing his equipment from the dam site on September 20 to which project officials responded by terminating the company's contract on September 28, 1965, because of default. This action almost completely halted work on the dam for the remainder of the year. In an unusual move, the contractor erected forms for the spillway bridge before leaving the job. Work on this phase of the project was eventually completed by the Olsen Construction Company.

During the fall of 1965, the Gibbons and Reed Construction Company received authorization to build a dike, which was located upstream from the existing zone 2 phase of the dam. This construction was to prevent the existing dam structure from being washed away during the spring runoff.

Also during the winter of 1965-66, the Bureau of Reclamation initiated a study concerning the surety of the Steenberg Construction Company and the desires of the bonding company. The bureau informed the bonding company that if they declined to complete the work under Specification No. DC-5935, bids for the completion of the Lost Creek Dam would probably be issued in the spring of 1966. Subsequently, on May 15, 1966, the LeGrand Johnson Construction Company of Logan, Utah, signed a contract for the completion of the dam. The contract to complete about 30 percent of the project came to $1,052,269. Though LeGrand Johnson had until December 1967 to complete the work, they finished in January 1967, with final payment being authorized by a letter from the chief engineer of the Bureau of Reclamation dated March 27, 1967.

The final major project on the Weber River was the construction of a new dam at East Canyon. The original East Canyon Dam had been constructed beginning in 1896 by the Davis and Weber Counties Canal Company. The dam site had been discovered in 1894 where the original earthen dam, built ninety-five feet high, could store 3,850 acre feet of water. The newly stored water proved so valuable during the growing season that the height of the dam was increased to 145 feet in 1902, which allowed a capacity of 13,800 acre feet. Samuel Fortier had been employed as the consulting engineer and engineer A. F. Parker was to prepare the plans for the new dam. The construction of the new concrete dam was completed during the summer of 1916 at a total cost of $175,054.74 to the Davis and Weber Counties Canal Company.

An enlargement of the East Canyon Reservoir was part of the original Weber Basin development plans. On April 6, 1964, the E-W Construction Company and L. D. Shilling Company was awarded the construction contract with a bid of $2,014,054. This new reservoir would have a capacity of 54,000 acre feet, and the new dam would be a double-curvature thin, concrete-arch structure approximately 260 feet high and 440 feet wide at the crest. The construction of the new dam, which was to be built just downstream from the older concrete dam, required involved negotiations with the owners of the old dam and reservoir. The Davis and Weber Counties Canal Company would retain annual water rights to the first 28,000 acre feet of water in the new reservoir.

During June of 1964, project planners decided to allow the contractor to use the old dam as a roadway in building the new dam. The use of the old dam contributed to the speedy completion of the new structure. During the same month, four subcontractors were engaged to push the project ahead—Miya Brothers Construction Company of Ogden, Utah; Sundwall and Company of Portland, Oregon; Tri-State Drilling Company of Lyons, California; and Ephrata Pre-Mix Company of Ephrata, Washington. The Ephrata Company began construction of a concrete batch plant immediately upstream from the right abutment of the existing dam. Sundwall and Company started excavations for the new dam foundations during July though they had to move ahead cautiously; to avoid vibrating the existing structure they could use only a limited amount of explosives. Another important construction element was the construction of a roadway across the then existent dam. Use of this roadway allowed easy access to equipment from both sides of the canyon. Finally, the Titan Steel Company of Salt Lake City was contracted to install steel supports for the roadway across the dam.

John Sundwall and Company, in their excavation of the thrust block and the spillway areas, used the pre-splitting method of drilling and blasting. In addition, accelerometer tests were made to record the effects of blasting on the existing dam, and close cooperation was maintained between the contractor and the inspectors to keep the vibration of the dam within specification limits. Work on the excavation moved slowly. The dam roadway was completed by mid-August, and work on rerouting the county road around the new dam and reservoir was an ongoing project that included the construction of a boat ramp and a trailer parking area.

Because of the concern for the original dam and the limited blasting, the contractor realized in the middle of October that he was approximately ten weeks behind his proposed construction program and requested an extension of time for this phase of the project. He explained that extraordinary care had to be taken in blasting operations for the new dam because in some instances the new excavations were less than ten feet from the old dam. The necessary extensions were granted.

The removal of a six- to ten-foot lift of material from the keyway was accomplished in six steps:

1. A Bureau of Reclamation survey party staked out lines and grades.
2. Pre-split holes around the periphery of the area were drilled to be blasted.
3. The pre-split holes were blasted after all other holes were deck loaded.
4. Holes for normal blasting were drilled within the pre-split area.
5. The holes for normal blasting were loaded with explosives, and then blasting occurred in one or more rounds, depending on the size of the blasting round and the calculated effect on the old dam.
6. The blasted rock was removed and the area cleaned up in preparation for a new six- to ten-foot lift.

Considerable variation in the effect of the blasting on the old dam, depending on the elevation of the blast and the locations in the keyways, was noted in ongoing accelerometer tests.

Another project undertaken during the summer and fall of 1964 and completed in November was the excavation of the old rockfill dam constructed by the Davis and Weber Counties Canal Company. During the same month unstable rock, located upstream from the left keyway, was removed at the suggestion of Neil Murdock, regional geologist for the Bureau of Reclamation. Breaching the old concrete dam then began during the fall of 1964 by drilling numerous holes through the dam within a five-foot circle marked on the upstream face of the dam. This process was completed by the end of the year.

Unlike many of the other Weber Basin Projects, work continued on the East Canyon Dam rather regularly throughout the winter of 1964-65 with further excavation on the dam keyway and grouting operations around the dam foundation. Concrete placement for the dam began during February as this was to be constructed of blocks of concrete placed at different times with the earliest concrete placed in blocks four and five. During the first months of 1965, 13,668 cubic yards, or about 40 percent of the total amount of concrete was placed (February, 758 cubic yards; March, 2,958 cubic yards; April, 4,746 cubic yards; and May, 5,206 cubic yards). The concrete was mixed on the site with the aggregate materials coming from the mouth of Weber Canyon and the cement from the Devil's Slide plant. Although the old concrete dam had been breached, some ten thousand acre feet of water was stored behind it during the spring of 1965 for release during the irrigation season.

During June the placement of another 5,979 cubic yards of concrete brought the new dam to the approximate level of the old dam. Summer weather allowed work on the dam to go forward rapidly with 5,870 cubic yards of concrete placed during July and 5,255 cubic yards during August. By August crews began the concrete work on the spillway and thrust block. Swift progress continued on the dam, achieved in part by the use of both a day and an afternoon shift of construction workers. The only real downside to the project was that the bidding schedule underestimated the amount of grouting needed around the dam. The schedule had listed ten-thousand sacks of cement for the entire process, but by the end of August, 28,718 sacks had been used.

By the end of September, with 2,141 cubic yards of concrete placed that month, concrete had been placed in all of the blocks of the dam. A shale seam protective wall was then positioned on the downstream side of the dam on the left wall of the canyon; this required 192 cubic yards of concrete. This seal, located on the same side of the dam as the spillway, would hopefully keep erosion and rock fall to a minimum under the spillway. Concrete work continued during October, although at a slower pace than earlier because the placements were smaller and more difficult to make, particularly in and around the spillway. The work on the dam in November included the placement of concrete in curbs and upper parapet walls with 556 cubic yards of concrete placed compared with 1,035.5 cubic yards placed during October. Also during October, Richard Cox of Orangeville, Utah, who had been awarded the contract to clear the bottom of the reservoir, completed 70 percent of his work, so that by the end of November 98.6 percent of the project was completed in 74.4 percent of the time.

William R. Croseclose, construction engineer on the project, tripped the gate on the last bucket of concrete during a special ceremony held on January 14, 1966. This was the last of the more than 36,000 cubic yards of concrete placed in the 210-foot high arch-type dam, which when completed was 440 feet long at the crest. James M. Ayers, superintendent of construction, Keith Jensen, president of the Weber Basin Water Conservancy District, and Wayne M. Winegar, district manager, were all involved in the special ceremony. Water storage in the newly completed reservoir started on December 21, 1965, and by the end of the year seven hundred acre feet had accumulated in the reservoir. The dam became fully operational by the end of January 1966.

With the East Canyon Dam, the last major project to be completed on the Weber Basin, running smoothly, the Bureau of Reclamation could turn its attention to some of numerous smaller projects awaiting completion. The construction of observation wells and pilot drains had been among the earliest of projects supervised by the bureau. Assistance in land acquisition had been an ongoing joint venture between the bureau and the Weber Basin Water Conservancy District.

The bureau assisted the district during 1956 in purchasing the rights to five thousand acre feet of water in Echo Reservoir on an annual basis. This agreement was made with Davis County. A particularly important contribution of the bureau was the improvement of existing irrigation laterals and the construction of new laterals. Laterals along Ricks Creek were completed during 1960 and the North Davis laterals were completed in 1962. Stream measuring controls and devices were installed both above and below Pineview Reservoir during 1958.

Throughout the construction period, officials of the Bureau of Reclamation and the Weber Basin Water Conservancy District continued negotiations with the public and with private concerns for the sale and exchange of water. In one case, a Utah State Board of Health ruling led to such an exchange. After the state board issued a statement saying that surface streams used by many cities and towns for culinary purpose were unsafe, numerous people from nearby the Weber Basin Project facilities asked to use the facilities to transport their stream water to the Weber Basin filtration plants for treatment and distribution. Such arrangements were made, particularly in Davis County.

Stone Creek, located on the north side of Bountiful, is an example of introduction or delivery into Weber Basin facilities. A contract for capturing Stone Creek water and placing it into the system (the Davis Aqueduct) was awarded to the R. W. Coleman Company in August of 1963, and this project was completed during November 1964. The Bureau of Reclamation also supervised construction of buildings and facilities for the Weber Basin Job Corps Conservation Center; much of this construction occurred during 1965 and 1966.

From 1967 through 1968 many projects undertaken by the Bureau of Reclamation on the Weber Basin Project reached completion. Of the eighty-five bureau employees involved in the project at the beginning of the year, only thirty-two remained by the end of 1968. Of the nine construction contracts in effect during 1967, seven were completed and the other two were to be completed in 1968. Projects for 1967 included equalizing reservoirs for Farmington and Woods Cross, laterals, access roads, and Gateway Canal repairs. Of the twelve projects contracted during 1968, eight were completed by year's end and four were finished by the middle of 1969.

In general the projects of 1967, 1968, and 1969 were the final elements of the undertaking that had been almost two decades under construction. Each major project was directed by a resident engineer who was responsible for supervising the construction work. The continuity of the project was maintained by Clinton D. Woods, who served as the project head from 1954 through 1965 (when Rex Greenhalgh became the project head) and Francis M. Warnick who worked as planning officer prior to the beginning of the project and then served as assistant project head (1952-1964) with responsibilities including contract negotiations, contact with irrigation companies including the selling of project water, and the development of right-of-ways for the project. The completion of the project was a success. It exemplified cooperation between a major governmental agency and private business and public interests—the Bureau of Reclamation and construction companies and private irrigation companies. And to manage the newly created reservoirs, canals and pipelines, a new entity had been born—the Weber Basin Water Conservancy District.

Chapter 8

The Weber Basin Water Conservancy District and Recent Water Use

With the passage of the enabling legislation for the Weber Basin Project during the late summer of 1949, and the move toward construction, the most necessary task facing the organizers was the formation of a conservancy district. Plans to accomplish this first materialized when passage of the Weber Basin legislation appeared likely. A particular concern was that the organization be able to contract with the Bureau of Reclamation for building the facilities on the river and to repay the federal government over an extended period of time.

The Davis-Weber Counties Municipal Water Development Association, with Ezra J. Fjeldsted as secretary, provided much of the force by urging cities and towns to join the association and then by pushing for organization of the conservancy district. Fjeldsted, in one of numerous letters to officials of Weber Basin towns, noted that towns and cities not involved with the conservancy district might pay "double" the regular rates for water that those who are members would pay.[1]

Fjeldsted arranged a meeting for June 9, 1949, at the Morgan County Court House for a presentation by Harold Ellison, Joe Johnson, and Fjeldsted himself to the people of Morgan County, but particularly to Morgan county commissioner James Palmer, Horald Clark, and George Florence. Engineer Win Templeton and Judge J. A. Howell stressed the need for Morgan County to join in the conservancy district. Howell noted that this would allow Morgan to participate in first claim on any excess water with the new conservancy district. George Florence, president of the Morgan County Water Users Association, voiced his organization's interest and promised to follow-up on the proposal at an early date. In order to organize the conservancy district with a financial base, Fjeldsted noted that each area was being assessed one-tenth of a mill of that area's 1948 assessed valuation. For Davis County in 1949 that amounted to $958.96.[2]

Advocates then arranged a meeting with Summit County water users for July 19, 1949, at Coalville. In addition to Ellison, Fjeldsted, and Johnson, DeLore Nichols, J. A. Howell, and C. J. Olsen, who was vice president of District No. 2 of the Utah Water Users Association, attended, as did twenty-two Summit County water users.

The Morgan County meeting had nine Morgan County water users present. Ellison told the Summit County group that the last legislature had passed amendments to the state conservancy law, allowing more than one county to join in the
formation of a conservancy district. Ellison pointed out that the Davis-Weber
Counties Municipal Water Development Association had sponsored the legislation and that he was most hopeful that Summit County would join the other three
counties in forming a broad program. The Summit County group expressed interest in becoming involved with the conservancy district.[3]

On August 19, a meeting of some sixty water users from the four-county area
was held in Layton. Harold Widdison presided at the meeting where participants
discussed organizing into a conservancy district. A motion passed the body assembled that a committee of eight be appointed (two for each county) to meet
with the Bureau of Reclamation to consider recommendations concerning
boundaries, divisions within the district, the number of directors, and the method
for choosing directors and then report back to the larger group. Members from
each county chose delegates to the committee from those in attendance from that
county. Harold Ellison and Joseph Johnson represented Davis County, with Ed
Sorenson and Maurice Boyden from Summit County, George Florence and
Horald Clark from Morgan County, and David Scott and Elmer Carver from Weber County. E. J. Fjeldsted was to serve as secretary for the group.

This ad hoc committee with Ellison as chairman began to meet with Bureau
of Reclamation officials E. O. Larson, F. W. Warnick, Hollis Hurst, and E. J.
Skeen. Skeen outlined the legal ramifications of the water conservancy district,
while Warnick detailed the engineering progress of the bureau on the project. As-
signments were made to contact various organizations to get their reaction to the
conservancy district plan and to present to them the ideas of the district in a very
favorable light. Organizations to be contacted included the canal companies,
Utah Power and Light Company, Union Pacific Railroad, Denver and Rio
Grande Western Railroad, Southern Pacific Railroad Company, Bamberger Electric Railroad, Mountain Fuel Supply Company, Utah Oil Transportation Company, the Ideal Cement Company, and mining interests at Park City. It was suggested at an August 23 meeting with the Bureau of Reclamation that all such
contacts and reactions be completed by the middle of September. Ellison's committee proceeded to contact the entities they had agreed upon and began to sell
the conservancy district, including its ability to tax, in a positive way. Most companies responded to the suggested district and more available water in a very
positive fashion. Fjeldsted as secretary for the committee did much of the legwork. He not only held several secretarial positions with water groups but also
served as the secretary-manager for the Ogden Chamber of Commerce. Members
of the four-county, eight-member committee, reported back in a September 22
meeting that irrigation and canal companies generally favored the new organization. Ellison's committee explained that the conservancy district could tax all ap-

propriate property within its boundaries up to a one mill limit, though a one-half mill limit was the maximum possible until construction was underway.

The special committee chaired by Harold Ellison met on October 6 at Layton when they unanimously approved the name "Weber Basin Conservancy District." They also discussed and agreed that the four county boundaries would serve as the boundaries for the district. The third question before Ellison's special research committee was how should the district be divided into divisions. David Scott made a motion to divide the district into five segments with the divisions and directors to be as follows:

Division	Directors
Davis County	3
West Weber County	3
East Weber County	1
Morgan County	2
Summit County	2

Maurice Boyden seconded the motion, which then passed unanimously.

The special committee then took these three recommendations to the larger group of thirty-five representatives from all four counties later in the evening. They too agreed that all three of the recommendations should be passed unanimously. Harold Ellison then noted that they would need to employ someone to coordinate all of the efforts concerning the district. E. J. Fjeldsted suggested that Fred M. Abbott of the Ogden Chamber of Commerce be approached for the position. The larger group then agreed that Ellison's eight member executive committee would do the hiring and set the terms of employment.

The conclusions of this meeting were forwarded to E. O. Larson who, in correspondence with Fjeldsted, proposed an October 25 meeting in the offices of the Bureau of Reclamation. In addition to Ellison's eight-member special committee and Secretary Fjeldsted, Fred Abbott, DeLore Nichols, and J. A. Howell attended the meeting. Larson, accompanied by four of his assistants, represented the bureau and offered a number of modifications, all of which were discussed at length and finally accepted by Ellison's special committee. The changes included: changing the boundaries of the district so that only that portion of the Summit County in the Weber Basin would be part of the district; leaving Park City out of the district so that it could join with the Central Utah District (part of Larson's recommendation for Park City related to the advantages that would come from draining the deep mines of the area of water); changing the name of the district to the Weber Basin Water Conservancy District; and, the issue discussed most fully, adjusting the number of directors for each area. Larson's suggestions, underlined by Bureau of Reclamation attorney J. S. McMasters, were to realign directors; since most of the repayment would come from the lower counties, they should have most representation. After lengthy discussions, Joseph Johnson proposed the following scheme, which was seconded by Edward Sorenson:

Division	Directors
Davis County	3
West Weber County (that part west of the Salt Lake Meridian)	3
East Weber County (that part east of the Salt Lake Meridian)	1
Morgan County	1
Summit County	1

All issues proposed by Larson were passed unanimously.[4] With the blessing of the Bureau of Reclamation, by the middle of November of 1949, all of the hurdles had been cleared for circulation of petitions for the formation of the district.

In order to proceed, Fred Abbott collected information concerning the value of irrigated land in the four county area for tax purposes. As of December 2, the value totaled $10,409,959 divided on the following basis: Davis County $5,343,508; Morgan County $523,826; Summit County $1,071,130; and Weber County $3,471,495. Although the land represented both city and county land, Abbott was interested in the total figure. As he, on behalf of the Ellison Committee, began to look at taxing within the one-half to one mill limit, Ellison wrote to state engineer Harold Linke to inform him that those he represented in Davis and Weber counties were opposed to raising the mill levy for any conservancy district and allowing them to tax more than one mill.[5]

During December, Ellison, Johnson, Fjeldsted, and others began to meet with community leaders from cities and towns in the proposed conservancy district. In particular, they wanted to explain the conservancy district concept and recruit supporters who would take petitions concerning the district to their areas and secure enough signatures so that the concept could be voted on.

The executive committee of the district in embryo chaired by Ellison also held a meeting with the directors of the Ogden River Water Users Association on December 29 who wanted to rewrite their repayment contract with the Bureau of Reclamation and extend the forty year contract to seventy years. Specifically, the association wanted the assistance of the soon-to-be conservancy district's executive committee in approaching the bureau. Ellison was empowered to appoint a subcommittee to work with the Ogden River Water Users Association; this action became a stepping-stone toward maintaining good relations between the conservancy district and water users presently active in the Weber Basin. Another step along the path of moving toward construction on the Weber Basin Project came in November 1949 as the Bureau of Reclamation established a Weber Basin Project office in Ogden.

One major obstacle to overcome before funding for the project would be granted by the federal government were the objections put forth by President Harry Truman in his three-page letter dated August 30, 1949. Truman noted he signed the bill with "reluctance" because it was not fully investigated before being approved. The bill, based on information gained in a preliminary Bureau of Reclamation report rather than a final study, had not been thoroughly reviewed by other interested agencies or by the Executive Office. Truman's specific reservations included the following:

1. Repayment of the project was to occur over sixty years rather than the usual practice of forty years under reclamation law.
2. Facilities to be constructed and used for recreational use amounted to $4,656,000. Recreational facilities were not authorized under current reclamation law and Truman was concerned about both the amount and precedent.
3. The interest rate on monies loaned by the federal government was too low.
4. The Army Corps of Engineers had yet to have adequate opportunity to review the project in reference to flood control.
5. The Department of Agriculture had not reviewed the report and thus were not able to express views concerning the agricultural and economic feasibility of the proposed plan.

Truman stated emphatically that until better preparations had been accomplished, including the ones he listed, he did not intend to have the project begin.[6]

The Bureau of Reclamation sought the help of Utah State Agricultural College employees in putting together a detailed report and analysis to answer questions concerning the project and get funding started on the project. At a January 9, 1950, meeting in Logan, eleven representatives from the USAC, one from the Bureau of Plant Industry, three from the Soil Conservation Service, including George D. Clyde, and four from the Bureau of Reclamation first met in the president's office and then in a follow-up meeting on campus. They established four committees to expedite the study and to recommend methods of cooperation between various agencies. The committees addressed land classification; repayment analysis; land development, including drainage, cleaning, land leveling, and farm distribution systems; and sample techniques for crop yield and pertinent information. Committee membership was assigned making certain that all agencies were well represented.

Over the next two years members of the four committees met together and did individual research on the agreed project areas. By 1952 they issued a forty-page report under the authorship of Walter U. Fuhriman, George T. Blanch, and Clyde E. Stewart. All three, economists by trade, used the information compiled over the previous two years by the four committees established in January of 1950. The report, *An Economic Analysis of the Agricultural Potentials of the Weber Basin Reclamation Project*, was published under the auspices of the Agricultural Experiment Station, Utah State Agricultural College. The well-coordinated and well-written study addressed the questions raised about the project, particularly those of President Truman, and concentrated almost exclusively on Wasatch Front agricultural areas. Most of the population as well as much of the repayment resources, the report said, would come from the Wasatch Front agricultural areas. Interestingly, although much of the anticipated revenue was expected from the use of irrigation water on agricultural cultivated crops, in the four decades since the study, large amounts of the agricultural land have become urban.

One of the interesting statistics concerned the proportion of survey farms reporting shortages of irrigation water on irrigated land in 1949 (Table 4).

Lands associated with fruit production were identified as being particularly productive when provided with additional water. One conclusion drawn from the research was that farms in the basin would become more "satisfactory" and that agriculture could contribute "substantial" amounts toward repayment of operating and construction costs. The economists did not forecast or suggest definite profits or charges for the agricultural lands within the project, but did suggest that water charges should be less than net returns to water at the prevailing price level.[7]

This detailed, tabled, and graphed report seemed to be just the kind of study Truman had asked for and the Bureau of Reclamation was interested in. It also provided a detailed agricultural analysis and inventory and helped to bridge the communication problems with Department of Agriculture people. Appended to the publication was a December 1951, Department of Interior Weber Basin Project multi-colored map. This map indicated two possible dams on the upper Weber (Magpie and Larrabee) that were later found unsuitable for building, but in large measure the project was ready to begin.

With one hurdle overcome, proponents of water development still had to form the basin area into a conservancy district. Fred Abbott, E. J. Fjeldsted, DeLore Nichols, and Joe Johnson continued to encourage individuals in the four-county area get county residents to sign the petitions so that organization could begin. Letters were written, phone calls made, and personal contacts were made to push the petition drive. One quotation from a letter is typical of the many sent: "If as you planned, the proper committee was selected and given a list of names to contact, and every man got busy it would in the whole be a small task in getting the required number of names."[8] The petitions could only be signed by property owners whose property was assessed at a value of $300 or more. The committee also had to secure from the incorporated towns and cities signatures of 5 percent of the landowners within each or one hundred signatures, whichever was lesser. In the unincorporated areas of the counties, either five hundred signatures or signatures of 10 percent of the landowners were required, whichever was lesser.[9]

TABLE 4
Shortages of Irrigation Water[10]

	Lake Plain	Foothill	Mountain Valley
Reporting late season shortage	21	65	69
Reporting shortage throughout season	79	35	31

Although stressing water development along the Weber River, representatives from the four-county area were also active in the Utah Water Users Association and the Utah Water and Power Board. On January 28, 1950, Harold Ellison as president of the Davis-Weber Counties Municipal Water Development Association, wired Oscar Chapman, secretary of the interior, urging the approval of dam projects on the Green River at Split Canyon and Echo Park. This was done in part because of a request from State Engineer Harold A. Linke.

Weber Basin leaders continued to petition their representatives in Congress to provide monies for the project for both planning and construction. Senator Arthur Watkins supported the successful effort to push an allocation of $350,000 for the planning phase through the Senate. Watkins noted that earmarking this amount for planning would be essential. "The President was able to save his face—that is, not put any money in for construction of the Weber Basin Project, and at the same time satisfy us out home because he put it in under some other item."[11] In addition, in a joint letter from Senator Elbert Thomas and Representatives Bosone and Granger, Johnson was informed that an additional $30,000 had been sent to the Salt Lake office of the Bureau of Reclamation to see it through 1950. The three also told Johnson that they were doing "everything possible to insure the earliest possible clearance so that appropriations might be requested for the actual construction of the Weber Basin Project which means so much to your territory."[12] Interestingly, the Thomas letter to Johnson did not mention the substantial monies that Watkins had alluded to and were earmarked for the project.

During February 1950, the subject of water rates surfaced in Ogden City. M. H. Harris, executive secretary of the Utah Taxpayers Association, informed Ogden Mayor Rulon White that because Ogden's water rates were very low, they should be raised to keep the water department on a "pay-as-you-go basis" and avoid expensive bonding for improvements and expansions. Harris noted that in an analysis of ninety-three cities with populations similar to Ogden (40,000 to

100,000), Ogden's water rates were very low. In the category of one thousand cubic feet consumption, only two cities had lower rates, and Ogden's rate of $.75 was only 39 percent of the average of $1.94 for the cities studied. In the bracket of ten thousand cubic feet of consumption, Ogden's rates were 80 percent of the average. It was Harris's contention that water rates should be raised, particularly in the lower consumption categories.[13]

Fjeldsted, especially interested in Harris's suggestions and study, corresponded with him concerning the matter. Fjeldsted was also instructed by Frances Warnick, area engineer for the Bureau of Reclamation, about the gathering of petitions for the organization of the conservancy district. Warnick explained that when ownership of property was listed in the name of both husband and wife, both signatures should be added to the petition so there would be no question as to the adequacy of signatures or the various petitions.

By March and April of 1950, most petitions for the conservancy district were completed. As officials, including J. A. Howell, evaluated the petitions, county water users were informed of problems or needed signatures. For example, petitions from six Davis County areas (Bountiful, Centerville, Layton, East Layton, Sunset, and South Weber) had just barely enough signatures to qualify, and Howell suggested obtaining more so the margin would not be so close.[14] County water users held meetings to suggest nominees for the board of the new conservancy district. During the same period, the Davis-Weber Municipal Water Development Association continued to meet, lobby for water use, and collect dues and assessments. During 1950, the association collected $14,505.52 from communities in the two counties and from Morgan County.

Senator Arthur Watkins continued his push for construction of the project. In a letter to President Truman, Watkins suggested that since 7,608 workers in Utah were drawing unemployment insurance and many of these were out-of-work miners, early construction on the Weber Basin Project would provide much-needed employment. Watkins suggested that out-of-work Park City miners would particularly benefit from construction, which were near their present residences. The early delivery, Watkins noted, would be of great assistance in the Weber Basin. Watkins finally noted that organization of the conservancy district was proceeding at a rapid pace. Truman's assistant, John R. Steelman, courteously but firmly answered Watkins's letter, reminding him that the conditions Truman had set forth when he signed the bill had not yet been met. This was particularly true in the area of preparatory work. Steelman suggested that Truman would not authorize construction funds until the work was done, and also since a large volume of work was already underway, consideration of funds for construction of the Weber Basin Project for the 1951 budget was very slight.[15]

Watkins continued to be dauntless as the Washington champion of the project. He received much support, pressure, and information from Nichols, Johnson, and Fjeldsted. Johnson and Nichols wrote to suggest that Davis County had five thousand acre feet of surplus water in the Echo Reservoir, which could be used if the proposed aqueduct were completed immediately. They also mentioned that Bountiful City had just denied the extension of the city limits, to allow for the construction of 1,100 new homes, because of insufficient water.[16] By this time Watkins was writing regularly to Johnson and referring to him as "Dear Friend Johnson." Watkins explained to Johnson that both the president and the Bureau of Reclamation seemed to be reticent on the project, but to "Rest assured that I shall do all in my power to expedite this work." Then in a personal, penned note, Watkins revealed that partisan politics were involved. "I expect a break in this attitude soon. They must do something to win this fall's elections and an early slant on the Weber could be the answer. They are not likely to let me announce it, however."[17]

Johnson responded by sending a two-page letter to all members of Utah's congressional delegation underlining facts and statistics, including the doubling of Davis County's population from 1940 to 1950 (15,774 to 30,771), the need for flood relief, the saving of irrigation water presently flowing unhindered into the Great Salt Lake, and the impact of the project on growth, particularly of defense industries.[18]

On April 17, 1950, a petition bearing 24,671 signatures of property holders from the four-county area was filed with the Second Judicial District of the State of Utah for the organization of the Weber Basin Water Conservancy District. A hearing on the petition was set for June 26, and on that day Judges John A. Hendricks and Charles G. Cowley established the district and appointed the directors shown in Table 5.

At its first regular meeting, set by the court, the board elected Ward Holbrook of Bountiful president of the district and chairman of the board. E. J. Fjeldsted received the appointment of temporary secretary while a search was conducted for a permanent manager-secretary. At this first meeting the board set a one-half mill tax to meet a $55,000 budget, to be collected on assessed valuations within counties by county assessors. The initial meeting was held in the law offices of Howell, Stine, and Olmstead, and thereafter moved to the chambers of the Ogden City Commission. It was determined that directors would be paid a salary of $500 yearly.

From a list of over a dozen applicants, which included DeLore Nichols, David Scott, and Joseph Johnson, the board selected E. J. Fjeldsted to serve as secretary-manager of the district with a salary of $10,000 and the opportunity to keep his position as secretary-manager of the Ogden Livestock Show. This board now became the major advocate for water development on the Weber River. With its multi-county taxing power and coordinating ability, it served to weld water interests on the Weber River together. It would help to provide services that smaller groups could only dream of. Nevertheless, until the very end of its organization, the district was opposed by various individuals, including some who had major water interests but saw the creation of the new district and subsequent construction of projects as a lessening of influence and power. Some opposition, for example, came from directors of the Davis and Weber Canal Company.[19]

TABLE 5
First Directors, Weber Basin Water Conservancy District

County	Director	Term
Davis	Ward C. Holbrook	3 years
Davis	Harold E. Ellison	2 years
Davis	LeRoy B. Smith	1 year
Weber	Lewis Spaulding	3 years
Weber	W. Rulon White	2 years
Weber	J. A. Howell	1 year
Weber (East)	D. D. McKay	1 year
Morgan	Horald G. Clark	2 years
Summit	Edward Sorenson	3 years

The board of the conservancy district became the major partner in the development of the Weber River with the Bureau of Reclamation. E. O. Larson and Francis Warnick met regularly with the board to give progress reports, identify problems, and seek assistance. On August 11, Larson explained the project in detail, noting various options that still remained open. Included in his discussion were responsibilities of the district, negotiations needed with the various water groups, particularly the Ogden River Water Users Association, the need to keep informed of the studies concerning the project, and the various aspects of the project and alternatives. One question that received much attention related to the proposed Perdue site on the upper Weber River. The Summit County people were interested in having part of the project benefit them directly, and were disappointed when Francis Warnick informed the conservancy board in October of 1950 that the site at Wanship was much more desirable than the Perdue site.[20] The cost would be lower in building a dam at Wanship, and the stability of the ground was much better there. Investigations discovered that there had been some rather serious faulting at the Perdue site. A final consideration was that the Perdue Dam would have to be built 200 feet high to impound 60,000 acre feet of water, while the Wanship Dam need be only 140 feet high to impound the same amount of water.

The Summit County people were unhappy with the decision and over the next two decades asked for some type of up-river storage facilities. By the summer of 1984, the board of the conservancy district decided that the time had come to honor the earlier commitment for such a facility. This would come with the enlargement of the Smith and Morehouse Reservoir.

The original Smith and Morehouse Dam had been built by using scrapers and other horse-drawn equipment. The original dam leaked, and some people felt that at the very least the dam needed to be repaired. Water demands had increased on the upper Weber, and the issue of further demands by Park City remained an unanswered question. The original dam held a capacity of 1,040 acre feet of water, and the dam designed in 1984 would hold a capacity of 8,350 acre feet. This total figure included 750 acre feet that was to be reserved as a conservation pool for fish and wildlife, with the balance of 6,560 acre feet available to Weber Basin for sale to individuals and groups anywhere below the dam on the river system. The new dam was to be raised to a height of eight-two feet.

Construction on the dam began in the autumn of 1984 with Gibbons and Reed Company of Salt Lake City as the general contractor. The dam was finally completed and ready for use in the early spring of 1988 at a cost of $11.7 million. There were a number of substantial difficulties related to the construction of the dam and its related facilities, including cost overruns. As the project proceeded, costs began to exceed the original bids, which were near the $6 million mark. While construction continued so did negotiations about costs, estimates, and overestimates. The construction company and project officials eventually agreed on cost figures for the dam, after lengthy discussions and negotiations, but without resorting to the courts for final settlement. The Smith and Morehouse project was generally a much smaller project when compared to other reservoirs in the system, but benefits to the upper basin, to campers, and to fish and wildlife conservation were substantial.[21]

As a result of their August 1950 board meeting, the directors sent a detailed three-page memorandum to the president of the United States and to the members of the Utah congressional delegation. By linking the Weber Basin Project to the Korean War effort, board members tried to expedite the approval of monies for construction.

The board continued to review in detail the progress of the various studies on the project, including the comprehensive report being prepared under the auspices of the Utah State Agricultural College. Win Templeton was selected as engineer for the district during December 1950.

Meetings of this board often included representatives from other water users groups working on similar problems. One common problem they encountered was flooding on the river. On December 1, 1950, representatives from the U.S. Army Corps of Engineers met with Weber Basin officials and representatives of the Weber River Water Users Association to discuss the use of Echo Reservoir for flood control. H. R. Reifsnyder and E. A. Sugg of the Corps of Engineers reported on the river's flooding from 1928 to 1947. They suggested that if the water in the reservoir was "manipulated" in the right way, serious flooding could be avoided and that the Weber River Water Users Association should control the outflow of the reservoir based on forecasts furnished by the Bureau of Reclamation.

D. D. Harris, manager of the Weber River Water Users Association, made it clear that his group was happy to comply with the suggestions of the Army Corps of Engineers, but they should not be held accountable for flood damages downriver caused by the water releases. The engineers believed that the capacity of the river below the dam could hold as much as three thousand second feet, but others at the meeting contended that the river could not be kept within its channel if there was much more than one thousand second feet.

Although flood control was an important aspect and benefit of the project, the problem of jurisdiction has remained fuzzy over the past three decades. This was particularly evident during the major flood years of 1983 and 1984, as each group—Army Corps of Engineers, Weber Basin Water Conservancy District, Bureau of Reclamation, and water users on the Ogden and Weber rivers—sought to absolve themselves of blame for floods. Nor were the groups eager to work with local government groups to spend resources to dredge the river and perform other needed flood relief acts. Part of the difficulty arose from the many agencies that have their fingers in the flood-control pie. Each seems willing to wait while another agency attacks the problem.

There is no doubt that the new and enlarged dams held back large amounts of water that would otherwise have cascaded downward in an uncontrolled fashion. Evidence of major flooding came during the spring of 1952, which washed out major roads and bridges and inundated farmland. The effects of the flooding on the Weber River were particularly graphic at the mouth of Weber Canyon near Uintah. Flood control, including such issues as water releases from reservoirs and river dredging, would be ongoing points of discussion and action by water users along the river centered on the Weber Basin Water Conservancy District Board. Related to flood control was the replacement of a large section of wooden pipeline in Ogden Canyon during the spring of 1984. A massive earthslide brought on by extremely wet weather and water-saturated soil crushed the old wooden pipeline. The Ogden River Water Users Association, with minimal inconvenience to water users downriver, replaced the wooden pipe with steel pipe.

On August 28, 1951, a public hearing was held at the auditorium of Weber College in Ogden concerning a proposed plan and budget for flood control along the river. The plan presented by the Army Corps of Engineers suggested that if channel capacities on the river were properly maintained and the proposed reservoirs on the river system were built that flooding on the river could be controlled. The corps estimated that it would cost $500,000 to improve the channel and protect banks from erosion from Morgan through the city of Ogden along the Weber River. Of the initial $500,000, $360,000 was to come from the federal government and the remainder was to be provided from other sources. They further recommended that, following this one-time river improvement for flood control, local governments allocate an annual sum of $20,000 per year be spent to maintain the river against flooding.[22]

Through 1951, the board of the conservancy district spent long hours with representatives of the Bureau of Reclamation concerning the Weber Basin plan discussing the location of reservoirs and all possible alternatives, including the use of ground water. By the end of the year, the board members agreed to support the final Definite Plan of the bureau and even endorsed the two most controversial features: the enlargement of Pineview Dam rather than building Magpie and the construction of Wanship Dam rather than at the Perdue site.

On November 23, 1951, engineer Tom Clark of the Bureau of Reclamation presented a plan concerning recreation development and benefits for the project. The program outlined by the National Park Service included amounts to be credited to the project as nonreimbursable items as listed in Table 6.

The board accepted the proposal and suggested that the bureau make it part of the final plan. When approved this would become the first nationally sponsored reclamation project to include funds for recreational facilities.

Another area of ongoing discussion for the conservancy district board with the Bureau of Reclamation and the United States Geological Survey (USGS) was the issue of ground water in the Weber Basin. USGS officials suggested in 1952 that a great deal had been done to study ground water, but at least two more years were needed to make a complete study of Davis and Weber counties. They also noted that the district needed to furnish at least $3,000 in order to complete the survey. Officials of the survey made it clear to district officials that the project was not a priority, but if the district wanted the survey and set aside the funds, the USGS could obtain matching funds. At any rate, it seemed certain that construction on the project would be underway for at least two years before a final underground water survey could be completed.[23]

On August 8, 1952, the Weber Basin Water Conservancy District Board met for a luncheon meeting with the commissioner of the Bureau of Reclamation, Michael W. Strauss, and regional director, E. O. Larson. Strauss informed the group that the bureau had nearly gained approval for the definite plan report for the project and that the first phase of the project should be bid on during September. The project Strauss spoke of was the Gateway Tunnel and he was confident that construction would begin by December. Members of the board were delighted with this pronouncement for initial construction on the project had been long awaited. Strauss stressed that although the project now has a "foot in the door," constant vigilance and close cooperation with Utah's congressional delegation was very much needed to guarantee that the project receive its needed annual appropriations. Strauss reassured the group that his department had recognized for a number of years the importance of the Weber Basin Project and they were hopeful of its timely completion.

Strauss's advice did not fall on deaf ears; over the past several years advocates of the Weber Basin Project had learned the importance of applying constant pressure on members of Utah's congressional delegation, and they were also aware of the many legislative pitfalls that could delay approval of construction funds

TABLE 6
Recreational Allocations

Reservoir	Amount
Wanship Reservoir	$ 634,400
East Canyon Reservoir	227,900
Willard Bay Reservoir	2,045,000
Pineview Reservoir	2,130,000
Total	$5,037,300

Members of water boards and water companies corresponded regularly with politicians and met with them as the occasion arose. Newly elected politicians like Governor J. Bracken Lee and Senator Wallace Bennett each received a full updated briefing on water developments and the proposed Weber Basin Project. When the Appropriations Committee of the United States Senate refused to approve construction funds for the project for the 1951-52 fiscal year, Utah's congressional delegation was bombarded with letters and telegrams. Joseph Johnson, president of the Utah Water Users Association, District No. 2, and president of the Davis County Water Users Association, wired this message to Senators Wallace F. Bennett and Arthur V. Watkins:

> This District Association comprising area of Weber Basin Project stunned to learn Senate Appropriations Committee denied request for starting funds. Weber and Davis Counties concentrated population in critical position as far as ample water is concerned. Tremendous increased population this area has been supplied with water only because of above normal precipitation past eleven years. We surviving on borrowed time and serious consequences will result should there be dry cycle prior to building of Weber Project that would make us self-sufficient so far as water is concerned. Request you do all in your power to acquaint Senate with serious situation and grant us necessary relief.

In his telegrams to Utah's senators, Harold Ellison indicated, "One dry year would reduce supply on Wasatch Front by 60% and detrimentally affect total operations this area and extensive program of our military installations which are so important to national preparedness and defense."[24] Letters, telegrams, and resolutions sent to Utah's representatives again and again requested construction monies, but none became available until fiscal year 1952-53.

The Weber Basin Water Conservancy District Board lobbied for construction funds, but also began to arrange to sell the water that would be delivered by the soon-to-be developed facilities of the Weber Basin Project. E. J. Fjeldsted as district manager and Ward Holbrook as district president started contacting the numerous irrigation, canal, and ditch companies to discuss their water desires and needs. They contacted municipalities and other water users such as the Ogden Arsenal and sold water on a commitment basis.

The conservancy district had become the facilitator for the project on the local level and worked as the mediator with the Bureau of Reclamation on the project, which would later be managed by the conservancy district. During the winter of 1952, E. J. Fjeldsted testified before the Subcommittee on Appropriations for the Department of Interior to assist in securing funds for the project. This testimony was given at the invitation of Congressman Walter Granger.

To the relief of the Water Basin Project supporters, Congress appropriated $1,350,000 for beginning construction on this project, which was to take place during the fall of 1952. Congresswoman Reva Beck Bosone presented a two-page analysis of the passage of construction money in *Intermountain Industry*. She noted that reclamation money had been difficult to get from Congress, particularly because of a coalition of Republicans and Dixiecrats that controlled the purse-strings in the House of Representatives over the past two years. Truman had also issued a policy of "no new starts except for defense or emergency" just as the Weber Basin Project was "a-borning." After writing that the House of Representatives had often been far too restrictive in originating projects related to new money, Bosone added insightfully,

> it was I who realized that Congress would never go along with the Bureau of Reclamation request for $4,031,000 to start the Weber Basin project; it was I who insisted that the Bureau of Reclamation scale down the figure to the bare minimum necessary to do the work which could be done during the fiscal year and, as a result of my efforts the low but sufficient figure of $1,350,000 was brought forth....

Senator Carl Hayden, a staunch friend of reclamation, was chairman of the Subcommittee on Interior Appropriations. He again proved his friendship to Utah and the West when he did what his compatriot on the House side would not do—wrote in several millions of dollars for nine new reclamation projects in addition to Weber.[25]

Telegrams of congratulations were sent to Utah's congressional delegation concerning the authorization of construction funds: "We're thrilled with your results," "Please accept our congratulations and high praise for outstanding work." Congressman Granger boasted that this was the largest single reclamation project ever approved for Utah and pledged his continued support for the project.[26] Helpful in advancing the Weber Basin Project through Congress had been the combined efforts and thousands of hours of congressional representatives, Weber Basin water users, Bureau of Reclamation officials, and many others, including the committee members who had analyzed the agricultural needs and potential in the project area from Utah State Agricultural College and the Department of Agriculture. Estimates show that USDA and the Utah Agricultural Experiment Station alone had contributed $71,450 to the study for the project without counting the contributions of the Bureau of Reclamation.[27]

The Davis-Weber Municipal Water Development Association proved to be one of the most active groups lobbying for water development in the Weber Basin. Its genesis has been traced earlier in this history. By 1951, with the Weber Basin Project well on its way to being funded, questions arose concerning the need of this organization continuing to operate. On August 6, Harold Ellison, president of the organization asked members of the association to consider its validity, and recommend continued involvement or dissolution. The association represented twenty-six cities and towns from the two-county area with representatives from the county commissions as well as token representation from Morgan and Summit counties. Each municipality appointed two representatives to attend meetings, but by late 1951, interest in the association was declining, partly because of the growth and involvement of the Weber Basin Water Conservancy District. This association continued to operate sporadically through the 1950s, including making an analysis of culinary water rates in the two-county area as of 1957.

The study indicated that the present minimum water charge for the towns in the two counties was $1.95, and as a result of the study each municipality was urged to set its rates at $2.50. The association suggested that the amount of money derived from the new minimum rate would allow all cities to maintain their annual payments to the Weber Basin Water Conservancy District. This payment to the district in 1957-58 amounted to $237,952 while income from current culinary water rates had been $474,419. By 1969 the payment to the district rose to $535,392, while the new proposed minimum water rate would produce $796,596.[28]

Through the 1950s, dinner meetings of the two county municipal water development associations were often held at Ma's and Pa's Restaurant in Roy. On October 12, 1954, they scheduled a tour of the Weber Basin Project for civic leaders in Weber and Davis County, with a luncheon scheduled at the Daughters of Utah Pioneers Hall, Coalville. To better acquaint themselves with development on the river, thirty-four participants visited the Gateway Tunnel, the Wanship Reservoir site, as well as other features of the project. Though Harold Ellison continued to operate as president of the organization with E. J. Fjeldsted as secretary, they could not revive the ailing organization. It had done its duty during 1948 and

1949 in lobbying for the Weber Basin Project and served as a catalyst to bring together the conservancy district.

In 1971, E. J. Fjeldsted, DeLore Nichols, and Boyd Storey, as the Trustees for the Davis-Weber Counties Municipal Water Development Association, gave a grant of $4,000 to Weber State College to provide seed money to begin a history of water development on the Weber River because many of the participants in the development of the Weber River over the last several decades were interested in chronicling water use and development.

A major issue in the construction of facilities on the river was the repayment method and means. Numerous studies had investigated the ability of the basin to repay the federal government for the facilities to be constructed. In fact, the Weber Basin Water Conservancy District had been conceived of to manage the new facilities and the orchestrate repayment. As early as October 1951, discussions were held between the district and the Bureau of Reclamation concerning repayment, and they decided that final negotiations would have to be delayed until President Truman approved the Definite Plan report and until the conditions set forth by him in his letter of August 30, 1949, were met. These conditions were met by the summer of 1952, with the first construction funds of $1.35 million approved for the project during late June and early July of that year. This initial sum of money, approved by the Senate, House, and President Truman guaranteed the project's beginning. With initial construction funds secured, the regional office of the Bureau of Reclamation prepared a draft of the contract for repayment of the funds. The draft was then reviewed by the commissioner of the bureau and over the next several months reviewed by officials of bureau and the conservancy district. The twenty-nine page contract was signed on December 12, 1952, by E. O. Larson, regional director of the Bureau of Reclamation, representing the secretary of interior, and by W. R. White, president of the Weber Basin Water Conservancy District, with E. J. Fjeldsted as witness.

The contract, unique in several respects, allowed for a repayment period of sixty years rather than the usual forty years. Recreational opportunities were among the benefits to be derived from the proposed storage facilities. This was a first for a water project in the United States and in contracting with the United States through the Department of the Interior. The contract, with an overall construction obligation of $57,694,000, was the largest single repayment contract entered into by the Bureau of Reclamation to that date. The contract specified storage facilities, diversion dams, pumping plants, drains, wells, and power plants. The contract required the district to acquire all of the lands and easements needed for construction, operation, and maintenance of the project works. The contract also contained a provision concerning the establishment of development units in the district by the secretary of the interior for assistance in repayment.

The district was given permanent and exclusive use of all project water with the stipulation that the United States would exercise control for flooding, recreational use, and fish and wildlife purposes. The district was directed to make water allotments and contracts, and supervise the distribution of project water. The title to all of the project works acquired or constructed by the United States was to remain in the name of the United States, even though the project was to be eventually paid for and the operation and maintenance of the project completely performed by the district.

The contract also noted that the United States would not be liable for water shortage resulting from "drought, inaccuracy in distribution, hostile diversion, or prior or superior claims." The board of directors approved the contract on November 18, 1952, and the secretary of the interior approved it on November 19.

One section of the contract referred to the issuance of district bonds to finance the construction of filtration plants, regulatory reservoirs, distribution systems, and other facilities to make project water available to the consumers. In order to be able to issue the needed bonds, the district needed to receive the approval of the voters in the district. Through the fall of 1952, preparations were made for the special bond election to be held December 6. The district developed an election brochure to encourage positive voting in the election. The brochure was 10 1/2 inches by 14 inches in size and was folded in thirds. One side of the brochure provided a six-color map of the Weber Basin Project, including proposed developments and lands to be benefitted by the project. The other side of the brochure was carefully put together to elicit "yes" votes on election day. It featured "Uncle Sam" pointing his finger and saying, "Water for the Weber Basin, Its up to YOU, I've done my part." A chart indicating precipitation in northern Utah suggested "Another Drought is Overdue" and that the Weber Basin Project would help to alleviate the impact of a drought. Fifteen specific questions and answers filled up the remainder of the brochure including: What is the Weber Basin Project? What will it do? How much will it cost? and Who will put up the money? The answers indicated that the estimated project costs of $70,385,000 were to be made available by the federal government, and of that, $12,691,000 would be "a gift of the nation in return for national benefits from flood control, recreation, and improvements for fish and wildlife." Repayment of the project costs were to be shouldered by farmers, and municipal and industrial water users whose costs would vary from $2.00 to $3.50 an acre foot. The brochure finally noted that the $6,500,000 to be raised by issuance of revenue bonds would be repaid by project water users who benefitted from the construction of treatment plants, supply lines, distribution systems, and other facilities to be constructed separately from the federal Weber Basin Project. The brochure underlined the idea, "You can't afford not to vote 'Yes' on December 6, 1952!"[29]

The brochure and publicity was aimed at allowing the district to issue the needed bonds through an election with positive results. In order to better inform citizens of the basin, copies of the proposed contract between the conservancy district and the United States were provided for inspection at prominent places in

Completed Gateway Canal carrying water near Peterson, Morgan County.
(1957 photograph by Mel Davis. Courtesy United States Bureau of Reclamation.)

One of the seven protective stair devices in Gateway Canal that allows deer or
livestock, which have gotten into the water of the canal, to exit from the canal
before being carried into the siphon by the water. (1957 photograph by Stan
Rasmussen. Courtesy United States Bureau of Reclamation.)

Spillway chute at the lower end of the Gateway Canal near the Gateway Tunnel. The spillway transfers excess water back into the Weber River. At this site is also located the Gateway Power Plant. (1957 photograph by Stan Rasmussen. Courtesy United States Bureau of Reclamation.)

View of section of Gateway Tunnel through Wasatch Mountains of Weber Canyon, with structural steel supports. (1953 photograph. Courtesy United States Bureau of Reclamation.)

William R. Wallace, chairman, Utah Water and Power Board, pushing the plunger that ignites the first round of explosives for the excavation of the Gateway Tunnel. This formally instituted the Weber Basin Project. Also in the photo *left to right:* T. A. Clark, area engineer, Bureau of Reclamation; Colonel H. H. Needham, commanding officer, Ogden Arsenal; Leroy B. Smith, director; Horald G. Clark, director, and W. R. White, president, Weber Basin Water Conservancy District; and George R. Putnam, vice president, Utah Construction Company. (January 9, 1953, photograph. Courtesy United States Bureau of Reclamation.)

Echo Dam celebration honoring the final payment on June 6, 1966 by the Weber River Water Users Association to the Bureau of Reclamation. Ralph A. Richards, director of the Weber River Water Users association was master of ceremonies. (1966 photograph by Mel Davis. Courtesy United States Bureau of Reclamation.)

Reclamation Commissioner Floyd E. Dominy on the left congratulates H. J. Barnes, vice president of the Weber River Water Users Association, for the final payment in the form of a five-foot check on the contract for construction of Echo Dam. (June 1966 photograph by Mel Davis. Courtesy United States Bureau of Reclamation.)

View of Echo Dam and Reservoir of the Weber Basin Project near Coalville, Summit County. (1958 photograph by Stan Rasmussen. Courtesy United States Bureau of Reclamation.)

Progress photo of dam embankment from right abutment above spill-way of Wanship Dam, Weber Basin Project. (1956 photograph by J. R. Hinchcliff. Courtesy United States Bureau of Reclamation.)

Ezra Clark, president of the Weber Basin Water Conservancy District, on right, receiving the key from David Crandall, regional director of the Bureau of Reclamation, that indicates control of the completed Wanship Dam has been turned over to the Conservancy District by the Bureau of Reclamation. Others in the group are Board of Directors of the Weber Basin District, *left to right*: Blaine Fisher, Elmer Carver, Ralph Richards, Clifford Linford, Jack Olsen, Francis Simpson, and Wayne Winegar, manager of the district. (1968 photograph. Courtesy United States Bureau of Reclamation.)

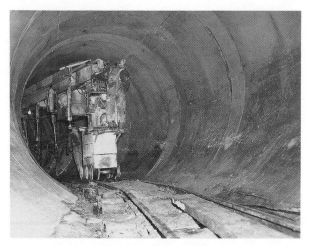

Jumbo machinery removing steel forms from newly poured concrete of the Outlet Works Tunnel of the Wanship Dam. (1955 photograph by Alton W. Timmins. Courtesy United States Bureau of Reclamation.)

Harold E. Ellison, president, Weber Basin Water Conservancy District, speaking at the dedication of Wanship Dam. Also in photo *left to right:* E. O. Larson, director of Region Four, Bureau of Reclamation; Utah Senator Arthur V. Watkins; Governor George D. Clyde; W. A. Dexheimer, commissioner, Bureau of Reclamation; and David W. Loertscher, commissioner of Summit County. (Photograph by D. B. Carr. Courtesy United States Bureau of Reclamation.)

Aerial view of Wanship Dam and Rockport Reservoir, Weber Basin Project, Summit County. This reservoir covered the town of Rockport. (1958 photograph by Stan Rasmussen. Courtesy United States Bureau of Reclamation.)

Boating and water skiing recreation activities on Rockport Reservoir behind the Wanship Dam, Weber Basin Project. Reservoir covered the town of Rockport, Summit County. (1957 photograph by Stan Rasmussen. Courtesy United States Bureau of Reclamation.)

Young Tommy Martin with a trout he caught in Wanship Dam Reservoir. One of the contributions the Weber Basin Project has made to its communities is the recreational facilities for boating, water skiing, fishing, and swimming. (1960 photograph by Stan Rasmussen. Courtesy United States Bureau of Reclamation.)

Looking east across John Braegger's farm of pasture, alfalfa and grain land, before Willard Bay Dam of the Weber Basin Project was started and affected this area. (1957 photograph by W. J. Eldredge. Courtesy United States Bureau of Reclamation.)

View of raft with drilling equipment on the bay of the Great Salt Lake where the Willard Bay Dike, Box Elder County, would be located. Tests of the lake bottom were taken to determine if it would support the Willard Bay Dam of the Weber Basin Project. (1951 photograph by E. C. Christensen. Courtesy United States Bureau of Reclamation.)

View from Sauerman dragline tail tower, showing series of canyons excavated for Willard Dam by successive moves of the dragline. (1960 photograph by Stan Rasmussen. Courtesy United States Bureau of Reclamation.)

Willard Dam-North Drain Ditch-Barrow area "B," looking east towards mountains. (1957 photograph by R. R. Robison. Courtesy United States Bureau of Reclamation.)

Willard Dam, Weber Basin Project. Photo showing the water at the by-pass channel of Pumping Plant No.1 on November 30, 1964. (Photograph by T. R. Smith. Courtesy United States Bureau of Reclamation.)

Clinton D. Woods, Ralph A. Richards, and Ezra J. Fjelsted inspecting newly placed riprap at Willard Bay Reservoir. (May 14, 1964, photograph. Courtesy *Salt Lake Tribune*.)

Aerial view looking northeast at Willard Dam, Weber Basin Project. (1963 photograph by the *Ogden Standard Examiner.*)

Causey Dam, built on Causey Creek a tributary of the upper Ogden River, was constructed between 1963 to 1966
by R. A. Heintz Construction Company for $3,836,419. It
is a 200-foot high, earth and rock dam with a water storage capacity of 6,900 acre feet. (1966 photograph by R.
R. Robison. Courtesy United States Bureau of Reclamation.)

Lost Creek Dam, constructed 1964 to January 1967, is located on Lost
Creek, a tributary of the Weber River above Devil's Slide. It is an earth
and rock dam with a height of 220 feet, and has a storage capacity of
20,000 acre feet.

Smith and Morehouse Dam, constructed 1984-1988 by Gibbons and Reed Construction Company for the Weber Basin Water Conservancy District as an enlargement of an earlier dam which had a capacity of 1,040 acre feet. The new dam has an 8,350 acre feet capacity. The dam cost $12,000,000 to construct and was financed by the Weber Basin Water Conservancy District and Utah State Division of Water Resources. (Photograph by Grant Salter. Courtesy Weber Basin Water Conservancy District.)

The Davis Aqueduct and facilities valued at over $16 million were transferred from the Bureau of Reclamation to the Weber Basin Water Conservancy District in ceremonies at the district headquarters in East Layton on February 4, 1966. The headquarters building has been upgraded and additional facilities added at this site since that time. *Left to right*: Richard Evans and Glen Flint, Davis County commissioners; Wayne M. Winegar, secretary-manager, Weber Basin Conservancy District; Golden L. Stewart, Bountiful Subconservancy District; Frances V. Simpson, Joseph F. Breeze, S. J. Olsen, and Ralph A. Richards, directors, Weber Basin Water Conservancy District; Rex L. Greenhalgh, Bureau of Reclamation; Davis L. Crandall, regional director, Bureau of Reclamation; J. Clifton Linford, director, Weber Basin; Bob Brown, Bureau of Reclamation; E. J. Fjeldsted, retired secretary-manager, Weber Basin; and Horald G. Clark, retired director, Weber Basin. (1966 photograph by Mel Davis. Courtesy United States Bureau of Reclamation.)

Weber Basin Water Conservancy District board members in 1993. The Board of Directors are appointed under Utah water law statutes, which state that for the Weber Basin District there are nine members: three from the lower Weber County area, one from the upper Ogden River area of Weber County, three from Davis County, one from Morgan County, and one from Summit County. County commissions involved nominate three candidates for the board position, and the governor of Utah makes the final appointment from among the nominees for a four-year term. The board sets the policy for the district. *Front row, left to right*: Ivan Flint, manager; Ted Christensen; R. Howard Cox; Charles F. Black, president; *second row*: Scott Peterson, Stephen Osguthorpe, Max Rigby, Norman Montgomery, Wayne Gibson, and Evan Whitesides. (Photograph courtesy Weber Basin Water Conservancy District.)

Ivan Flint, general manager of the Weber Basin Water Conservancy District in 1993. The general manager is appointed on an annual basis, and is responsible for the general management of the district operations and implementing the policy determined by the board of directors. It includes making decisions on when to store or when to release waters from the reservoirs and for carrying out flood control measures. (Photograph courtesy Weber Basin Water Conservancy District.)

each of the counties, often in post offices. Overall, twenty-seven sites displayed the contract: eight in Davis County, four in Morgan County, five in Summit County, and ten in Weber County.[30]

For the election of Saturday December 6, special election precincts, which were usually a combination of several regular voting precincts, were established in Davis County, with two in Bountiful, one each in Centerville, Farmington, Kaysville, Layton, Clearfield, and Sunset. One precinct was set in Morgan County in Morgan at the county court house while two precincts were established in Summit County—one at Coalville and the other at Kamas. Weber County designated nine voting places, with one each in South Ogden, Plain City, Huntsville, Roy, and North Ogden and four in Ogden City. The ballot for the election included two propositions:

Proposition One
> Shall the Weber Basin Water Conservancy District enter into a contract with the United States of America for the construction of the Weber Basin Project works, and the acquisition of a water supply for the District, at a cost to the District of not to exceed $57,694,000 to be paid in annual installments over a period of 60 years without interest, except interest at 6% in case of delinquency, on the terms and conditions of the draft of contract referred to in the Resolution by which this election is called?

YES ( )
NO ( )

Proposition Two
> Shall the Board of Directors of the Weber Basin Water Conservancy District, Utah, be authorized to issue and sell bonds of said district in the amount of Six Million Five Hundred Thousand Dollars ($6,500,000) bearing interest at not exceeding six per cent (6%) per annum, for the purpose of paying the cost of acquiring and constructing waterworks and facilities for said district, comprised principally of water treatment plants and distribution facilities, and expenses preliminary and incidental thereto, said bonds to be payable from and, to the extent provided by the Board of Directors, secured by pledge of the revenue derived by said district, including ad valorem taxes, assessments, and revenues received by it for water sold and services rendered?

For the issue of Bonds YES ( )
NO ( )

As the day for the special election drew closer, the conservancy district board held a meeting to consider the feelings of a delegation from Summit County concerning membership in the conservancy district. To many members of the Summit County group the benefits offered by the district seemed to be very limited, and some suggested that Summit County should withdraw from the district. They contended that the county was losing money in taxes paid to the district with little return, and that land presently being used for farming would soon be inundated by the Wanship Dam, reducing farm land in the county. Finally, the county's economy was not in a slump, particularly in relationship to the mining decline at Park City. Summit County attorney Morgan Lewis, Jr., took particular interest in the method Summit County might use to withdraw from the district.

District president Rulon White responded to the concerns of the Summit County citizens by saying the board could not make a decision concerning Summit County without a detailed study of the impact on the entire basin and project. Summit County Commissioner Maurice Boyden and Weber Basin Water Conservancy District Director Edward Sorensen from Summit County were particularly involved in the discussion. They decided to hold a meeting in Summit County on November 25 to consider the matter with the residents of the county.

The board of directors of the district prepared a special statement for the meeting to urge residents of Summit County to stay affiliated with the conservancy district. The board committed themselves to "lend its best efforts to have a small reservoir feature for the Upper Weber Valley Users added as a part of the Weber Basin Project."[31] The Weber Basin Board and the Bureau of Reclamation suggested that a small reservoir site might be located at the Middle or Lower Larrabee sites and that test drilling would be scheduled to ascertain the feasibility of such a site. As noted earlier, both Perdue and Larrabee were ruled out as drilling proceeded.

The dissatisfaction of some elements with the Weber Basin Project and the conservancy district was evidenced by the vote of the special election held in December, as shown in Table 7.

Though both propositions passed by a more than four to one vote, the people in Kamas in upper Summit County approved Proposition One but disapproved Proposition Two. The strongest vote against both proposals came from those who voted at the Summit County Court House at Coalville. There the vote was seven to one against both proposals. Overwhelming support of the project came from the populous areas of Davis and Weber counties, whereas the Ogden Valley residents voting at Huntsville defeated both proposals. There is no doubt that rural residents of the basin did not favor the project or bond issue. Many did not favor new dams such as Wanship or reservoir expansion as at Pineview and many farmers opposed taking good agricultural land for a reservoir that would not directly benefit them. At any rate, a majority of voters in the district approved both the project and the bond issue—the project now had new life breathed into it.

The voting turn out was largest at Bountiful where the residents hoped to benefit directly from the project. The voting by county is shown in Table 8.

TABLE 7
Voting Results, Special Election, December 6, 1952

| | Proposition One | | Proposition Two | |
Place	For	Against	For	Against
South Bountiful Elementary School	235	2	215	6
American Legion Hall, Bountiful	640	15	608	29
Town Hall, Centerville	174	3	171	1
Davis Co. Court House, Farmington	167	1	155	5
City Hall, Kaysville	225	22	200	32
Town Hall, Layton	230	22	221	27
Town Hall, Clearfield	217	2	208	2
Town Hall, Sunset	192	7	187	9
Morgan Co. Court House, Morgan	106	140	92	144
Summit Co. Court House, Coalville	47	357	42	352
City Hall, Kamas	85	77	77	81
Pingree School, Ogden	161	6	147	10
Mound Fort School, Ogden	281	33	265	41
Lorin Farr School, Ogden	252	12	247	13
Madison School, Ogden	398	18	392	20
Weber County Infirmary, Roy	175	5	174	6
Recreation Hall, Plain City	209	47	210	47
LDS Church House, North Ogden	102	18	100	17
Soil Cons. Bldg., Huntsville	47	55	37	57
Johnson-Perrin Floor Co., Ogden	241	10	231	10
Totals	4184	852	3979	909

The conservancy district board spent the next month determining which fiscal agent should be retained to handle the $6,500,000 bond. It agreed at the January 9, 1953, meeting that J. A. Burrows and George Danton be employed by contract at a maximum service fee of $17,000. The board would pay the fee in two installments, 50 percent at the issuance of the first bonds and the remainder as the bonds were issued and sold. The board also decided to allow its president to serve two one-year terms.[32]

TABLE 8
Voting by County

County	Proposition One		Proposition Two	
	For	Against	For	Against
Davis County	2080	74	1965	111
Morgan County	106	140	92	144
Summit County	132	434	119	433
Weber County	1866	204	1803	221

A great deal of time was spent investigating land purchases and right-of-ways, and on these procedures the Bureau of Reclamation worked closely with the conservancy district. Most property agreements were settled out of court. The Daughters of the Utah Pioneers in the Wanship area petitioned the district to obtain an old log cabin, which had been the first school in the Rockport area. The district had paid $700 for the school and wanted the Daughters of the Pioneers to have the building but could not, by law, give it away. Bureau of Reclamation officials informed the conservancy district board that to establish the log cabin as a building of historical value, an act of Congress would be required.[33]

Clinton D. Woods, director for the Weber Basin Project for the Bureau of Reclamation often met with the conservancy district board and kept them well informed on the progress of the project. Francis Warnick, the project planning engineer for the bureau also met on a regular basis with the conservancy district board and on occasion served as a mediator. One instance occurred September 13, 1954, when representatives of the conservancy district and the Ogden River Water Users Association met to discuss how much credit would be given to the association for their current investment in the Pineview Dam when the dam was to be raised. It would take several years of meetings before a final agreement would be reached.

The conservancy district then had to iron out its relationship with Salt Lake City. During the early part of 1955, the Salt Lake City Corporation made application before the state engineer for 365 second feet of water from the Weber River. Salt Lake City had filed on this water as early as 1930. The Bureau of Reclamation indicated that it would cost between $23 and $25 million to deliver the water at an acre-foot cost of $300 per year. The conservancy district board, however, notified the state engineer of their opposition to any water from the Weber River going to the city of Salt Lake.[34]

On May 27, 1955, the board of directors of the conservancy district did approve the purchase of five thousand shares of stock in the Weber River Water Users Association owned by Davis County. The board unanimously agreed upon the purchase price of $146,311.67. Davis County also requested that the conservancy district repay its $2,500, which had been originally spent for drilling work on some of the original dam sites (Perdue). Harold Ellison opposed repaying the money to Davis County as that county had not been assessed annual dues to the Davis-Weber Counties Municipal Water Development Association because of the earlier monetary contribution. To clearly illustrate his position, Director Ellison submitted in summary in the form of a letter discussing the historical background of the situation to Davis County. Ellison concluded that Davis County had been treated fairly in terms of money assessed and should not receive reimbursement from the conservancy district.

With several issues resolved, the board determined that the date of February 23, 1956, should be set for opening the bids on the $5,400,000 worth of bonds to be issued by the Weber Basin Water Conservancy District through its fiscal agent, J. A. Burrows. The district published notice of the sale in *The Bond Buyer* in New York City on January 31, and the sealed bids were opened in the Ogden City Council room in the municipal building in Ogden. Of the four bids, F. S. Smithers and Co., located at 1 Wall Street, New York, submitted the lowest. The Smither's bid included a net effective interest rate of 3.38866 percent over a period of forty years. In a special meeting held February 23, the board evaluated the bids and granted the issuance of $5,400,000 in bonds to the Smithers Company. President D. D. McKay of the conservancy district and Secretary-Manager E. J. Fjeldsted, finalized the agreement with Smithers through a series of meetings in New York City during April of 1956.

During this same time, lengthy negotiations were conducted with Bountiful City to acquire the needed land for a filtration plant. After condemnation proceedings had begun, a compromise figure of $2,800 an acre was reached for the eleven-acre site. This agreement would allow the building of what would be known as Filtration Plant Number 4. The Board began to contract with municipalities, irrigation companies, and industrial water users for the delivery of Weber Basin Water, which was now becoming available through the construction of the projected facilities. The conservancy district board also agreed, during a meeting held on July 26, 1957, to move the headquarters of the district from downtown Ogden to the site of Treatment Plant Number 3 at the junction of Hill Field Road and U.S. Highway 89. The reasons for this move included the central location and fact that the district would now own its building and other facilities.

Officials from the Bureau of Reclamation met regularly with the Weber Basin Water Conservancy Board and other water users along the river concerning the Weber Basin Project. By late summer 1958, project director Clinton Woods reported to the conservancy district board that the estimated total cost for the project had jumped nearly $10 million over the past five years. Woods determined that the increase resulted from the number and changes in types of lateral water systems that water users were insisting upon. Woods noted, too, that increased costs at this stage would eliminate other hoped-for parts of the project, such as providing additional storage on the upper reaches of the Weber River.

By 1961 it was apparent that the Weber Basin Project could not be built within the original specified monetary limit of 70,385,000, with the Weber Basin Conservancy District having a maximum repayment obligation of $57,694,000. The difference was to be covered by the federal government for recreation and fish and wildlife protection and development. By the spring of 1961, the conservancy district board put together another special election for the voters of the district to approve an increase of costs to $97,500,000 for the total project and to $81,656,000 as the amount repayable by the conservancy district. Secretary-Manager Fjeldsted reported to the board that he had been working with local newspaper people to have them carry "factual" stories about the needed increase. In addition to newspaper stories, the district sponsored public meetings in the counties to disseminate information and urge the public to vote "yes" on the single proposition on the ballot. Once again, copies of the contract, this time amended, were placed on public display at twenty-two places throughout the district: two places in Summit County, two in Morgan County, and nine each in Weber and Davis counties. The special election was set for Saturday, June 24, with the polling places being at the same locations as the sites for inspecting the proposed amendatory contract. The ballot was arranged in the following fashion:

Sample Ballot
Special Election
June 24, 1961
Proposition

Shall the Weber Basin Water Conservancy District enter into a contract with the United States of America, amendatory to that contract heretofore entered into between the said District and the United States of America, dated December 12, 1952, for the construction of the Weber Basin Project and the acquisition of a water supply, which amendatory contract provides for the cost of the Project works to the District to be an amount not exceeding $81,656,000.00, to be paid in annual installments over a period of sixty (60) years, without interest, except interest at the rate of Six (6%) percent per annum in case of delinquency, on the terms and conditions of the draft of amendatory contract referred to in the resolution by which this election is called?

YES ( )
NO   ( )

If the voter desires to vote in favor of the proposition he shall mark a cross (X) in the square opposite the word "Yes".

If the voter desires to vote against the proposition, he shall mark a cross (X) in the square opposite the word "No".

On the reverse side of the ballot shall be printed the following:

"Official Ballot For

_____—Election Precinct of the Weber Basin
Water Conservancy District, Utah, Special Election,
Saturday, June 24, 1961

_____

Secretary, Board of Directors of the Weber Basin Water Conservancy District, Utah."

The results of the special election favored the proposition by about a three to one margin. The voters of both Summit and Morgan counties again opposed the increase, which was particularly evident in Summit County, where only ten voters supported the proposal while 101 voted against it (Table 9).

With the approval for repayment assured, all systems were kept in high gear, including construction on the project monitored by the Bureau of Reclamation, sales of water by the conservancy district, and construction of filtration plants and delivery systems by the district. The district had originally planned to build four filtration plants and the needed delivery systems at an estimated cost of $6,500,000. As investigation took place over financing the proposed plants, Ogden City requested that it be allowed to build and operate its own filtration plant below Pineview Dam. The district granted the request, which allowed it to prepare its proposed bond issue in the amount of $5,400,000.

The monies gained from bond sales were used mainly for construction of the three filtration plants. Plant Number 4 was constructed east of Bountiful with a capacity of treating 6.5 million gallons of water on a daily basis for municipal use. Filtration Plant Number 3, located on the same site as the district headquarters, also has a 6.5 million gallon daily capacity. Filtration Plant Number 2, located southeast of Ogden, can treat 13 million gallons daily. Construction on these filtration plants was completed during 1957.

The first sale of municipal water that had been treated was made to Clearfield City on October 26, 1953. In order to make certain that the petition procedure under which Clearfield had agreed to purchase water from the board of directors of the conservancy district was legal, a test case was entered into the courts. In this case, *Barlow v. Clearfield City, et al.*, [1 (2d) 419, 268 P. 2d 682], the Supreme Court of the State of Utah ruled that the approval of the petition by the board of directors of the conservancy district created a binding and enforceable contract for the purchase of water on the terms specified in the petition. The

TABLE 9
Voting Returns, Special Election, June 24, 1961

Place	Votes For	Votes Against
Davis County		
South Bountiful Elementary School	15	1
American Legion Hall	170	30
Marian Boulton, Bountiful	96	22
Centerville Elementary School	33	9
Davis County Courthouse, Farmington	28	5
Kaysville City Hall	85	14
Layton City Hall	33	17
Clearfield Town Hall	54	6
Sunset Town Hall	43	11
Morgan County		
Morgan County Courthouse, Morgan	29	48
Devil's Slide, Church House	12	3
Summit County		
Summit County Courthouse, Coalville	3	77
City Hall, Kamas	7	24
Weber County		
Lake View School, Roy	72	36
Burch Creek School, Ogden	137	29
Edna McPhie, Ogden	38	6
Phoebe Lund, Ogden	118	18
Edna B. Wright, Ogden	68	8
Flora Black, Ogden	60	2
Plain City Town Hall	115	38
Orba Brown, Ogden	45	10
Huntsville School, Huntsville	67	1
Total	1,328	415

The court also upheld the right of the district to levy taxes on all taxable property in the city or town in an amount sufficient to pay for the water sold by the district. The decision of the court further stated that the contract between the city and the district was not unreasonable and did not constitute a loan of the credit of the city to the district. This decision would serve as an important piece of bedrock in municipal water sale by conservancy districts in the state of Utah.

Before the Weber Basin Water Conservancy District could deliver water, its only source of revenue was the annual ad valorem tax. Between the time that the district was organized and actual construction on the project began, the district could only levy a tax of one-half mill on an annual basis on the assessed valuation of the property within the district. This taxing policy was regulated by the Conservancy Law passed by the state of Utah. After the beginning of construction, the district's levy was increased to one mill annually. The conservancy district operated until 1957 within a budget based on the one mill levy, and by 1957 water sales added to the district's budget. Since then, the budget has expanded each year to deal with administrative expenses, the filtration plants and pipeline operations, salaries of personnel, annual payment to the national government, payment to bond holders, the continuing purchase of rights-of-way, and operation and maintenance expenses necessary for maintaining the facilities built during the 1950s, 1960s and later.

The multilayered system of water use and control begun in pioneer days is still evident in the decades following the creation of the Weber Basin Water Conservancy District. Local ditch and irrigation companies continue to operate, but often purchase water from the district. The Bountiful-Mill Creek Irrigation Company was the first such organization to purchase water from the district, and on March 31, 1954, that company made its initial purchase under the direction of Mark T. Hatch, president of the company. Many companies have made arrangements for "wheeling" their own water through the facilities of the conservancy district.

The Ogden River Water Users Association continues their involvement along the Ogden River, including their partnership in Pineview Reservoir with the conservancy district. The Weber River Water Users Association still maintains Echo Reservoir with some assistance from the conservancy district. Both water users groups meet annually to discuss problems and recommend a river commissioner who submits an annual report concerning his area of responsibility to the state engineer. That report includes stream flow and storage records, water exchanges, and an assessment of general conditions along the river. Annual reports allow a basis for comparison, year to year, and provide historical statistics, which inform future water judgements. River commissioners and their assistants seem to be cast somewhat in the mold of earlier community and/or church appointed ditch riders.

The 1983 annual report on the Weber River noted that 5,400 acre feet of water was diverted annually through the Weber-Provo Canal in the Kamas Valley to the Provo River. This water was diverted so that those subscribers on the Provo River who own 5,400 shares in the Echo Reservoir may use the water on the

Provo drainage. Provo water users, through a variety of agreements, received 9,954 acre feet in 1983. All of this was delivered through the Weber-Provo Canal to the Provo River, originally built under the direction of the Bureau of Reclamation. The commissioner's report also said that in 1969, the Deseret Land and Livestock Company constructed a reservoir on Heiner Creek, which is a tributary to Echo Creek.

To recognize the accomplishment of building the Echo Reservoir and Dam, the Weber River Water Users Association sponsored a last payment ceremony on June 11, 1966. The official program of the day put the total cost of the project at $2,875,871.83, resulting in annual payments $88,192.90. The subscribers for Echo water included eighty-five individual water users, fifty-nine irrigation companies, and other organized water groups. Floyd E. Dominy, commissioner of the Bureau of Reclamation, delivered the major address, after which H. J. Barnes, vice president of the Weber River Water Users Association, presented the last payment. A free lunch and a tour of the dam followed the official ceremony. Many officials of the Weber Basin Conservancy District attended the Echo ceremonies.

Originally, the judges of the Second District Court appointed the nine directors for the Weber Basin Water Conservancy District with staggered terms so that three directors would be appointed on an annual basis. The judges have often reappointed those whose terms have expired, but the 1984 session of the Utah state legislature passed a law changing the method of appointing directors. Some suggested the new format would ensure that the directors would be more representative of the people in their areas. The new appointment system, which began during the summer of 1984, had the county commissioners nominate individual(s) who resided in the particular county; the prospective director was then officially appointed by the governor of the state.

Since its beginning, the conservancy district has had four secretary-managers. Ezra J. Fjeldsted served first from 1950 to 1965 and was followed by Wayne Winegar, who held the position from 1965 to 1980. Winegar was succeeded by Keith Jensen who directed the operations of the district from 1980 to 1985. Ivan Flint has managed the district from 1985 until the present (1994). Each secretary-manager has had new and different challenges. Fjeldsted spent much of his time helping put together the new district as well as working with the Bureau of Reclamation to have the projects built. A major effort of the Winegar years was to sell water, to use that revenue to assist in paying off the project. By the time Keith Jensen became manager, much of the project water had been sold, and so Jensen needed to look toward new water supplies and maintain a water system that was two and three decades old. Ivan Flint's era has included the completion of the Smith and Morehouse project, the drilling of new wells, much maintenance and replacement of old and worn out facilities, and a good deal of strategic planning.

During construction of new facilities during the 1950s and 1960s, the Weber Basin Water Conservancy District gradually took over operation and maintenance of these facilities. By 1969 the conservancy district assumed full responsibility for operating and maintaining all of the newly constructed dams, canals, aqueducts, etc. The directors of the conservancy district noted in a statement in their Second Annual Report that water development, particularly the Weber Basin Project, was a cooperative effort:

It would be manifestly unfair to indicate that the progress made in the preliminary work of the Project is a result of the sole efforts of the board of directors of the Conservancy District. Little could have been accomplished had it not been for the splendid cooperation of all of the citizens throughout the District. Special recognition should be given to the cooperating organizations and agencies, such as the Davis-Weber Counties Municipal Water Development Association, the Utah Water Users Association, the Utah Weber and Power Board, the State Engineer, District No. 2 of the Utah Water Users Association comprising Davis, Morgan, Summit, and Weber counties and the individual water associations of these same four counties, the Weber River Water Users Association, the Davis and Weber Canals Company, the Ogden River Water Users Association, and many others.

Especially is appreciation extended to the county commissioners of Davis, Morgan, Summit, and Weber counties in recognition of their unselfish and untiring cooperation, all to the end that continuous progress may be made toward the realization of the Weber Basin Project. Extra responsibility has been placed on these four counties as a result of the creation of the Conservancy District, and the board is not unmindful of this in publicly acknowledging their unstinted cooperation.

The directors are grateful for the outstanding services of all these groups, and solicit their continued efforts.

The Board of Directors of the Weber Basin Water Conservancy District are fully determined that the Weber Basin Project is economically feasible and sound, and that this area cannot fully develop until the project is underway in progressive steps or sufficient advancement to keep abreast, and even ahead, of the demands of the potential consumers of the waters to be developed. They dedicate their best efforts to the accomplishment of this worthwhile service to the basin area.[35]

The conservancy district has issued annual reports concerning water use and water sales since the district first began to sell water in 1958 during its first full year of operation. The annual reports describe operation and maintenance development during each year and document water flow, water storage, and water use in a variety of ways. The seriousness of the 1984 and 1977 droughts as well as the slight drought of 1971 is evident in the water use tables. Electrical power production at both Wanship and Gateway Power plants as well as water used for such production is reported on a daily, monthly, and annual basis.

The Weber Basin Project has stood the test of economic feasibility because of its multiple-use features. Originally planned to benefit agricultural users as well as municipal water users, over the past two decades municipal and industrial water use has grown very rapidly. There is no way that the urban growth along the Wasatch Front could have been maintained without the water available from storage in the new reservoirs and delivery through facilities such as the aqueducts and Gateway Tunnel. Recreational use on all of the facilities has grown steadily. One newspaper headline indicated recreational use at Willard Bay: "Willard Bay Area Drawing Capacity Crowds." The June 30, 1982, article noted that crowds were so large, restrictions had to be placed on many of the facilities in order to accommodate even part of those who wanted to enjoy them. Some visitors were turned away while others waited three or four hours to be admitted to certain areas.[36] Not all facilities host such crowds, but outdoor sun lovers have enjoyed fishing, boating, water-skiing, swimming, and all of the other outdoor kinds of recreation at facilities on the Weber Basin.

A 1976 study showed that East Canyon, Lost Creek, and Willard all had inadequate facilities for recreation. According to a letter from secretary-manager Wayne Winegar, this had the potential to cause conflicts between water users and recreationists:

> Willard Bay was planned with the entire water contents designated to be used for irrigation, and low quality municipal and industrial use.
>
> Since construction of the reservoir its use for recreational purposes has increased. The State Parks and Recreation Division, managers of the recreation facilities, are currently pursuing extensive recreational development. The continued use of Willard Reservoir for fishing has grown immensely, making it one of the State's most popular reservoirs for fishing. The fact that there was no conservation pool purchased by the Bureau of Sport Fisheries means the reservoir can almost be entirely drained if the water user's [sic] need the water. It is hoped that determinations will be negotiated in the near future and water will be purchased from the District to maintain fish and aid in securing recreational stability.
>
> At present the Conservancy District has not released much water from Willard Reservoir; however, in the future this will not likely be

the case. When large quantities of water are withdrawn in the future, recreation and fishing activities will be adversely affected.[37]

Recreational use of Weber Basin waters grew at a pace during the 1960s, 1970s, and 1980s that was not foreseen during the early developments of the project. Other water use during the same period grew in similar fashion. The tenuous but needed marriage between recreation and other users had started, and it would become a stronger relationship as time passed. The need for recreational use pushed in the original Weber Basin Bill was borne out by the public, who voted with boats, water skis, picnic baskets, and fishing poles.

Dams, canals, and other facilities that have been built over the last half-century are now beginning to need both repair and, in some cases, replacement. A particularly important project is the replacement of a one-mile stretch of the Gateway Canal, which has been constantly bothered by land-slippage. During 1983 and 1984, a section of the canal slipped downhill about three feet. If the canal were to cease operation and all water stopped going through the Gateway Tunnel, many cities, particularly in Davis County, would not have enough water to last a full day. After several years of studying the problem, the decision was made in 1993 to move ahead with drilling a tunnel through about one mile of the mountain near the west end of the canal and to install a large pipe through which this section of the canal will run. This solution was decided upon rather than installing a syphon. Another major construction effort for the district will be a settling basin, which will be constructed near Stoddard to collect the massive amounts of mud and silt that tends to flow into the Gateway Canal during the periods of high water. This settling basin will be continually cleaned out and will keep the Gateway Canal water much cleaner.

The cost of these two major projects and some other clean-up and repair work on the system will total $30 million. A portion of the total will be borrowed from the Bureau of Reclamation in a Rehabilitation and Betterment Loan (R & B Loan). The bureau will also assist Weber Basin in engineering plans for the repairs and developments.[38]

Water needs in the basin continue to increase, and there is only a limited amount of water available, even though planners for the Weber Basin Project suggested that water needs would be handled well into the twenty-first century. Developments along the Wasatch Front as well as in Ogden Valley and Summit County require more water. This water will be used primarily for urban development. Weber Basin has begun, particularly in the last decade, to develop all of the wells that were originally identified when the project began. As these twenty wells are developed, along with the upstream reservoirs, the water needs of the Weber Basin should be covered until about 2010. The wells that have been recently drilled and those in the planning stages are being drilled to culinary standards rather than irrigation standards. In general, the wells are being used as back-up to upstream reservoirs.

Growth and water use will then need to be planned and managed until another source of water, including the Bear River, can be brought to the Weber Basin and even to the Salt Lake Valley. There are as many as seven proposed dams for the Bear River. Some have proposed moving Bear River water by canal or otherwise to Willard Bay and then moving it through the present Weber Basin systems.

Water use and distribution along the Weber River system has never been an easy task. The future promises to increase management difficulties just as certain as it promises water rate increases. The extension of the Layton Canal and use of water in Willard Bay for irrigation are certain indicators that the Weber Basin Project of the 1940s and 1950s is not only reaching maturity but is being fully used. Careful management by all water-affected groups and ongoing cooperation can bring the Weber River to be even more of a servant to the dry land that it flows through.

Perhaps water development on the river is most apparent from a historical perspective. That perspective can be gained by comparing current practices that force water through the Gateway Tunnel to the Wasatch Front with Miles Goodyear's practice of watering his garden with buckets of irrigation water that passed through his hand-carved canals or by comparing the early efforts of the Brown family and other early Mormon pioneers at ditch irrigation with the year-round water use provided by reservoirs, ditches, and canals of modern vintage. Water under pressure is now used to sprinkle thousands of lawns, gardens, and fields where irrigation was never thought possible. Increased development cannot take place without water, and in such development wisdom and foresight will be essential.

 **Notes**

INTRODUCTION

1. William E. Smythe, *The Conquest of Arid America* (New York: The MacMillan Co., 1911; reprint, Seattle: University of Washington Press, 1970), 261.
2. Remarks made by Brigham Young in the bowery, Great Salt Lake City, June 8, 1856, *Journal of Discourses*, ed. Orson Pratt, vol. 3 (Liverpool: Orson Pratt, 1856), 339–40.
3. Robert G. Dunbar, *Forging New Rights in Western Waters* (Lincoln: University of Nebraska Press, 1983), 82.
4. Many works on Utah water history have concentrated on early times. They include Leonard J. Arrington, *Great Basin Kingdom: An Economic History of the Latter-day Saints, 1830–1900* (Linclon: University of Nebraska Press, 1966); Leonard J. Arrington, Feramorz Fox, and Dean L. May, *Building the City of God: Community and Cooperation among the Mormons* (Salt Lake City: Deseret Book Co., 1976); G. Lowery Nelson, *The Mormon Village: A Pattern and Technique of Land Settlement* (Salt Lake City: University of Utah Press, 1952); and Charles S. Peterson, "Imprint Agricultural Systems on the Utah Landscape," in Richard H. Jackson, ed., *The Mormon Role in the Settlement of the West* (Provo: Brigham Young University Press, 1978).

   Some other authors have dealt with various Utah water laws, such as George Thomas, who traces the laws from settlement times to 1919 in his work, *The Development of Institutions under Irrigation with Special Reference to Early Utah Conditions* (New York: Macmillan Co., 1920). Elwood Mead, in *Irrigation Institutions: A Discussion of the Economic and Legal Questions Created by Growth of Irrigated Agriculture in the West* (New York: Macmillan Co., 1907), traces Utah water law up to 1903; Charles Brough does the same to 1898 in *Irrigation in Utah* (Baltimore: Johns Hopkins Press, 1898). Robert G. Dunbar, in parts of his book, *Forging New Rights*, looks at Utah's water laws and makes comparisons with other western states.

   Some works have studied various Utah water projects, such as David Merrill's *An Historical Mitigation Study of the Strawberry Valley Project, Utah* (Salt Lake City: MESA Corporation for the Bureau of Reclamation, Upper Colorado Region, 1982). William Peterson, "History of Federal Reclamation in Utah by Projects," in *Utah: A Centennial History*, by Wain Sutton, vol. 1 (New York: Lewis Historical Publishing Co., 1949), and the *Utah Historical Quarterly* 39 (Summer 1971), edited by Charles

S. Peterson contain articles on Newton Reservoir, the Weber River Project, and the Strawberry Valley Project. James Hudson has written about *Irrigation Water Use in Utah Valley, Utah* (Chicago: University of Chicago, Department of Geography Research Paper No. 79, 1962).

5. Thomas, *Development of Institutions under Irrigation*, 45.
6. Ibid., 50, 121; Mead, *Irrigation Institutions*, 221.
7. Mead, *Irrigation Institutions*, 232.
8. Brough, *Irrigation in Utah*, 8. The act further stated that "a vote of two-thirds in favor would officially establish the water district. The district had the power to levy a direct tax on district residents for the purpose of constructing and improving on the district projects. In 1898 an amendment changed the powers for levying the taxes and restricted the levies to only the farmland irrigated by the system which meant that only the lands receiving water were taxable. This was the main change to the original law until it was repealed in 1897."
9. U. S. Congress, House, *Congressional Globe*, 39th Congress, 1st Session, 1866, 4049–54; William K. Wyant, *Westward in Eden: The Public Lands and the Conservation Movement* (Berkeley: University of California Press, 1982), 170–72.
10. For example, in Ogden in 1890, William "Coin" Harvey, who would go on to national prominence as an advocate of free silver, staged in Ogden on July 4, 1890, an economic "Carnival" that was nationally advertised to bring people from throughout the United States to partake of the great land and economic "boom," which he thought was on its way. The spectacular fizzled, but it did bring a flurry of interest in land speculation and mining to the area. It also demonstrated that Utah was caught up in the new economic way and would continue to be involved in that development. Along with that economic change came a political change which brought the non-Mormon leadership into power beginning with the election of Fred J. Kiesel as mayor of Ogden and George M. Scott as mayor of Salt Lake City. Mormon polygamy had created a great deal of political interest in Utah both nationally and locally. Federal law kept many Mormons disenfranchised from voting and holding office, and the "Gentile" leadership was capable of Americanizing the Utah economy much quicker.
11. Mead, *Irrigation Institutions*, 333.
12. Thomas Alexander, as stated in a communication with David Merrill and cited in Merrill, *An Historical Mitigation Study*, 5–6.
13. Richard M. Alston, *Commercial Irrigation Enterprise, The Fear of Water Monopoly and the Genesis of Market Distortion in the Nineteenth Century American West* (New York: Arno Press, 1978), 68–70.
14. Brough, *Irrigation in Utah*, 63–73.
15. William C. Darrah, "John Wesley Powell and the Understanding of the West," *Utah Historical Quarterly* 37 (spring 1969): 150–51; Wallace Stegner, *Beyond the Hundredth Meridian: John Wesley Powell and the Second Opening of the West* (Cambridge, Mass.: Riverside Press, 1962), 217–20, 223–28; Donald Worster, *Rivers of Empire, Water Aridity, and the Growth of the American West* (New York: Pantheon Books, 1985), 138–39.
16. Especially supportive of Powell's position are Philip L. Fradkin, *A River No More: The Colorado River and the West* (Tucson: University of Arizona Press, 1981), 24–

25, and Worster, *Rivers of Empire*, 138–39. Fradkin wrote that "Powell preached the uniqueness of arid lands and their need for special institutions. He used common sense and proposed that instead of the rectangular grid survey useful to the east on flat, equally watered lands, the arid West should be divided into watersheds, such as the Colorado River Basin. The West has paid dearly for not following that suggestion: witness the bitter intrastate water feuds. Powell knew that the West did not have an unlimited amount of land that could be irrigated or an inexhaustible supply of water, two false impressions spread widely by various boosters."

17. Samuel P. Hays, *Conservation and the Gospel of Efficiency: The Progressive Conservation Movement, 1890–1920* (New York: Antheneum, 1980), 8–10.
18. Smythe, *Conquest of Arid America*, 267–73.
19. William E. Warne, *The Bureau of Reclamation* (New York: Praeger, 1973), 13–20, 27.
20. Merrill, *An Historic Mitigation Study*, 15–17.
21. Dunbar, *Forging New Rights*, 117.
22. Thomas, *The Development of Institutions under Irrigation*, 197; Merrill, *An Historic Mitigation Study*, 17.
23. Barry Worth, Bureau of Reclamation, interview by Richard C. Roberts, 5 February, 1988.
24. The full citation of the key works referred to are as follows: Leonard J. Arrington, *Great Basin Kingdom: An Economic History of the Latter-day Saints, 1830–1900* (Lincoln: University of Nebraska Press, 1966); Leonard J. Arrington, Feramorz Y. Fox, and Dean May, *Building the City of God* (Salt Lake City: Deseret Book Co., 1976); G. Lowery Nelson, *The Mormon Village: A Pattern and Technique of Land Settlement* (Salt Lake City: University of Utah Press, 1952); Newburn I. Butt, *The Soil as One Factor in Early Mormon Colonization* (master's thesis, Brigham Young University, 1938); Charles S. Peterson, "Imprint Agricultural Systems on the Utah Landscape," in Richard H. Jackson, ed., *The Mormon Role in the Settlement of the West* (Provo: Brigham Young University Press, 1978); Elwood Mead, *Irrigation Institutions: A Discussion of the Economic and Legal Questions Created by the Growth of Irrigated Agriculture in the West* (New York: Macmillan Co., 1907); Charles H. Brough, *Irrigation in Utah* (Baltimore: Johns Hopkins Press, 1898); George Thomas, *The Development of Institutions under Irrigation with Special Reference to Early Utah Conditions* (New York: Macmillan Co., 1920); William E. Smythe, *The Conquest of Arid America* (Seattle: University of Washington Press, 1970); Robert G. Dunbar, *Forging New Rights in Western Waters* (Lincoln: University of Nebraska Press, 1983); Robert J. Hinton, "Water Laws, Past and Future," *Irrigation Age* 2 (November 1, 1891); Wallace Stegner, *Beyond the Hundredth Meridian: John Wesley Powell and the Second Opening of the West* (Cambridge, Mass.: Riverside Press, 1962); Thomas G. Alexander, "John Wesley Powell, the Irrigation Survey and the Inauguration of the Second Phase of Irrigation Development in Utah," *Utah Historical Quarterly* 37 (Spring 1969); William Culp Darrah, "John Wesley Powell and an Understanding of the West," *Utah Historical Quarterly* 37 (Spring 1969); Samuel P. Hays, *Conservation and the Gospel of Efficiency: The Progressive Conservation Movement, 1899–1920* (New York: Antheneum, 1980); Donald Worster, *Rivers of Empire: Water Aridity and the Growth of the American West* (New York: Pantheon

Books, 1985); Philip L. Fradkin, *A River No More: The Colorado River and the West* (Tucson: University of Arizona Press, 1981); William K. Wyant, *Westward in Eden: The Public Works and the Conservation Movement* (Berkeley: University of California Press, 1982); Samuel G. Houghton, *A Trace of Desert Waters: The Great Basin Story* (Salt Lake City, Howe Brothers, 1986); William E. Warne, *The Bureau of Reclamation* (New York: Praeger, 1973); David Merrill, *An Historical Mitigation Study of the Strawberry Valley Project, Utah* (Salt Lake City: MESA Corp. for the Bureau of Reclamation, Upper Colorado Region, 1982); Karl A. Wittfogel, *Oriental Despotism: A Comparative Study of Total Power* (New Haven: Yale University Press, 1967); Richard M. Alston, *Commercial Irrigation Enterprise, The Fear of Water Monopoly and the Genesis of Market Distortion in the Nineteenth Century American West* (New York: Arno Press, 1978); Keith D. Wilde, *Defining Efficient Water Resource Management in the Weber Drainage Basin, Utah* (Ph.D. diss., Utah State University, 1976).

CHAPTER ONE

1. *Deseret News*, July 23, 1936.
2. William Clayton, *William Clayton's Journal* (Salt Lake City: Deseret News Press, 1921), 323, 341–42.
3. Leonard J. Arrington and Dean May, "'A Different Mode of Life': Irrigation and Society in Nineteenth Century Utah," *Agricultural History* 49 (Jan. 1975): 7–8; Leonard J. Arrington, *Great Basin Kingdom: An Economic History of the Latter-day Saints, 1830–1900* (Lincoln: University of Nebraska Press, 1966), 53. Charles S. Peterson's position on this matter was determined by conversations with him over a period of time.
4. Howard Stansbury, *Exploration and Survey of the Great Salt Lake of Utah* (Philadelphia, 1855), 128–29.
5. Kate B. Carter, comp., "Pioneer Water Supply," in *Heart Throbs of the West* (Salt Lake City: Daughters of the Utah Pioneers, Central Co., 1948), 155–60.
6. Newbern I. Butt, "The Pioneer Spirit in Irrigation," *The Irrigation Farmer*, April 5, 1945, 2; Arrington, *Great Basin Kingdom*, 53.
7. William Peterson, "History of Agriculture in Utah," chap. 2 in *Utah: A Centennial History*, by Wain Sutton, vol. 1 (New York: Lewis Historical Publishing Co., 1949), 50–53; Brigham Young, *Journal of Discourses*, ed. Orson Pratt, vol. 3 (Liverpool: Orson Pratt, 1856), 328–30.
8. B. H. Roberts, *A Comprehensive History of the Church of Jesus Christ of Latter-day Saints: Century 1*, vol. 3 (Provo: Brigham Young University Press, 1965), 269; Arrington, *Great Basin Kingdon*, 52.
9. Charles H. Brough, *Irrigation in Utah* (Baltimore: Johns Hopkins Press, 1898), 19; George W. Rollins, "Land Policies of the United States as Applied to Utah to 1910," *Utah Historical Quarterly* 20 (July 1952): 242; Newbern I. Butt, *The Soil as One Factor in Early Mormon Colonization* (master's thesis, Brigham Young University, 1938), 145; Peterson, "History of Agriculture in Utah," 57.
10. Howard Egan, *Pioneering the West* (Richmond, Utah: Howard R. Egan Estate, 1917),

126–29. Names of the streams involved were: City Creek, Red Butte Creek, and Canyon Creek (later called Parley's Creek).

11. Leonard J. Arrington, Feramorz Y. Fox, and Dean May, *Building the City of God* (Salt Lake City: Deseret Book Co., 1976), 51; "First General Epistle," *Millennial Star* 2 (1849): 228–30; Edward Tullidge, *History of Salt Lake City and Its Founders* (Salt Lake City, 1886), 57–58; Carter, "Pioneer Water Supply," 167; Peterson, "History of Agriculture in Utah," 49.

12. Leonard J. Arrington and Davis Bitton, *The Mormon Experience: A History of the Latter-day Saints* (New York: Alfred A. Knopf, 1979), 115–16; Carter, "Pioneer Water Supply," 167.

13. Arrington, *Great Basin Kingdom*, 53; Arrington and May, "'A Different Mode of Life,'" 8–9; George Thomas, *The Development of Institutions under Irrigation, with Special Reference to Early Utah Conditions* (New York: Macmillan Co., 1920), 23.

14. Brough, *Irrigation in Utah*, 36–41.

15. "An Ordinance, Granting the Water of North Mill Creek Kanyon, and the Waters of the Next Kanyon North to Heber C. Kimball, Jan. 9, 1851," in Dale L. Morgan, *The State of Deseret* (Logan: Utah State University Press, 1987), 163.

16. "An Ordinance to Incorporate Great Salt Lake City, January 9, 1851" and "An Ordinance to Incorporate Ogden City, February 6, 1851," both in Morgan, *State of Deseret*, 166, 182.

17. "An Act Defining the Powers of the County Court, February 4, 1852," section 39 in Thomas, *Development of Institutions under Irrigation*, 45.

18. Thomas, *Development of Institutions under Irrigation*, 53–55, 117–24.

19. Ibid.

20. Donald J. Pisani, "Enterprise and Equity: A Critique of Western Water Law in the Nineteenth Century," *Western Historical Quarterly* 18 (Jan. 1987): 28.

21. Andrew Love Neff, *History of Utah, 1847 to 1869*, ed. and comp. Leland H. Creer (Salt Lake City: University of Utah, 1940), 255–61; Edward W. Clyde, "Law on Water Rights in Utah" in *Utah: A Centennial History*, by Wain Sutton, 98–105; Brough, *Irrigation in Utah*, 36–41; Pisani, "Enterprise and Equity," 20–23, 26–29.

22. Elwood Mead, *Irrigation Institutions: A Discussion of the Economic and Legal Questions Created by the Growth of Irrigated Agriculture in the West* (New York: MacMillian Co., 1907), 66, 228–33; Brough, *Irrigation in Utah*, 43; Roy P. Teele, "A General Discussion of Irrigation in Utah," in *Report of Irrigation Investigations in Utah, under the Direction of Elwood Mead, Chief of Irrigation Investigations, Assisted by R. P. Teele, A. P. Stover, A. F. Doremus, J. D. Stannard, Frank Adams, and G. L. Swendsen* (Washington, D. C.: Government Printing Office, 1903), 24; Robert G. Dunbar, *Forging New Rights in Western Waters* (Lincoln: University of Nebraska Press, 1983), 82.

23. Thomas, *Development of Institutions under Irrigation*, 45–55.

24. Clyde, "Law on Water Rights in Utah," 98–105.

25. David Merrill, Donald L. Snyder, and Jay Anderson, *An Historical Mitigation Study of the Strawberry Valley Project, Utah* (Salt Lake City: MESA Corp. for the Bureau of Reclamation, Upper Colorado Region, 1982), 5–6.

26. *Wilford Woodruff's Journal*, vol. 3 (Midvale, Utah: Signature Books, 1983), 236; Roberts, *A Comprehensive History of the Church of Jesus Christ of Latter-day Saints*, 292.

CHAPTER TWO

1. William J. Critchlow III and Richard W. Sadler, "Miles Goodyear's Fort Buenaventura," n.d., manuscript in authors' possession, Ogden, Utah, 22–36; Charles Kelly and Maurice L. Howe, *Miles Goodyear, First Citizen of Utah: Trapper, Trader and California Pioneer* (Salt Lake City: Western Printing Co., 1937), 50–53, 62–65, 70–75, 84–87, 142–47; Dale Morgan, "Miles Goodyear and the Founding of Ogden," *Utah Historical Quarterly* 21 (July 1953): 214–15; Milton R. Hunter, ed., *Beneath Ben Lomond's Peak: A History of Weber County, 1824–1900* (Salt Lake City: Deseret News Press, 1945), 57–64; Edward W. Tullidge, *Tullidge's Histories: Containing the History of All the Northern, Eastern, and Western Counties of Utah, also the Counties of Southern Idaho* (Salt Lake City: Juvenile Instructor Press, 1889), 2–9; Utah Historical Records Survey Project, *History of Ogden* (Ogden: Odgen City Commission, 1940), 11–19.

2. Utah Historical Records Survey Project, *History of Ogden*, 11–19.

3. Hunter, *Beneath Ben Lomond's Peak*, 67–69, 287; Utah Historical Records Project, *History of Ogden*, 19, 29; Gladys Brown White, "Captain James Brown," n.d., manuscript in Weber County Library, Ogden; Tullidge, *Tullidge's Histories*, 10; *Ogden Standard Examiner*, Dec. 23, 1931, p. 1; Weber County Property Records, "Tabulation of Water Claims, Weber River System, Weber County," book 2, Weber County Recorder's Office, Ogden, 22.

4. The main tributaries of the Weber River are Smith and Morehouse Creek, Beaver Creek, Silver Creek, Chalk Creek, Echo Canyon, Lost Creek, East Canyon Creek, and Cottonwood Creek. The main tributaries of the Ogden River are the North Fork, Middle Fork, South Fork, and Wheeler Canyon.

   Those streams beginning on the northern end of the Wasatch front of the Weber-Ogden Basin area and working southward are Rice Creek, North Ogden Canyon, Cold Water Canyon, One Horse Canyon, Garner Canyon, Jump Off Canyon, Taylor Canyon, Waterfall Canyon, Strong's Canyon, Beus Canyon, Burch Creek, Dry Canyon, Spring Creek, Broad Hollow, Corbett Creek, Hobb's Canyon, Kay's Creek (with North, Middle, and South forks), Holmes Creek, Adam's Canyon (with a branch of Holmes Creek), Webb Canyon (with another branch of Holmes Creek), Baer Creek (named Baer Creek on the maps, but should be Bair after John Bair), Shepard Creek, Farmington Canyon, Rudd Creek, Steed Creek, Davis Creek, Ford Canyon (with Ricks Creek), Barnard Creek, Parrish Creek, Centerville Canyon (with Deuel Creek), Ward Canyon (with Stone Creek), Holbrook Canyon (with Barton Creek), Mill Creek, North Canyon, and Hooper Canyon.

5. Thomas Bullock's notes in "Journal History of the Church," Sept. 2–3, 1849, Mormon Church Archives, Salt Lake City, Utah; T. Earl Pardoe, *Lorin Farr, Pioneer* (Provo: Brigham Young University Press, 1953), 106, 118–19; Utah Historical Records Survey Project, *History of Ogden*, 21; *The Deseret News* vol. 1, Sept. 7, 1850, 95.

6. "Journal History of the Church," Aug. 28, 1850, Mormon Church Archives, Salt Lake City.

7. Photo of Weber Canal at 25th Street, Ogden, in Pardoe, *Lorin Farr*, 137; *Weber County Court Record*, vol. A., Weber County Recorder's Office, 2; Hyrum A. Shupe, "School Once Stood Right in Center of Washington Avenue," *Ogden Standard Ex-*

aminer, Dec. 3, 1933; William W. Terry, "Ditches and Canals Dug 1849–1858," 1980, manuscript, Ogden, 9; John Farr, *My Yesterdays* (Salt Lake City: Granite Publishing Co., 1957), 41–42; Aaron Jackson, "Tragic Handcart Journey Described by Subscriber," *Ogden Standard Examiner*, in *Ogden Standard Examiner* scrapbook, Weber County Library; Hyrum Belnap, "Early Reminiscences of Hooper, Utah," n.d., manuscript dictated to Flora Belnap; Hunter, *Beneath Ben Lomond's Peak*, 76–77; Leo Haefeli and Frank J. Cannon, comps., *Directory of Ogden City and Weber County, 1883* (Ogden: Ogden Herald Publishing Co., 1883), 31; *Ogden City Council Minute Books From 1851 to 1869* in Ogden City Recorder's Office, passim for year 1852, 8. Hunter, *Beneath Ben Lomond's Peak*, 77.

9. Farr, *My Yesterdays*, 41–42.

10. Pardoe, *Lorin Farr*, 137; Shupe, "School Once Stood"; Terry, "Ditches and Canals," 9; Hunter, *Beneath Ben Lomond's Peak*, 76–77; Haefeli and Cannon, *Directory*, 31; *Weber County Court Record*, vol. A, 2; *Ogden City Council Minute Books From 1851 to 1869*, passim for year 1854; Farr, *My Yesterdays*, 41–42; *Sanborn Maps of Ogden* (New York: Sanborn Map and Pub. Co., 1884).

11. Isaac Newton Goodale, "Journal: 1850–1857," typed copy in Weber County Library, Ogden, 91–115; "Death of a Pioneer: Three Quarters of a Century Spent on This Earth, I. N. Goodale Leaves Many Sad Friends," *Ogden Standard Examiner*, April 26, 1890; Electa Goodale and Martha Goodale Buswell, "Isaac Newton Goodale," n. d., in possession of author; Irving W. Jones and Emma Anderson Jones, "The Ogden Bench Canal," n. d., in possession of author; Gordon Jones, interview by Richard Roberts, Sept. 5, 1982, Ogden. Mr. Jones and the author walked a portion of the Ogden Bench Canal, which is still easily identifiable in several places. Raymond E. Wardleigh, interview by Richard Roberts, Sept. 5, 1982, Ogden. The Wardleigh family, living at 950 East 23rd Street, Ogden, were long-time users and owners of shares in the Ogden Bench Canal. Much of the old water lines used to irrigate the Wardleigh lot are still in place.

12. Utah Historical Records Survey Project, *History of Ogden*, 56.

13. Ibid., 56–57; "A Stroll through Ogden," *Tullidge's Quarterly Magazine of Utah* 1 (1881): 475; Haefeli and Cannon, *Directory*, 111–12.

14. Haefeli and Cannon, *Directory*, 111–12 ; *Sanborn Maps of Ogden*.

15. James Macbeth, "Plumbing Scarce in Ogden as Macbeth Opened Shop," *Ogden Standard Examiner*, in *Ogden Standard Examiner* scrapbook, Weber County Library.

16. Haefeli and Cannon, *Directory*, 111–12.

17. "Stroll through Ogden," 476.

18. Utah Historical Records Survey Project, *History of Ogden*, 56–57; *Ogden City Council Minute Book*, vol. B, 24, 88, 301; vol. C, 141; vol. F, 162.

19. William W. Terry, *Weber County is Worth Knowing* (Ogden: William W. Terry and Weber County School District, n.d.), 39–51. Today the old route of the Mill Creek can be seen west of Harrison Boulevard and south of the Farr Orchard packing shed at 1235 Canyon Road. The diagonal orchard road running south-easterly and north-westerly is the old Mill Creek route. This creek bed was used until a pipeline was developed that brought water from the canyon to the Pioneer Electric Plant and then ran back into Mill Creek west of the electric plant. The tail race is still visible today.

20. Terry, *Weber County is Worth Knowing*, 39–52; Hunter, *Beneath Ben Lomond's Peak*, 86–87, 156.

21. Floyd Woodfield and Clara Woodfield, eds., *A History of North Ogden* (Ogden: Empire Printing, 1986), 202; Floyd Woodfield, "Notes on the BiCentennial History of North Ogden," (North Ogden: held by author, 1983); John Q. Blaylock, "History of North Ogden," n.d., manuscript, Ogden Genealogical Library, 60.

22. F. Woodfield and C. Woodfield, *A History of North Ogden*, 203–4. Those credited with digging the North Ogden Canal are: Levi Wheeler; Calvin Wheeler; Samuel Dean; David J. Evans; John Cardon (and his wife, probably Anne Regula Furrer Cardon, using a pick and shovel); the Humphries; Rose, Jenkins, John, and Josiah Price; Nathaniel, John, James, and Robert Montgomery; William James; David Levy; Isaac Riddle; John Riddle; John and Henry Mowers; Simeon Cragun; the Browns; George Norvill; Joseph Bidwell; the Georges; Thomas and Crandall Dunn, Newman G. Blodgett and sons; James Barker; Johnathan and Solomon Campbell; William Roylance; David Moore; Abraham Chadwick; Thomas Chapman; Jared Campbell; William and George Marler; Charles Sansom; James Lofthouse; Alfred, Robert, and Richard Berrett; John Cole; Thomas Painter; Wesley Rose; Henry Holmes; the Mathewses; Edward Austin; John Davis; David R. Jones; James Rice; Benjamin Cazier; William Hall; Isaac Bosenbart; James Monroe; John Campbell; John and Isaac White; Bailey Lake; Edward Wade; William Hill; the Ferrins; Charles H. Rhees; Thadeus and Gideon Alvord; and others.

23. F. Woodfield and C. Woodfield, *A History of North Odgen*, 201–11; Woodfield, "Notes"; Leonard H. Judkins, "Two Decades in North Ogden," n.d., manuscript in possession of Floyd Woodfield, 15; Terry, *Weber County is Worth Knowing*, 50.

24. Hunter, *Beneath Ben Lomond's Peak*, 190–92; Terry, *Weber County is Worth Knowing*, 51; *Ogden City Council Minutes Record, Book A, from 1851 to 1869*, 48.

25. Brian L. Taylor, comp., "History of Farr West," 1980, manuscript, Ogden Genealogical Library.

26. Luman A. Shurtliff, *His Personal History, 1807–1884* (Harrisville, Utah, n.d.), 93–94.

27. Ibid., 101–3.

28. Ibid.

29. Robert O'Brien, *Marriott: The J. Willard Story* (Salt Lake City: Deseret Book Co., 1981), 34, 105, 119, 22–122 passim.

30. Hunter, *Beneath Ben Lomond's Peak*, 157–58; Norman Bingham, Lillian Belnap, and Lester Scoville, comps., *Sketch of the Life of Erastus Bingham and Family* (n.d., n.p.), 30–31; Edward Tullidge, *Tullidge's Quarterly Magazine of Utah* 2 (1883): 632–34. 31. Utah Historical Records Survey Project, *Preliminary Inventory of the County Archives of Utah: No. 29, Weber County* (Ogden: Utah Historical Records Survey Project, 1940), 10; *Ogden City Council Minutes Record, Book A, from 1851–1869*, 9–10, 18–19; Hunter, *Beneath Ben Lomond's Peak*, 161, 162, 164.

32. Timothy Kendell, "A Brief History of Uintah, Weber County, Utah," n.d., manuscript in Ogden Genealogical Library, Ogden, 3–7; Andrew Jenson, *Encyclopedia History of the Church of Jesus Christ of Latter-day Saints* (Salt Lake City: Deseret News Pub. Co., 1941), 896; Hunter, *Beneath Ben Lomond's Peak*, 166–69.

33. Leila Heslop, comp., *History of the West Weber Relief Society, 1868–1968* (n.d.,

n.p.), 96–101, 103; Arvil Hipwell and Ida Mae Deem Hipwell, comps., *The Big Bend of the River: History of West Weber, 1859–1976* (Ogden: Empire Printing Co., 1976), 403–6; Hunter, *Beneath Ben Lomond's Peak*, 198–202.

34. Emma Anderson, Marian Berrett, and Rosalie Clark, comps., *A Brief History of Taylor in Story and Pictures: About 1822–1875* (n.p., n.d.), 1–5; Jenson, *Encyclopedia History*, 862; Hunter, *Beneath Ben Lomond's Peak*, 206.
35. Hunter, *Beneath Ben Lomond's Peak*, 206–8; Jenson, *Encyclopedia History*, 955.
36. Lyman Cook and Dorothy Cook, eds., *History of Plain City: March 17th 1859 to Present* (n.p., n.d.), 7.
37. Jensen, *Encyclopedia History*, 659–60; Hunter, *Beneath Ben Lomond's Peak*, 211–14, 217–18; L. Cook and D. Cook, *History of Plain City*, 1–2, 5–8; Edward Tullidge, *Tullidge's Quarterly Magazine of Utah* 2 (1883): 53–58; Tullidge, *Tullidge's Histories*, 51–52. By 1889 some eleven miles of canal had been made at a cost of $1,000 per mile. From the time of first settlement of Plain City to 1890, about $20,000 (in value of that period) had been expended on improvements on the irrigation system.
38. Jensen, *Encyclopedia History*, 922–23; Utah Historical Records Survey Project, *Inventory of the County Archives of Utah*, 27; Hunter, *Beneath Ben Lomond's Peak*, 223–26.
39. Belnap, "Early Reminiscences of Hooper, Utah."
40. John D. Hooper, "Memoirs and Life of John D. Hooper" (n.p., n.d.), 16.
41. Hunter, *Beneath Ben Lomond's Peak*, 226–30; Belnap, "Early Reminiscences of Hooper, Utah"; Roger Rawson and Kay Rawson, *A Bicentennial History: Hooper-Kanesville* (n.p., 1976); Hooper, *Memoirs and Life of John D. Hooper*, 16; A. Hipwell and I. Hipwell, *Big Bend of the River*, 403–5; John M. Belnap, *History of Hooper, Utah: Land of Beautiful Sunsets* (n.p., n.d.), 37–43. Belnap, on page 203, lists nine people of Hooper who were drowned in the river over the years. He also includes a poem written in 1869 by John Thompson titled "Muscrat Springs" in which a stanza tells of the importance of the Weber River to the community: "And down from the Weber, the water will flow; / It will spread o'er the land with it's [sic] life giving power, / Will revive and support both the fruit and the flower, / Deck the land in full bloom and all nature will sing, / In her floral array by the Muscrat Springs."
42. Emma Russell, *Footprints of Roy: 1873–1979* (Roy, Utah: Emma M. Russell, 1979), 9.
43. Ibid., 9, 10, 221; Hunter, *Beneath Ben Lomond's Peak*, 233–34. 44; LaVerna Newey, *Remember My Valley: A History of Ogden Canyon, Huntsville, Liberty, and Eden, Utah, from 1825–1976* (Salt Lake City: Hawkes Pub. Co., 1977), 52.
45. Ibid., 258.
46. Ibid., 51–52, 258–59; Hunter, *Beneath Ben Lomond's Peak*, 242–45; "Huntsville Ward Dedicatory Services: Short History of Huntsville," May 3, 1959, photocopy, Ogden Genealogical Library; Andrew Jensen, comp., "Huntsville Ward," in *Souvenir, Homecoming Week* (Huntsville, Utah, 1917), 5–21, 29.
47. Hunter, *Beneath Ben Lomond's Peak*, 255.
48. Newey, *Remember My Valley*, 32–34, 72–73, 258–59; Hunter, *Beneath Ben Lomond's Peak*, 253–58; Melba Colvin and Ren Colvin, *Over 100 Years in Eden: History of Eden Ward, Ogden, Utah, 1877–1977* (n.p., n.d.).
49. Newey, *Remember My Valley*, 258.

50. Mary Chard McKee, *Early History of Liberty and the People* (n.p., n.d.), 9–10.
51. Newey, *Remember My Valley*, 34–35, 258; McKee, *Early History of Liberty*, 1–10; Parley J. Clark, *History of Liberty* (n.p., n.d.).
52. Newey, *Remember My Valley*, 82.
53. Ibid., 39–40, 42, 82–85.

CHAPTER THREE

1. William Robb Purrington, "The History of South Davis County, from 1847–1870" (master's thesis, University of Utah, 1959), 1–3.
2. Ibid., 2.
3. Aubrey L. Haines, ed., *Osborne Russell's Journal of a Trapper* (Lincoln: University of Nebraska Press, 1970), 120.
4. J. Roderic Korns, ed., "West from Fort Bridger," *Utah Historical Quarterly* 19 (1951): 134–35.
5. John W. Gunnison, *The Mormons or Latter–day Saints in the Valley of the Great Salt Lake, A History of Their Rise and Progress, Peculiar Doctrines, Present Conditions and Projects* (Philadelphia, 1852), 15–18.
6. Leslie T. Foy, *The City Bountiful: Utah's Second Settlement from Pioneers to Present* (Bountiful, Utah: Horizon Publishers, 1975), 47, 64–66, 73–74; Annie Call Carr, "Bountiful," in *East of Antelope Island*, comp. Daughters of the Utah Pioneers, Davis County Chapter, 4th ed. (Salt Lake City: Publishers Press, 1971), 49.
7. Foy, *The City Bountiful*, p. 69; Carr, "Bountiful," 53; Davis County Probate Court Records, Davis County Recorder's Office, Farmington, Utah, minutes in passim since 1852.
8. Foy, *The City Bountiful*, 70.
9. Ibid., 69.
10. Carr, "Bountiful," 52–53.
11. Truman Woodrow Barlow, interview by Richard C. Roberts, Burley, Idaho, Feb. 5, 1983.
12. Charles R. Mabey, *Our Father's House: Joseph Thomas Mabey Family History* (Salt Lake City: Beverly Craftsman, 1947), 151–52, 236–38.
13. Ibid.
14. Carr, "Bountiful," 53.
15. Arlene H. Eakle, Adelia Baird, and Georgia Weber, *Woods Cross: Patterns and Profiles of a City* (Woods Cross, Utah: Woods Cross City Council, 1976), 48.
16. Leo J. Muir, comp. and ed., *The Life Story of Truman Heap Barlow, Including His Forebears, Immediate Kinfolk and Posterity* (Los Angeles: Paul D. Bailey-Westernlore Press, 1961), 29–30; Priscilla Muir Hatch, "West Bountiful," in *East of Antelope Island*, 192–201.
17. Hatch, "West Bountiful," 192–201; Muir, *Life Story of Truman Heap Barlow*, 29–30; Eakle, Baird, and Weber, *Woods Cross*, 4–5, 12, 16–17, 19, 25, 49.
18. Edith Folsom Hatch, "South Bountiful," in *East of Antelope Island*, 145–46.
19. Ibid., 158–60.
20. Mabey, *Our Father's House*, 161.

21. Glen M. Leonard, "A History of Farmington, Utah" (master's thesis, University of Utah, 1966), 27.

22. Ibid.

23. Ibid., 28–29; Marva Udy Earl, "History of James Udy," n.d., Daughter of the Utah Pioneers, Miller Camp Files, Daughters of the Utah Pioneer Museum Library, Salt Lake City, 3.

24. Leonard, *History of Farmington*, 113–16.

25. Purrington, "History of South Davis County," 63.

26. Ibid., 63–65; Joseph Holbrook, "Diary Excerpt," in *East of Antelope Island*, 363.

27. Mabel S. Randall, "Centerville," in *East of Antelope Island*, 60–65, 69–71; Mary Ellen Smoot and Marilyn Sheriff, *The City In-Between: History of Centerville, Utah, Including Biographies and Autobiographies of Some of Its Original Settlers* (Bountiful: Carr Printing Co., 1975), 5–7; David F. Smith, *My Native Village: A Brief History of Centerville, Utah* (n.p.: David F. Smith, 1943), 9–13; Purrington, "History of South Davis County," 10, 13, 15–17.

28. Pearl Green and LaVaun Burgin, "History of the Old Grist Mill of Centerville" (Centerville, Utah, 1967); "Structure Erected in Early 1850's [Old Mill]," *Centerville Newsetter* 1, no. 6 (May 1944).

29. John M. Whittaker, *Daily Journal of John M. Whittaker: From the Founding of Centerville in 1842 [sic] to January 1, 1883*, 1883, Special Collections, University of Utah Marriott Library, Salt Lake City, 2; Ves Harrison, "Centerville Historical Society: Antiquated Pipe," *Davis County Clipper*, Aug. 3, 1983; "Centerville Deuel Creek Irrigation Company Records," 1900–1983, records in possession of Vernon Carr, secretary, Centerville, Utah, passim.

30. Ibid.

31. "Kaysville," in *East of Antelope Island*, 108.

32. Ibid., 102–8; Kenneth H. Sheffield and Sherman B. Sheffield, "Early History of Kaysville Ward, Utah," prepared for Centennial Celebration of Kays Ward held Aug. 24–25, 1951, Weber State University Library Documents, Ogden; Wayne Wahlquist, "Settlement Processes in the Mormon Core Area, 1847–1890" (Ph.D. diss., University of Nebraska, 1974), 231–42. Wahlquist says that the Joseph Barton and Government Survey plat maps of Kaysville locate Kays Fort inaccurately. Ivy Blood Hill, ed., *William Blood Biography* (Logan, Utah: J. P. Smith and Son, 1962), 17; Carol Ivins Collett, *Kaysville—Our Town: A History* (Salt Lake City: Moench Letter Service, 1976), 5–7, 16–17, 38–45, 83–85, 102, 108–9, 134.

33. Sarah J. Humphrey Adams, "Layton: A History of Pioneer Days of Layton," in *East of Antelope Island*, 128, 118–40.

34. Frank D. Adams and Bonnie Adams Kesler, comps. and eds., *Biography of Hyrum Adams and Annie Laurie Penrod Adams, Layton, Utah* (Salt Lake City: Paragon Printing Co., 1953), 3–5, 10, 13, 24, 34, 75; Frank Adams, comp. and ed., *Ancestors and Descendants of Elias Adams: The Pioneer, 600–1930* (Kaysville, Utah: Inland Printing Co., n.d.), 96–97, 111–14, 138–39.

35. Sarah K. Adams and Irene Doney Higley, "West Layton," in *East of Antelope Island*, 203.

36. Ibid., 202–5.

37. Janice Page Dawson, "Three Sections of East Layton History" (paper presented in senior seminar, Department of History, Weber State College, Ogden, June 2, 1980).

38. Leonard Bowman, "South Weber," in *East of Antelope Island*, 172–73; Lee Bell, "Irrigation and Farming," in *South Weber: The Autobiography of One Utah Community* (Salt Lake City: K/P Graphics, Utah Division, 1990), 147–56.

39. Doneta Gatherum, "West Point Pioneer Settlement Related to Founding of Hooper," *Salt Lake Tribune*, Nov. 19, 1981; Cathy Free, "Visiting West Point Not an Activity for Thrill Seekers," *Salt Lake Tribune*, Mar. 3, 1983.

40. Ethel Steed Mitchell, "Clinton: Davis County, Utah, from 1870 to 1900," in *East of Antelope Island*, 79–83; Mary Bowman and Bowen Hadlock, "Story of the First Settler of Clinton," in *East of Antelope Island*, 84–86.

41. Eva M. Warren, "Syracuse," in *East of Antelope Island*, 187–91.

42. Briant S. Jacobs, "Clearfield," in *East of Antelope Island*, 74–79.

43. Mary Easthope, "History of Sunset," 1959, manuscript, Ogden Genealogical Library.

CHAPTER FOUR

1. There are several accounts that deal with the settlement era of Box Elder County. Among those studied that had somewhat varied accounts are the following: Andrew Jenson, *Encyclopedia History of the Church of Jesus Christ of Latter-day Saints* (Salt Lake City: Deseret News Pub. Co., 1941), 83–84; The Historical Records Survey, Division of Women's Professional Projects, Works Progress Administration, *Inventory of the County Archives of Utah No. 2* Box Elder County (Ogden: The Historical Records Survey, 1938), 1–9; Box Elder Chapter Sons of Utah Pioneers, *Box Elder Lore of the Nineteenth Century* (Brigham City, Utah: Box Elder Chapter of the Sons of the Utah Pioneers, 1951), 81–85, 138–43; Box Elder County Daughters of the Pioneers, *History of Box Elder County* (n.p., [1937?]), 2–3, 40, 59–60, 63–69, 125–28, 257–60, 264–66, 269–70, 273–76, 326–30, 347–48; Wayne L. Wahlquist, "Settlement Processes in the Mormon Core Area 1847–1890" (Ph.D. diss., University of Nebraska, 1974), 197–230; Veara S. Fife and Chloe N. Petersen, *Directory to Commemorate the Settling of Brigham City* (Golden Spike Chapter, Utah Genealogical Society, 1976), 172.

2. Box Elder County Daughters of the Pioneers, *History of Box Elder County*, 2.

3. Ibid.

4. Wahlquist, "Settlement Processes," 216.

5. Ibid.; Fife and Petersen, *Settling of Brigham City*, 172.

6. Wahlquist, "Settlement Processes," 200–1.

7. Box Elder County Daughters of the Pioneers, *History of Box Elder County*, 265–66.

8. Ibid., 270–76.

9. Ruth Johnson and Glen F. Harding, *Barnard White: Convert to the L.D.S. Church, 1854; Utah Pioneer, Bishop and Business Leader; Pioneer in Paradise, Cache County, Utah; Prominent in Weber and Box Elder Counties, Utah* (Provo: Brigham Young University Press, 1967), 163.

10. Ibid., 163–66.

11. Box Elder County Daughters of the Pioneers, *History of Box Elder County*, 276.

12. Ibid., 63–66; 125–28.
13. Ibid., 67–68.
14. Ibid., 326–30; Box Elder Chapter Sons of Utah Pioneers, *Box Elder Lore of the Ninetheenth Century*, 138–43.
15. The Fine Arts Study Group, an Affiliation of the National Federation of Women's Clubs, comps., *Mountains Conquered: The Story of Morgan with Biographies* (Morgan, Utah: Morgan County News Publishers, 1959), 5–7, 9–11, 13–49, 62–65; Morgan, Utah North Stake, *Morgan Stake 1877–1981* (Salt Lake City: Publishers Press, 1988), 1–3; Edward W. Tullidge, *Tullidge's Histories: Containing the History of All the Northern, Eastern, and Western Counties of Utah, also the Counties of Southern Idaho* (Salt Lake City: Juvenile Instructor Press, 1889), 109–22.
16. Quoted from Huldah Cordelia Thurston Smith's journal in Jeanine Fry Ricketts, "Father Thomas Jefferson Thurston," in *By Their Fruits: A History and Genealogy of the Fry Family* (n.p. n.d.), 147.
17. Lincoln Jensen, "Line Creek Irrigation Company," Jan. 1982, manuscript, Morgan County Historical Society, Morgan, Utah (hereafter MCHSL); C. Alfred Bohman, "Water Resource History, Peterson Area," Mar. 16, 1981, manuscript, MCHSL; C. Alfred Bohman, "History of Peterson Irrigation Water at Peterson," Mar. 1980, manuscript, MCHSL.
18. Lincoln Jensen, "Mecham-Madsen-Nelson Ditch," n.d., manuscript, MCHSL; Lincoln Jensen, "The Smith Creek Irrigation Company," n.d., manuscript, MCHSL.
19. Bohman, "Water Resource History, Peterson Area," 1–3.
20. Fine Arts Study Group, *Mountains Conquered*, 20.
21. Bohman, "Water Resources History, Peterson Area," 4–6; Bohman, "History of Peterson Irrigation Water at Peterson," 1; Bohman, "History of Culinary Water Supply for the Town of Peterson, Morgan County," Mar. 16, 1981, manuscript, MCHSL, 1–2.
22. Fine Arts Study Group, *Mountains Conquered*, 24.
23. Grace Bowen Kilbourn, *History of the Old Porterville Church, 1864–1948* (Morgan, 1981), 8, 15, 16, 19–21, 25–27, 50, 138; Elsie Delia Adams Florence, comp., *Henry Samuel Florence* (Provo: Brigham Young University Press, n.d.), 3–5, 8, 12, 21, 50, 52, 55.
24. Kilbourn, *History of the Old Porterville Church*, 8, 15, 16, 19–21, 25–27, 50, 138; Linden Porter, "Porterville Ditch," Jan. 1984, manuscript, MCHSL.
25. Louise B. Waldron, "West Richville Irrigation and Canal Company," 1983, manuscript, MCHSL, 1. See also: Tullidge, *Tullidge's Histories*, 115–16; Fine Arts Study Group, *Mountains Conquered*, 30–32.
26. Fine Arts Study Group, *Mountains Conquered*, 32.
27. Waldron, "West Richville Irrigation and Canal Company," 2–3.
28. Fine Arts Study Group, *Mountains Conquered*, 34–35; "The South Round Valley Canal Company," n.d., manuscript, MCHSL; Jay D. Stannard, "Irrigation in the Weber Valley," in *Report of Irrigation Investigation in Utah*, by Elwood Mead (Washington, D.C.: Government Printing Office, 1903), 174–75.
29. Lincoln Jensen, "History of Enterprise Field Ditch," n.d., manuscript, MCHSL; Fine Arts Study Group, *Mountains Conquered*, 36–37, 39–40.
30. Fine Arts Study Group, *Mountains Conquered*, 46.

31. Ibid., 47–48; Tullidge, *Tullidge's Histories*, 117–18.
32. Joseph H. Francis, "Weber Canal," Mar. 1984, manuscript, MCHSL, 1–6; Joseph H. Francis, comp., *Our Heritage* (Morgan: Samuel and Esther Francis Family Organization, 1984), 69–73; Fine Arts Study Group, *Mountains Conquered*, 40–46; Tullidge, *Tullidge's Histories*, 107–112.
33. John A. Compton, "Welch Field Ditch," Mar. 1980, manuscript, MCHSL.
34. Fine Arts Study Group, *Mountains Conquered*, 42; Francis, "Weber Canal," 2–6; Francis, *Our Heritage*, 69–73.
35. Morgan, Utah North Stake, *Morgan Stake 1877–1981*, 48–49; Francis, *Our Heritage*, 199–204.
36. Fine Arts Study Group, *Mountains Conquered*, 44–45.
37. Ibid., 45–46.
38. Marie Ross Peterson and Mary H. Pearson, comps., *Echoes of Yesterday: Summit County Centennial History* (n.p.: Daughters of Summit County, 1947), 12.
39. E. G. Beckwith, "Report of Explorations and Surveys for a Railroad Route to the Pacific Ocean," in *Report of the Secretary of War on the Several Pacific Railroad Explorations*, 33rd Cong., 1st sess., 1853, H. Doc., vol. 1, p. 12.
40. Tullidge, *Tullidge's Histories*, 125.
41. Fannie J. Richins and Maxine R. Wright, comps., *Henefer: Our Valley Home* (Salt Lake City: Utah Printing Co., n. d.), 151.
42. Ibid.
43. Ibid., 152.
44. Ibid., 92–93.
45. Ibid., 93.
46. Amelda Richins, "History of Echo, Utah," in *Echoes of Yesterday*, 76.
47. Jessie R. Stevens, "History of Peoa," in *Echoes of Yesterday*, 215–16.
48. Ibid.
49. Tullidge, *Tullidge's Histories*, 129; Stannard, "Irrigation in the Weber Valley," 172–73.
50. Tullidge, *Tullidge's Histories*, 126–27; Peterson and Pearson, *Echoes of Yesterday*, 187–96; Stannard, "Irrigation in Weber Valley," 173.
51. Tullidge, *Tullidge's Histories*, 138.
52. Ibid.
53. Ibid.
54. Ibid., 139; J. Kenneth Davies, *George Beard: Mormon Pioneer Artist with a Camera* (Provo: n.p., 1975), 21–22.
55. Davies, *George Beard*, 22.
56. Tullidge, *Tullidge's Histories*, 138.
57. Stannard, "Irrigation in Weber Valley," 173–74; Thomas E. Moore, "Coalville City Water Works," in *Echoes of Yesterday*, 129–30.
58. Hazel Wright, "Irrigation," in *Echoes of Yesterday*, 172–73; Elizabeth Brown, "History of Amos Sargent," in *Echoes of Yesterday*, 186.
59. J. Kenneth Davies, *Thomas Rhoads: The Wealthiest Mormon Gold Miner* (Provo: J. Kenneth Davies, 1980), 92; Laurua E. Bloggard, "History of Kamas," in *Echoes of Yesterday*, 266–67, 272–73, 283; Tullidge, *Tullidge's Histories*, 129–30.
60. Ralph A. Richards, interview by Richard W. Sadler and Steve Readdick, Oakley,

Utah, Apr. 9, 1973; Gwen Gibbons "History of Marion," in *Echoes of Yesterday*, 250–52, 259–61; Grace H. Lemon and Pearl W. Atkinson, "Francis," in *Echoes of Yesterday*, 283–85, 290–91.

61. Gibbons, "History of Marion," 260.
62. Mrs. Charles Gibbons and Mrs. T. Edward Brown, "Rockport, Summit County, Utah," in *Echoes of Yesterday*, 199–208; Tullidge, *Tullidge's Histories*, 127–28. Tullidge spells the first name as Enoch.
63. Leora Franson, "Irrigation in Oakley," in *Echoes of Yesterday*, 246.
64. Richards, interview.
65. Peterson and Pearson, *Echoes of Yesterday*, 33–34.
66. Franson, "Irrigation in Oakley," 229–31, 242–43.

CHAPTER FIVE

1. George Thomas, *The Development of Institutions under Irrigation with Special Reference to Early Utah Conditions* (New York: Macmillan Co., 1920), 138–39.
2. Ibid., 138. See also: Charles H. Brough, *The Economic History of Irrigation in Utah* (Baltimore: Johns Hopkins Press, 1898), 55–62; Elwood Mead, *Irrigation Institutions: A Discussion of the Economic and Legal Questions Created by the Growth of Irrigated Agriculture in the West* (Washington, D. C.: Government Printing Office, 1903), 82–87, 222–27; Roy P. Teele, "Irrigation Investigations in Utah: General Discussion of Irrigation in Utah," in *Report of Irrigation Investigations in Utah*, by Elwood Mead (Washington, D. C.: Government Printing Office, 1903), 22–24, 29–30; Jay D. Stannard, "Irrigation in the Weber Valley," in *Report of Irrigation Investigations in Utah*, 176–79; Robert G. Dunbar, *Forging New Rights in Western Waters* (Lincoln: University of Nebraska Press, 1983), 16–17, 28–31.
3. Stannard, "Irrigation in the Weber Valley," 176–79.
4. Thomas, *Development of Institutions under Irrigation*, 150–51.
5. Ibid., 209–12.
6. Ibid., 218.
7. Thomas Alexander, as stated in a communication with David Merrill and cited in his book, *An Historical Mitigation Study of the Strawberry Valley Project Utah* (Salt Lake City: MESA Corp. for the Bureau of Reclamation, Upper Colorado Region, 1982), 5–6.
8. Stannard, "Irrigation in the Weber Valley," 179; Mead, *Irrigation Institution*, 223–27; Teele, "Irrigation Investigations in Utah," 24–25.
9. William E. Warne, *The Bureau of Reclamation* (New York: Praeger Publishers, 1973), 3–13; William Peterson, "History of Agriculture in Utah," in *Utah: A Centennial History*, by Wain Sutton, vol. 1 (New York: Lewis Historical Publishing Co., 1949), 82–98; Thomas, *Development of Institutions under Irrigation*, 248–62; Dunbar, *Forging New Rights*, 39, 50–52; Samuel P. Hayes, *Conservation and the Gospel of Efficiency: The Progressive Conservation Movement, 1890–1920* (1959; reprint, New York: Atheneum, 1980), 5–22.
10. Teele, "Irrigation Investigations in Utah," 29.
11. Dunbar, *Forging New Rights*, 118.

12. Ibid., 117–19, 125. See also: *Message of the Governor of Utah to the Fifth Session of the State Legislature of Utah, January 13, 1903* (Salt Lake City, 1903), 11; *Salt Lake Tribune*, Oct. 2, 3, 4, 1903; Teele, "Irrigation Investigations in Utah," 29–30; Thomas, *Development of Institutions under Irrigation*, 196–202.

13. Thomas, *Development of Institutions under Irrigation*, 274–85.

14. J. L. Robson, "Weber County Report on Irrigation Claims on Ogden River," n.d., manuscript, Ogden River Water Users Office, Ogden. This document is especially significant because it lists the various dates, names, and claimants for water rights on the Ogden River between 1848 and 1889. D. D. McKay, "Report: Irrigation in Weber County" n.d., manuscript, Ogden River Water Users Office, 11.

15. McKay, "Irrigation in Weber County," 11.

16. *In the District Court of the Second Judicial District of the State of Utah, in and for Weber County, Plain City Irrigation Company, A Corporation, North Ogden Irrigation Company, A Corporation, the State of Utah, and all other claimants to the right to the use of water of the Weber River System, Defendants, No. 7487, Findings of Tort, Conclusions of Law and Judgement and Decree*, in possession of author. This document is especially significant for listing the claims, dates, claimants, and water allotment from 1850 to 1937 on the Weber River.

17. McKay, "Irrigation in Weber County," 11.

18. LaMar C. Kap, "File on John Kap, Sr.," n.d., Office of LaMar C. Kap, Weber State University, Ogden; *Ogden Standard Examiner*, May 20, 27, 28; June 2; July 7, 8, 10, 11, 1936.

19. Edward Tullidge, *Tullidge's Histories: Containing the History of All the Northern, Eastern, and Western Counties of Utah, also the Counties of Southern Idaho* (Salt Lake City: Juvenile Instructor Press, 1889), 65. See also: Stannard, "Irrigation in the Weber Valley," 175; Daughters of the Utah Pioneer, Davis County Chapter, comps., *East of Antelope Island* 4th ed. (Salt Lake City: Publishers Press, 1971), 129.

20. Horace Steed, interview by Steve Readdick, Kaysville, Utah, Jan. 23, 1973. See also: Samuel Fortier, "Report of Samuel Fortier, Consulting Engineer for the Davis and Weber Counties Canal Company," in *The Davis and Weber Counties Canal Company: Engineer's Report* (Ogden: n.p., 1910); D. E. Harris, interview by Jerry Mower and Steve Readdick, Ogden, Jan. 12, 1973; David Jay, "The Development of Irrigation and Reclamation on the Weber River" (paper written for history course at Weber State College, Ogden, May, 1974); Malcolm MacKenzie, "The Development of Irrigation and Reclamation on the Weber River During the 20th Century" (paper written for history course at Weber State College, May, 1974); David Schoenfeld, "Irrigation and Canal Companies in Davis County 1890–1974" (paper written for history course at Weber State College, May, 1974); D. Earl Harris, "Brief History of the Davis and Weber Counties Canal Company," 1965, manuscript.

21. Octave Frederick Ursenbach, "Vocation—Employment," excerpts from biography of Ursenback, Morgan County Historical Society Library, 101–4.

22. Stannard, "Irrigation in the Weber Valley," 196; Fortier, "Report of Samuel Fortier"; The Fine Arts Study Group, An Affiliation of the National Federation of Women's Clubs, comps., *Mountains Conquered: The Story of Morgan with Biographies* (Morgan: Morgan County News Publishers, 1959), 28–29; Robin Tippets, "Northern Utah Earthen Dam a Century Old," *Ogden Standard Examiner*, Oct. 4, 1973, p. 9B; *Speci-*

fications for a Storage Reservoir Dam, on East Canyon Creek, Morgan County, Utah to be Constructed by Davis and Weber Counties Canal Company (Ogden: Hestmark and Wilcox Printers and Bookbinders, n. d.); Lincoln Jensen, "History of the East Canyon Reservoir," Mar. 1984, manuscript, Morgan County Historical Society Library.

23. *Ogden Standard Examiner*, Oct. 19, 1896; Leonard J. Arrington and Lowell Dittmer, "Reclamation in Three Layers: The Ogden River Project, 1934–1965," *Pacific Historical Review* 35 (Feb. 1966): 17–18.

24. Leslie T. Foy, *The City Bountiful: Utah's Second Settlement from Pioneers to Present* (Bountiful, Utah: Horizon Publishers, 1975), 212.

25. Ibid., 213.

26. Oscar Wood and DeLore Nichols, interview by Richard W. Sadler, Oct. 11, 1973, Weber State University, 15–27.

27. Ibid.

28. Foy, *The City Bountiful*, 211–14; Wood and Nichols, interview; Thoral Tuttle, interview by Richard C. Roberts, Bountiful, Aug. 16, 1984; George L. Florence, interview by Richard C. Roberts, Centerville, Utah, Aug. 16, 1984.

29. Stephen A. Merrill, "Reclamation and the Economic Development of Northern Utah: The Weber River Project," *Utah Historical Quarterly* 39 (summer 1971): 257–64. See also: Jensen, "History of the Echo Dam and Reservoir"; "History of the Echo Dam Project" Morgan County Historical Society Library; U. S. Department of Interior, Bureau of Reclamation, *Report of the Commissioner, 1930* (Ogden: Bureau of Reclamation, 1930), 39; U. S. Department of Interior, Bureau of Reclamation, "Final Report on Design and Construction of Echo Dam and Reservoir, Salt Lake Basin Project" (Ogden: Bureau of Reclamation, 1934), 210; Weber River Water Users Association, "Echo Dam: Last Payment Ceremony, June 11, 1966," Morgan County Historical Society Library, copy; "Articles of Incorporation of the Weber River Water Users Association," 1926, manuscript, Office of Weber River Water Users Association, Ogden.

30. Arrington and Dittmer, "Reclamation in Three Layers," 27.

31. Arlie Campbell, interview by Jerry Mower, Pleasant View, Utah, Jan. 10, 1973, History Department, Weber State University, 4–9.

32. McKay, "Irrigation in Weber County"; Arrington and Dittmer, "Reclamation in Three Layers," 15–27; Arlie Campbell, interview, 4–9; *Ogden Standard Examiner*, Nov. 19, 1916 and July 13, 20, 1919; Bureau of Reclamation, "Ogden River Project," in *Report: Utah, Weber and Box Elder Counties: Region 4 Bureau of Reclamation* (Ogden: Project Headquarters, n.d.), 1–2; Rulon H. Sorensen, "Rehabilitating Ogden's Culinary Water System," n.d. manuscript, Office of Director of Public Works, Ogden.

33. A. Russell Croft, *History of Development of the Davis County Experimental Watershed* (Ogden: Wasatch National Forest Publication, 1981), 10.

34. Ibid., 13.

35. Ibid., 41–42.

36. Croft, *Development of the Davis County Experimental Watershed*; A. Russell Croft, *Rainstorm Debris Floods: A Problem in Public Welfare* (Tucson: University of Arizona, Agricultural Experiment Station, 1967); "Davis Experimental Area Looks for

More Water," *Davis County Clipper*, Oct. 23, 1964; "The Floods of 1923 in Northern Utah," *Bulletin of the University of Utah* 25, no. 3 (Mar. 1925); "Torrential Floods in Northern Utah, 1930," Agricultural Experiment Station, Utah State Agricultural College, circular 92, *Quarterly*, 1931; "Historical Record of 1930 Flood," Sept. 4, 1930, Weber Basin File, History Department, Weber State University.

CHAPTER SIX

1. *Weber Basin Project History*, vol. 1 (Ogden: Weber Basin Water Users Association Office, 1952), 6.
2. *Salt Lake Tribune*, April 25, 1937.
3. *Ogden Standard-Examiner*, Dec. 6, 1939.
4. *Ogden Standard-Examiner*, Aug. 16, 24, 26, 27, 1940.
5. *Salt Lake Tribune*, Dec. 7, 1940.
6. Postcard, Feb. 19, 1945, Weber Basin File, Weber State University, Ogden, Utah (hereafter, WBF).
7. DeLore Nichols, interview by Richard Sadler, Jerry Mower, and Steve Readdick, Nov. 29, 1972, Weber State University Library (hereafter WSUL).
8. Robert G. Harding to Warwick C. Lamoreaux, April 11, 1945, WBF.
9. R. Bruce Major, Davis County clerk, to Ed H. Watson, Aug. 7, 1945, WBF.
10. Robert G. Harding to Davis County Water Users Association, Mar. 6, 1945, WBF.
11. Ibid.
12. DeLore Nichols to Davis County Water Users Association, Mar. 21, 1946, WBF.
13. D. W. Thorne, Utah State Agricultural College, to DeLore Nichols, Mar. 27, 1946, WBF.
14. DeLore Nichols, interview by Richard W. Sadler, Nov. 29, 1972, WSUL. Most information concerning the May 3, 1946, meeting comes from this interview.
15. Nichols also noted "I went to school with Ollie, his face was as red as . . . you know . . . as red as a red rose." Nichols, interview. Also, Joseph W. Johnson, interview by Steve Readdick, Nov. 22, 1972, WSUL.
16. DeLore Nichols, interview by Richard Sadler, Jerry Mower, and Steve Readdick, Nov. 29, 1972, WSUL.
17. Francis Warnick, interview with Richard Sadler, Dec. 1, 1981, WSUL.
18. Francis Warnick, interview by Jerry Mower and Steve Readdick, Dec. 20, 1972, WSUL.
19. DeLore Nichols, interview by Richard Sadler, July 31, 1972, WSUL.
20. Johnson, interview.
21. Ibid.
22. The plan is outlined in a letter from Thomas to Nichols, July 13, 1946, WBF.
23. Merrill Parkin, secretary, South Davis Water Users Protective Association, to Ed H. Watson, state engineer, Aug. 5, 1946, WBF.
24. Ed H. Watson to DeLore Nichols, July 12, 1946, WBF.
25. DeLore Nichols to Ed H. Watson, Sept. 21, 1946, WBF.
26. Note Report of Davis County Water Users Association, Oct. 3, 1946, WBF.
27. Table, 1946, Utah Irrigation Companies, Davis County, WBF.

28. Arthur V. Watkins to Joseph W. Johnson, Jan. 5, 1947, WBF.
29. Joseph W. Johnson to William R. Wallace, Feb. 5, 1947, WBF.
30. DeLore Nichols to directors, Davis County Water Users Association, Feb. 27, 1947, WBF.
31. Elizabeth Tillotson, *A History of Ogden* (Ogden: J. O. Woody Printing, 1962).
32. E. O. Larson to DeLore Nichols, Mar. 7, 1947, WBF.
33. Ibid.
34. Note correspondence between William Dawson and Joseph W. Johnson, April and May 1947, WBF.
35. Joseph W. Johnson to William Dawson, Nov. 5, 1947, WBF.
36. Minutes, Davis County Water Users Association meeting, Nov. 25, 1947, WBF.
37. Petition, Nov. 29, 1947, WBF.
38. William Dawson to Joseph W. Johnson, Dec. 20, 1947, WBF.
39. DeLore Nichols to William Dawson, Dec. 29, 1947, WBF.
40. *Salt Lake Tribune*, Oct. 2, 1947; *The Weekly Reflex*, Oct. 2, 1947, WBF.
41. Joseph W. Johnson to Orson Christensen, Oct. 30, 1947, WBF.
42. Joseph W. Johnson to E. J. Fjelsted, Oct. 30, 1947, WBF.
43. Note letters from E. J. Fjelsted concerning the meeting, Dec. 1947, WBF.
44. Articles of association and bylaws of Davis and Weber Counties Municipal Water Development Association, copy in WBF.
45. Joseph W. Johnson to Elbert Thomas, Arthur Watkins, and W. K. Granger, letters, Jan. 3, 1948, WBF.
46. Joseph W. Johnson to William Dawson, Jan. 3, 1948, WBF.
47. Michael Strauss to Joseph W. Johnson, Jan. 22, 1948, WBF.
48. William Dawson to Joseph W. Johnson, Jan. 23, 1948, WBF.
49. Minutes, Davis County Water Users Association, special meeting, Feb. 3, 1948, WBF.
50. H. R. 5327, 80th Cong., 2d sess., Feb. 9, 1948, copy in WBF.
51. Warnick, interview by Mower and Readdick, Dec. 20, 1972, WBF.
52. Warnick, interview by Sadler, Dec. 1, 1981, WBF.
53. Johnson, interview by Readdick, Nov. 22, 1972, WBF.
54. Minutes, Executive Committee, Davis and Weber Counties Municipal Water Development Association, April 27, 1948, WBF.
55. Minutes, Executive Committee, Weber and Davis Metropolitan Water District, May 11, 1948, WBF.
56. E. O. Larson to Joseph W. Johnson, May 21, 1948, WBF.
57. E. O. Larson to Joseph W. Johnson, June 2, 1948, WBF.
58. Joseph W. Johnson to E. O. Larson, June 5, 1948, WBF.
59. Harold R. Howard to E. J. Fjelsted, Aug. 2, 1948, WBF.
60. R. A. Wheeler to Arthur V. Watkins, Aug. 9, 1948, WBF.
61. Howard Widdison and Ernest Ekins, Weber Planning Commission, to the Honorable Board of County Commissioners, Aug. 9, 1948, WBF.
62. Minutes, Executive Committee, Davis and Weber Counties Municipal Water Development Association, Aug. 12, 1948, WBF.
63. Reva Beck Bosone to Joseph W. Johnson, Oct. 14, 1948, WBF.
64. Rationale for Weber Basin Project, 1950, WBF.

65. Municipal water rates in Utah's cities and towns, 1948, WBF.
66. "Ogden's Water Supply," survey by Win Templeton, May 10, 1948, WBF.
67. Win Templeton, *Davis and Weber Counties Water Development* (Salt Lake City: n.p., 1949), 29-30, WBF.
68. It is important to note that even after three years of intensive investigation, the planned dam sites would change: Perdue to Wanship, Jeremy to an enlargement of East Canyon, and Magpie to Causey Reservoir.
69. Note Johnson's letter to newly elected Governor J. Bracken Lee, Feb. 21, 1949, WBF; see also minutes, Davis and Weber Counties Municipal Water Development Association, Mar. 7, 1949, WBF.
70. W. K. Granger to E. J. Fjelsted, April 7, 1949, WBF.
71. Reva Beck Bosone to Joseph W. Johnson and DeLore Nichols, April 24, 1949, WBF.
72. Minutes, Davis and Weber Counties Municipal Water Development Association, April 26, 1949; E. J. Fjelsted to J. A. Howell, newsletter, May 6, 1949, both in WBF.
73. W. K. Granger to E. J. Fjelsted and Joseph W. Johnson, telegrams, May 9, 1949, WBF.
74. W. K. Granger to Joseph W. Johnson, July 21, 1949, WBF. Much of the information concerning the passage of the Weber Basin Bill comes from Francis Warnick, interview by Richard Sadler, Dec. 1, 1981, WSUL.
75. *Congressional Record*, 81st Cong., 2d sess. Aug. 15, 1949. On this topic note the numerous pieces of correspondence during Aug. 1949 in the WBF.
76. Reva Beck Bosone to Joseph W. Johnson, Aug. 15, 1949, WBF.
77. Arthur Watkins to E. J. Fjelsted, Aug. 24, 1949, and Arthur Watkins to Harry Truman, Aug. 24, 1949, WBF.
78. Elbert Thomas to E. J. Fjelsted, Sept. 9, 1949, WBF.

CHAPTER SEVEN

1. As construction began on Weber Basin projects in 1952, an annual project history was compiled under the auspices of the Bureau of Reclamation. The yearly project history includes specifics on constructions, bids, photographs, and other documents. Copies of these construction histories are available at the Weber Basin Water Conservancy District Office, Layton, Utah. This chapter is written from the information in those detailed but unpublished histories.

CHAPTER EIGHT

1. E. J. Fjeldsted to John Dye, president, Uintah town board, May 10, 1949, Weber Basin File, Weber State University Library, Ogden, Utah (hereafter WBF).
2. E. J. Fjeldsted to Hyrum C. Brough, Davis County clerk, July 14, 1949, WBF.
3. Minutes, Davis and Weber Counties Municipal Water Development Association and Summit County Water Users, meeting at Coalville, July 19, 1949, WBF.
4. Minutes, Ellison Committee with Bureau of Reclamation, meeting, Oct. 25, 1949, WBF. See also Francis Warnick to E. O. Larson, Nov. 14, 1949, WBF.
5. Harold Ellison to Harold A. Linke, Dec. 2, 1949.

6. Harry S. Truman, "Statement, Aug. 30, 1949" in *Weber Basin Project History*, vol. 1 (Ogden: Weber Basin Water Users Association Office, 1952).

7. Walter U. Fuhriman, George T. Blanch, and Clyde E. Stewart, *An Economic Analysis of the Agricultural Potentials of the Weber Basin Reclamation Project, Utah* (Logan: Utah State Agricultural College, 1952), 32–33.

8. Fred M. Abbott to George W. Florence, Jan. 10, 1950, WBF.

9. E. J. Fjeldsted to Alma Ellis, Jan. 25, 1950, WBF.

10. Fuhriman, Blanch, and Stewart, *Economic Analysis of the Agricultural Potentials*, 23.

11. Arthur V. Watkins to Joseph W. Johnson, Feb. 6, 1950. WBF.

12. Elbert D. Thomas, W. K. Granger, and Reva Beck Bosone to Joseph W. Johnson, Feb. 3, 1950.

13. M. H. Harris to Rulon White and members of the Ogden City Commission, Feb. 7, 1950, WBF.

14. J. A. Howell to Joseph W. Johnson, Harold Ellison, and DeLore Nichols, Mar. 31, 1950.

15. Arthur V. Watkins to President Harry S. Truman, Apr. 22, 1950; John R. Steelman to Senator Arthur V. Watkins, May 9, 1950, WBF.

16. Joseph W. Johnson and DeLore Nichols to Arthur Watkins, Apr. 27, 1950, WBF.

17. Arthur Watkins to Joseph W. Johnson, May 29, 1950, WBF.

18. Joseph W. Johnson to W. K. Granger, June 10, 1950, WBF.

19. On this topic, note oral interviews with Joseph W. Johnson, Weber State University Library.

20. The Perdue site, located about five miles above Oakley on the Weber River, and the Larrabee site, located about twelve miles above Oakley, were both originally considered as dam sites. Perdue was ruled out because of extensive faulting, and Larrabee was later excluded because it sat in part on a glacial moraine.

21. Ivan Flint, interview with Richard Sadler, Oct. 20, 1993. Information concerning this project gained in conversation with Weber Basin engineer Mark Anderson by Richard Sadler, July 10, 1984.

22. U.S. Army Corps of Engineers, Notice of Public Hearing, Aug. 9, 1951, WBF.

23. Important to note on this topic is Keith D. Wilde, "Defining Efficient Water Resource Management in the Weber Basin Drainage, Utah" (Ph.D. diss., Utah State University, 1976). Wilde's thesis includes the idea that underground water capabilities were not thoroughly investigated and could have been used more extensively rather than building costly dams. In taking all of the historical evidence under consideration, it is difficult to take Wilde's charges seriously.

24. Joseph W. Johnson to Wallace F. Bennett and Arthur V. Watkins, telegrams, July 5, 1951; Harold Ellison to Wallace F. Bennett and Arthur V. Watkins, telegrams, July 5, 1951, WBF.

25. Reva Beck Bosone, "The Capitol Reports," in *Intermountain Industry*, Aug. 1952, 11, 33.

26. Harold Ellison to W. K. Granger, Reva Beck Bosone, Arthur Watkins, and Wallace Bennett, telegrams, July 9, 1952, WBF; W. K. Granger to Harold Ellison, July 18, 1952, WBF.

27. United States Department of Agriculture, Office of the Secretary, *Appendix to the*

Report on the Ability of Irrigation Water Users on the Weber Basin Project Utah to Repay Project Construction Costs and Other Water Changes (Washington, D. C.: Government Printing Office, July 20, 1952), 80.

28. Municipal Water Rate Study, Davis and Weber Counties Municipal Water Development Association, 1957, WBF.

29. Voting brochure, Dec. 6, 1952 election, WBF.

30. Minutes, Weber Basin Water Conservancy District Board, Nov. 18, 1952, Weber Basin Water Conservancy District Office, Ogden (hereafter WBWCD).

31. Minutes, Board of Directors, Weber Basin Water Conservancy District, Nov. 21, 1952, WBWCD.

32. Rulon White, interview by Jerry Mower.

33. Minutes, Weber Basin Water Conservancy District Board, Mar. 15, 1954, WBWCD.

34. Minutes, Board of Directors, Weber Basin Water Conservancy District, Feb. 25, 1955, WBWCD.

35. Weber Basin Water Conservancy District, Second Annual Report, June 1, 1951 to May 31, 1952, 9, WBWCD.

36. *Ogden Standard-Examiner*, June 30, 1982, sec. D, p. 1.

37. Wayne Winegar, "May 9, 1974 letter," in *Improvement of Planning for Post-Development Water Resource Management: A Study of the Weber Basin Project* by Gary E. Madsen and Wayne Andrews, Institute for Social Sciences Research on Natural Resources, Utah State University, Logan, Research Monograph No. 6, Sept. 1976, 67–68.

38. Flint, interview.

Appendix A

 Weber Basin Water Conservancy District Directors and Managers

Black, Charles F. Jr.	1987–present	Davis County
Bohman, Frank W.	1979–1985	Morgan County
Breeze, Joseph F.	1964–1970	Weber County
Burbidge, George S.	1989–1992	Davis County
Carver, Charles D.	1970–1985	Weber County
Carver, Elmer	1952–1978	Weber County
Christensen, Ted	1988–present	Weber County
Clark, Horald G.	1950–1964	Morgan County
Clark, Ezra T.	1965–1989	Davis County
Cox, R. Howard	1985–present	Weber County
Ellison, Harold	1950–1961	Davis County
Fisher, Blaine	1964–1987	Davis County
Francis, Joseph	1964–1965	Morgan County
Fuller, Haynes	1984–1988	Upper Weber County
Gibson, Wayne	1984–present	Weber County
Holbrook, Ward	1950–1959	Davis County
Howell, J. A.	1950–1952	Weber County
Jenkins, T. Bruce	1971–1980	Weber County
Jensen, Keith	1961–1980	Upper Weber County
Linford, J. Clifton	1965–1982	Davis County
Mabey, Rendell N.	1959–1965	Davis County
McFarland, Bruce	1980–1984	Weber County
McKay, D. D.	1950–1961	Upper Weber County
Montgomery, Norman	1990–present	Upper Weber County
Olsen, S. Jack	1965–1979	Morgan County
Osguthorpe, Steve	1989–present	Summit County
Peterson, Scott	1991–present	Morgan County
Richards, Ralph	1958–1983	Summit County

Rigby, Max	1992–present	Davis County
Simpson, Francis V.	1956–1971	Weber County
Smith, LeRoy	1950–1964	Davis County
Sorensen, Edward	1950–1958	Summit County
Spaulding, Lewis	1950–1956	Weber County
Storey, Boyd K.	1988–1990	Upper Weber County
Storey, Boyd K.	1980–1984	Upper Weber County
Thackeray, Jack H.	1985–1991	Morgan County
White, W. Rulon	1950–1964	Weber County
White, W. Robert	1979–1988	Weber County
Whitesides, Evan	1982–present	Davis County
Winegar, Wayne	1961–1965	Davis County
Wright, Dennis K.	1983–1989	Summit County

MANAGERS

Fjeldsted, Ezra J.	1950–1965	Ogden
Winegar, Wayne	1965–1980	Kaysville
Jensen, Keith	1980–1985	Huntsville
Flint, Ivan W.	1985–present	Kaysville

Appendix B

Water Companies in the Weber Basin Area

BOX ELDER COUNTY

BOX ELDER CREEK WATER USERS ASSOCIATION, Brigham City, Utah; Irrigated Acreage: 442.39 acres (1990); Incorporation Date: NA
BRIGHAM CITY NORTH FIELD WATER COMPANY, Brigham City, Utah; Irrigated Acreage: 664.56 acres (1990); Incorporation Date: 1851
COLD SPRINGS DAM AND IRRIGATION COMPANY, Brigham City, Utah; Irrigated Acreage: 698.90 acres (1990); Incorporation Date: 1861
HARPER IRRIGATION COMPANY, Brigham City, Utah; Irrigated Acreage: 461.60 acres (1990); Incorporation Date: 1881
HILLCREST WATER USER CORPORATION, Perry, Utah; Irrigated Acreage: 10.00 acres (1990); Incorporation Date: NA
MANTUA IRRIGATION COMPANY, Willard and Mantua, Utah; Irrigated Acreage: 274.93 acres (1990); Incorporation Date: NA
NORTH STRING IRRIGATION COMPANY, Brigham City, Utah; Irrigated Acreage: 263.95 acres (1990); Incorporation Date: NA
NORTH WILLARD IRRIGATION COMPANY, Willard and Brigham City, Utah; Irrigated Acreage: 194.35 (1990); Incorporation Date: NA
PERRY IRRIGATION COMPANY, Perry, Utah; Irrigated Acreage: 395.75 acres (1990); Incorporation Date: NA
SOUTH WILLARD WATER COMPANY, INC., Willard, Utah; Irrigated Acreage: 14.0 acres (1990); Incorporation Date: NA
THREE MILE CREEK IRRIGATION AND WATER COMPANY, Perry, Utah; Irrigated Acreage: 277.75 acres (1990); Incorporation Date: NA
WILLARD WATER COMPANY, Willard, Utah; Irrigated Acreage: 400.0 acres (1990); Incorporation Date: NA

DAVIS COUNTY

BAMBROUGH IRRIGATION COMPANY, South Weber, Utah; Irrigated Acreage: 306 (1990); Incorporation Date: 9/16/1955
BARNARD CREEK IRRIGATION COMPANY, Centerville, Utah; Irrigated Acreage: 9,600 (1963); Incorporation Date: 4/17/1912
BARTON CREEK IRRIGATION COMPANY, Bountiful, Utah; Irrigated Acreage: 7,100 (1963); Incorporation Date: 3/13/1901

BOUNTIFUL CITY MILL CREEK IRRIGATION COMPANY, Bountiful, Utah; Irrigated Acreage: 11,000 (1963); Incorporation Date: 4/3/1896
BOUNTIFUL IRRIGATION COMPANY, Bountiful, Utah; Irrigated Acreage: NA; Incorporation Date: NA
BOUNTIFUL WATER SUBCONSERVANCY DISTRICT, Bountiful, Utah; Irrigated Acreage: NA; Incorporation Date: NA
CENTERVILLE DEUEL CREEK IRRIGATION COMPANY, Centerville, Utah; Irrigated Acreage: 1,000 (1963); Incorporation Date: 5/7/1900
CENTRAL CLEARFIELD GARDEN IRRIGATION COMPANY, INC., Clearfield, Utah; Irrigated Acreage: NA; Incorporation Date: NA
CENTRAL IRRIGATION COMPANY, West Point and Clinton, Utah; Irrigated Acreage: NA; Incorporation Date: NA
CLARK WATER COMPANY, Bountiful, Clinton, and Farmington, Utah; Irrigated Acreage: NA; Incorporation Date: NA
CLEARFIELD DITCH COMPANY, Clearfield, Utah; Irrigated Acreage: NA; Incorporation Date: NA
CLEARFIELD IRRIGATION COMPANY, Clearfield, Utah; Irrigated Acreage: 4,500 (1963); Incorporation Date: NA
CLINTON DITCH COMPANY, Clinton, Utah; Irrigated Acreage: 800 (1963); Incorporation date: 6/16/1937
CLINTON PIPE LINE & WATER COMPANY INC., Clinton, Utah; Irrigated Acreage: NA; Incorporation Date: NA
DAVIS AND WEBER COUNTIES CANAL COMPANY, Ogden, Layton, Roy, and So. Weber, Utah; Irrigated Acreage: NA; Incorporation Date: 12/29/1884
DAVIS CREEK IRRIGATION COMPANY OF FARMINGTON, Farmington, Utah; Irrigated Acreage: 150 (1963); Incorporation Date: 1/21/1954
FALULA WATER ASSOCIATION, Kaysville and Bountiful, Utah; Irrigated Acreage: NA; Incorporation Date: NA
FARMINGTON AREA PRESSURIZED IRRIGATION DISTRICT, Kaysville, Utah; Irrigated Acreage: NA; Incorporation Date: NA
FIFE DITCH CORP, Clinton, Utah; Irrigated Acreage: 800 (1963); Incorporation Date: NA
HAIGHT BENCH IRRIGATION & WATER COMPANY, Farmington and Bountiful, Utah; Irrigated Acreage: 950 (1974); Incorporation Date: 5/1/1899
HAIGHTS CREEK IRRIGATION COMPANY, Kaysville and Fruit Heights, Utah; Irrigated Acreage: 1,200 (1963); Incorporation Date: 5/1/1899
HIGH LINE DITCH COMPANY, Kaysville, Utah; Irrigated Acreage: NA; Incorporation Date: NA
HOLMES CREEK IRRIGATION COMPANY, INC., Layton and Kaysville, Utah; Irrigated Acreage: 4800 (1963); Incorporation Date: 2/15/1897
HOME & GARDEN IRRIGATION COMPANY, Clearfield, Utah; Irrigated Acreage: NA; Incorporation Date: NA
JONES DITCH COMPANY, Ogden, Utah; Irrigated Acreage: 75 (1963); Incorporation Date: NA
KAYS CREEK IRRIGATION COMPANY, Layton, Utah; Irrigated Acreage: 1,000 (1963); Incorporation Date: 4/24/1897
KAYSVILLE IRRIGATION COMPANY, Kaysville, Utah; Irrigated Acreage: 5,000 (1963); Incorporation Date: 4/18/1952
LAST CHANCE DITCH COMPANY, Layton, Utah; Irrigated Acreage: 425 (1974); Incorporation Date: 5/10/1955

LAYTON CANAL IRRIGATION COMPANY, Syracuse, Utah; Irrigated Acreage: NA; Incorporation Date: NA

MISTY MEADOWS IRRIGATION COMPANY, Sunset, Utah; Irrigated Acreage: NA; Incorporation Date: NA

MORGAN-HEINER DITCH, Layton, Utah; Irrigated Acreage: NA; Incorporation Date: NA

NEW SURVEY IRRIGATION COMPANY, Kaysville, Utah; Irrigated Acreage: 400 (1963); Incorporation Date: 7/15/1921

NORTH CANYON WATER COMPANY, Bountiful, Utah; Irrigated Acreage: 100 (1963); Incorporation Date: 3/27/1893

NORTH COTTONWOOD IRRIGATION & WATER COMPANY OF FARMINGTON Farmington, Utah; Irrigated Acreage: 2,000 (1963); Incorporation Date: 1/27/1945

OAK HAVEN MUTUAL WATER USERS ASSN. (CULINARY), Bountiful and Salt Lake City, Utah; Irrigated Acreage: NA; Incorporation Date: NA

OLD MADSEN DITCH, Layton, Utah; Irrigated Acreage: NA; Incorporation Date: NA

PARRISH CREEK IRRIGATION COMPANY OF CENTERVILLE, Centerville, Utah; Irrigated Acreage: NA; Incorporation Date: NA

PRETTY VALLEY WATER COMPANY, Farmington, Utah; Irrigated Acreage: NA; Incorporation Date: NA

RICKS CREEK IRRIGATION COMPANY, Centerville, Utah; Irrigated Acreage: 500 (1963); Incorporation Date: 6/11/1909

ROSEDALE WATER COMPANY, Salt Lake City, Centerville, and Bountiful, Utah; Irrigated Acreage: NA; Incorporation Date: NA

SHEPHERD CREEK IRRIGATION & WATER COMPANY, Farmington and Kaysville, Utah; Irrigated Acreage: 350 (1963); Incorporation Date: 1/29/1902

SILL ADAMS DITCH COMPANY, INC., Layton, Utah; Irrigated Acreage: NA; Incorporation Date: NA

SNAKE CREEK MUTUAL WATER COMPANY, Salt Lake City and Fruit Heights, Utah; Irrigated Acreage: NA; Incorporation Date: NA

SOUTH BOUNTIFUL JUNCTION WATER COMPANY, Irrigated Acreage: NA; Incorporation Date: NA

SOUTH DAVIS COUNTY WATER IMPROVEMENT DISTRICT, Bountiful, Utah; Irrigated Acreage: NA; Incorporation Date: NA

SPRING CREEK IRRIGATION AND WATER COMPANY OF FARMINGTON Salt Lake City and Farmington, Utah; Irrigated Acreage: NA; Incorporation Date: NA

STEED CREEK IRRIGATION AND WATER COMPANY, Farmington, Utah; Irrigated Acreage: 600 (1974); Incorporation Date: 6/24/1895

STEVENSON DITCH CORPORATION, Layton, Utah; Irrigated Acreage: 1,500 (1963); Incorporation Date: NA

STRAIGHT DITCH COMPANY, Layton and Kaysville, Utah; Irrigated Acreage: 2,200 (1963); Incorporation Date: 6/15/1911

VAL VERDA WATER COMPANY, Woods Cross, Utah; Irrigated Acreage: 35 (1963); Incorporation Date: 1/11/1936

WEBER BASIN WATER CONSERVANCY DISTRICT, Layton, Utah; Irrigated Acreage: NA; Incorporation Date: 6/26/1950

WEST BOUNTIFUL MILL CREEK IRRIGATION COMPANY, Bountiful, Utah; Irrigated Acreage: 500 (1963); Incorporation Date: 4/1/1905

WEST BRANCH IRRIGATION COMPANY, Syracuse, Utah; Irrigated Acreage: 2,904 (1963); Incorporation Date: 6/5/1913

WEST LAYTON IRRIGATION COMPANY, Layton, Utah; Irrigated Acreage: 2,500 (1963); Incorporation Date: 6/1/1911

MORGAN COUNTY

ANDERSON BOWMAN DITCH, Morgan, Utah; Irrigated Acreage: NA; Incorporation Date: NA

BENCH DITCH COMPANY, Morgan, Utah; Irrigated Acreage: NA; Incorporation Date: NA

CLARENCE THURSTON DITCH, Morgan, Utah; Irrigated Acreage: 14 (1990); Incorporation Date: NA

CREECHLEY MIKESELL DITCH, Morgan, Utah; Irrigated Acreage: 500 (1990); Incorporation Date: NA

CROYDON IRRIGATION COMPANY, Croydon, Utah; Irrigated Acreage: 424.5 (1990); Incorporation Date: NA

DEEP CREEK DITCH COMPANY, Morgan and Centerville, Utah; Irrigated Acreage: NA; Incorporation Date: NA

DEEP CREEK GARDEN DITCH, Morgan, Utah; Irrigated Acreage: NA; Incorporation Date: NA

EAST PORTERVILLE CANAL COMPANY, Morgan and Ogden, Utah; Irrigated Acreage: 353 (1990); Incorporation Date: 11/20/1926

EAST RICHVILLE DITCH COMPANY, Morgan, Utah; Irrigated Acreage: 312.5 (1990); Incorporation Date: NA

ENTERPRISE FIELD DITCH, Morgan, Utah; Irrigated Acreage: 281 (1990); Incorporation Date: NA

ENTERPRISE STODDARD IRRIGATION COMPANY, Morgan and Farmington, Utah; Irrigated Acreage: 535 (1990); Incorporation Date: 10/18/1926

ENTERPRISE WATER SYSTEM; Irrigated Acreage: NA; Incorporation Date: NA

GLEN THURSTON DITCH, Morgan, Utah; Irrigated Acreage: NA; Incorporation Date: NA

HEINER MORRIS DITCH, Morgan, Utah; Irrigated Acreage: 43.1 (1990); Incorporation Date: NA

HEINER PUMP, Morgan, Utah; Irrigated Acreage: NA; Incorporation Date: NA

HIGHTLANDS WATER COMPANY, INC., Morgan and Layton, Utah; Irrigated Acreage: NA; Incorporation Date: NA

H.L.J. DITCH COMPANY (HEINER-LOWE-JOHNSON); Irrigated Acreage: 136 (1990); Incorporation Date: NA

LINE CREEK IRRIGATION COMPANY, Morgan, Utah; Irrigated Acreage: 404.4 (1990); Incorporation Date: NA

LITTLETON AND MILTON IRRIGATION COMPANY, Morgan, Utah; Irrigated Acreage: 768 (1990); Incorporation Date: NA

LOWER RIVER DITCH COMPANY, Morgan, Utah; Irrigated Acreage: NA; Incorporation Date: NA

MADSEN OLSEN DITCH, Morgan, Utah; Irrigated Acreage: 45.2 (1990); Incorporation Date: NA

MEACHAM-NELSON-MADSEN DITCH COMPANY, Morgan, Utah; Irrigated Acreage: 183 (1990); Incorporation Date: NA

MONTE VERDE WATER ASSOCIATION, Mountain Green, Utah; Irrigated Acreage: NA; Incorporation Date: NA

MOUNTAIN GREEN SUBDIVISION WATER ASSOCIATION, Morgan, Utah; Irrigated Acreage: NA; Incorporation Date: NA

MUSSER DITCH COMPANY, Morgan, Utah; Irrigated Acreage: 180.6 (1990); Incorporation Date: NA

NORTH BENCH CANAL COMPANY, Morgan, Utah; Irrigated Acreage: 108 acres; Incorporation Date: NA

NORTH DALTON DITCH, Morgan, Utah; Irrigated Acreage: NA; Incorporation Date: NA

NORTH MORGAN EXTENSION DITCH COMPANY, Ogden, Stoddard, and Morgan, Utah; Irrigated Acreage: NA; Incorporation Date: 4/17/1897

NORTH MORGAN IRRIGATION COMPANY, Morgan, Utah; Irrigated Acreage: 629.6 (1990); Incorporation Date: 8/29/1912

NORTH MORGAN MILL RACE DITCH COMPANY; Irrigated Acreage: NA; Incorporation Date: 5/20/1936

NORTH ROUND VALLEY CANAL COMPANY, Morgan, Utah; Irrigated Acreage: 252 (1990); Incorporation Date: NA

NORTHWEST IRRIGATION DITCH COMPANY, Morgan, Utah; Irrigated Acreage: 406 (1990); Incorporation Date: 2/7/1962

OLD FORT DITCH COMPANY, Morgan, Utah; Irrigated Acreage: 28.6 (1990); Incorporation Date: NA

PARSONS HOMESTEAD WATER COMPANY, INCORPORATED, Morgan and Cedar City, Utah; Irrigated Acreage: NA; Incorporation Date: NA

PENTZ CROUCH DITCH COMPANY, Morgan, Utah; Irrigated Acreage: NA; Incorporation Date: NA

PENTZ & SMITH DITCH COMPANY, Morgan, Utah; Irrigated Acreage: 95.4 (1990); Incorporation Date: NA

PETERSON CREEK DITCH COMPANY, Morgan, Utah; Irrigated Acreage: NA; Incorporation Date: NA

PETERSON IRRIGATION COMPANY, Morgan, Utah; Irrigated Acreage: 521.4 (1990); Incorporation Date: NA

POULSEN-PREESE COMPANY, Morgan, Utah; Irrigated Acreage: NA; Incorporation Date: NA

RICHARDS DITCH IRRIGATION COMPANY, Sandy, Utah; Irrigated Areage: 205 (1990); Incorporation Date: NA

RICHVILLE IRRIGATION AND CANAL COMPANY, Morgan, Utah; Irrigated Acreage: NA; Incorporation Date: NA

RICHVILLE PIPELINE COMPANY (CULINARY), Morgan, Utah; Irrigated Acreage: NA; Incorporation Date: NA

SETTLEMENT DITCH, Morgan, Utah; Irrigated Acreage: 125.8 acres Incorporation Date: NA

SMITH CREEK IRRIATION COMPANY, Morgan, Utah; Irrigated Acreage: 76.9 acres; Incorporation Date: NA

SOUTH LINE CREEK DITCH, Morgan, Utah; Irrigated Acreage: NA; Incorporation Date: NA

SOUTH MORGAN WATER DITCH COMPANY, Morgan, Utah; Irrigated Acreage:365.8 (1990); Incorporation Date: 10/2/1924

SOUTH ROUND VALLEY CANAL COMPANY, Morgan, Utah; Irrigated Acreage: 145.2 (1990); Incorporation Date: 5/15/1928

SPENDLOVE DITCH, Morgan, Utah; Irrigated Acreage: 22.9 (1990); Incorporation Date: NA

SPENDLOVE-CRIDDLE DITCH, Morgan, Utah; Irrigated Acreage: 59.7 (1990); Incorporation Date: NA

THACKERY-CROUGH, Morgan, Utah; Irrigated Acreage: NA; Incorporation Date: NA

VERL POLL ENTERPIRSES ESTATES WATER ASSOCIATION, Morgan, Utah; Irrigated Acreage: NA; Incorporation Date: NA

WEBER CANAL COMPANY, Morgan, Utah; Irrigated Acreage: 333.8 (1990); Incorporation Date: NA

WELCH FIELD DITCH IRRIGATION COMPANY, Morgan, Utah; Irrigated Acreage: 159.5 (1990); Incorporation Date: 2/14/1961
WEST ENTERPIRSE WATER ASSOCIATION, Morgan, Utah; Irrigated Acreage: 670 (1990); Incorporation Date: 1/14/1934
WEST PORTERVILLE CANAL COMPANY, Morgan, Utah; Irrigated Acreage: 670 (1990); Incorporation Date: 4/14/1934
WEST PORTERVILLE IRRIGATION COMPANY, Morgan, Utah; Irrigation Acreage: 303.7 (1990); Incorporation Date: NA
WEST PORTERVILLE PIPE LINE COMPANY (CULINARY), Morgan, Utah; Irrigation Acreage: NA; Incorporation Date: NA
WHITTIER GORDON DITCH COMPANY, Morgan, Utah; Irrigated Acreage: NA; Incorporation Date: NA
WILKINSON WATER COMPANY, Morgan, Utah; Irrigated Acreage: NA; Incorporation Date: NA
WOOLEY DITCH, Morgan, Utah; Irrigated Acreage: NA; Incorporation Date: NA

SUMMIT COUNTY

ANDERTON BROTHERS DITCH COMPANY, Henefer, Utah; Irrigated Acreage: 87.8 (1990); Incorporation Date: NA
ASPEN MOUNTAIN CULINARY WATER COMPANY, Murray, Salt Lake and West Valley, Utah; Irrigated Acreage: NA; Incorporation Date: NA
ASPEN MOUNTAIN WATER COMPANY, Salt Lake and Sandy, Utah; Irrigated Acreage: NA; Incorporation Date: NA
ASPEN SPRING NONPROFIT PRIVATE WATER CORPORATION (CULINARY), Salt Lake City, Utah; Irrigated Acreage: NA; Incorporation Date: NA
ATKINSON WATER COMPANY, INC., Park City, Bloomington and Salt Lake City, Utah; Irrigated Acreage: 80 (1990); Incorporation Date: NA
BANNER DEMING DITCH COMPANY, Coalville, Utah; Irrigated Acreage: 147.4 (1990); Incorporation Date: NA
BEAVER AND SHINGLE CREEK IRRIGATION COMPANY, Kamas, Utah; Irrigated Acreage: 3305 (1990); Incorporation Date: 5/6/1911
BIG CANYON-BIRCH GROVE IRRIGATION COMPANY, Coalville, Utah; Irrigated Acreage: 109 (1990); Incorporation Date: NA
BIRCH & GARN DITCH COMPANY, Coalville, Utah; Irrigated Acreage: 39.2 (1990); Incorporation Date: NA
BOULDERVILLE DITCH COMPANY, Kamas and Oakley, Utah; Irrigated Acreage: 478 (1990); Incorporation Date: 5/22/1913
BOYER DITCH COMPANY, Coalville, Utah; Irrigated Acreage: 138.7 (1990); Incorporation Date: NA
BRADBURY MUTUAL WATER COMPANY, Hoytsville and Coalville, Utah; Irrigated Acreage: NA; Incorporation Date: NA
BROWN-WILLIAMS DITCH COMPANY, Coalville, Utah; Irrigated Acreage: 17.50 (1990); Incorporation Date: NA
CARMEN-RICHINS; Irrigated Acreage: NA; Incorporation Date: NA
CHALK CREEK HOYTSVILLE WATER USERS ASSOCIATION, Coalville and Salt Lake City, Utah; Irrigated Acreage: 2000 (1990); Incorporation Date: 1/13/1936
CHALK CREEK IRRIGATION COMPANY, Coalville, Utah; Irrigated Acreage: 156.9 (1990); Incorporation Date: 1/17/1935
CHALK CREEK WATER DISTRIBUTORS, Coalville, Utah; Irrigated Acreage: NA; Incorporation Date: NA

CITIZENS' WATER IMPROVEMENT ASSOCIATION OF SUMMIT PARK, Park City and Summit Park, Utah; Irrigated Acreage: NA; Incorporation Date: NA
COALVILLE CITY DITCH, Coalville, Utah; Irrigated Acreage: 119.2 (1990); Incorporation Date: NA
COALVILLE & HOYTSVILLE IRRIGATION COMPANY, Coalville, Utah; Irrigated Acreage: 230.3 (1990); Incorporation Date: 8/31/1937
COMMUNITY WATER COMPANY, Park City, Utah; Irrigated Acreage: NA; Incorporation Date: NA
COOL SPRINGS MUTUAL WATER COMPANY, Midvale, Salt Lake and West Valley, Utah; Irrigated Acreage: NA; Incorporation Date: NA
COTTONWOOD CREEK IRRIGATION COMPANY, Coalville, Utah; Irrigated Acreage: 43.1 (1990); Incorporation Date: NA
COTTONWOOD WATER PIPELINE COMPANY INC., Hoytsville and Wanship, Utah; Irrigated Acreage: NA; Incorporation Date: NA
CROSSROADS WATER COMPANY, Salt Lake City and Park City, Utah; Irrigated Acreage: NA; Incorporation Date: NA
DANIELS, WINTERS & MCMICHAEL DITCH COMPANY, Coalville, Utah; Irrigated Acreage: 255.2 (1990); Incorporation Date: NA
DEEP SPRINGS WATER COMPANY, INC., Summit Park and Kamas, Utah; Irrigated Acreage: NA; Incorporation Date: NA
EAST HOYTSVILLE IRRIGATION COMPANY NUMBER 2, Coalville, Utah; Irrigated Acreage: 600 (1990); Incorporation Date: 3/24/1936
EAST WANSHIP IRRIGATION COMPANY, Coalville, Utah; Irrigated Acreage: 298.2 (1990); Incorporation Date: NA
ECHO IRRIGATION (DITCH); COMPANY, Echo, Utah; Irrigated Acreage: 186 (1990); Incorporation Date: 3/25/1905
ECHO MUTUAL WATER COMPANY (CULLINARY), Echo, Utah; Irrigated Acreage: NA; Incorporation Date: NA
EDDINGTON DITCH, Croydon, Utah; Irrigated Acreage: 49.6 (1990); Incorporation Date: NA
ELKHORN RESERVOIR AND DITCH COMPANY, Coalville and Midvale, Utah; Irrigated Acreage: 332.6 (1990); Incorporation Date: 12/8/1953
EXTENSION MIDDLE CHALK CREEK IRRIGATION COMPANY, Coalville, Utah; Irrigated Acreage: 92.2 (1990); Incorporation Date: 5/5/1910
EXTENSION NORTH MORGAN CANAL COMPANY; Irrigated Acreage: 39.2 acres: NA; Incorporation Date: NA
FISH LAKE RESERVOIR COMPANY, Proa and Oakley, Utah; Irrigated Acreage: 4800 (1990); Incorporation Date: 7/6/1926
FLINDERS' F 7 MUTUAL WATER COMPANY, Park City, Utah; Irrigated Acreage: NA; Incorporation Date: NA
FORT CREEK IRRIGATION COMPANY, Peoa, Utah; Irrigated Acreage: 205.6 (1990); Incorporation Date: NA
FRANCIS WATER WORKS COMPANY, Francis, Utah; Irrigated Acreage: NA; Incorporation Date: 2/18/1947
GIBBONS & PACE DITCH, Coalville, Utah; Irrigated Acreage: 293 (1990); Incorporation Date: NA
GORGOZA MUTUAL WATER COMPANY, Park City and Salt Lake City, Utah; Irrigated Acreage: 200 (1990); Incorporation Date: NA
HARRIS DITCH COMPANY, Coalville, Utah; Irrigated Acreage: 72.6 (1990); Incorporation Date: NA

HENEFER IRRIGATION COMPANY #1, Henefer, Utah; Irrigated Acreage: 1157.5 (1990); Incorporation Date: 3/28/1904

HENEFER PIPELINE, Henefer, Utah; Irrigated Acreage: NA; Incorporation Date: NA

HENEFER UPPER DITCH COMPANY, Henefer, Utah; Irrigated Acreage: 273.7 (1990); Incorporation Date: NA

HIGH VALLEY WATER COMPANY, Park City, Utah; Irrigated Acreage: NA; Incorporation Date: NA

HILLIARD EAST FORK CANAL COMPANY, Bountiful, Utah and Evanston, Wyoming; Irrigated Acreage: NA; Incorporation Date: NA

HOBSON BULLOCK DITCH COMPANY, Coalville, Utah; Irrigated Acreage: 95.8 (1990); Incorporation Date: NA

HOOP LAKE RESERVOIR AND IRRIGATION COMPANY, Roy, Utah; McKinnon and Loetree, Wyoming; Irrigated Acreage: NA; Incorporation Date: NA

HOVORKA DITCH; Irrigated Acreage: NA; Incorporation Date: NA

HOYT DITCH COMPANY, Coalville, Utah ; Irrigated Acreage: 60.10 (1990); Incorporation Date: NA

HOYTSVILLE IRRIGATION COMPANY, Coalville, Utah; Irrigated Acreage: 599.3 (1990); Incorporation Date: NA

HOYTSVILLE PIPE WATER COMPANY, Coalville, Utah; Irrigated Acreage: NA; Incorporation Date: NA

HUFF CREEK IRRIGATION COMPANY, Coalville, Utah; Irrigated Acreage:131.5 (1990); Incorporation Date: NA

ISLAND DITCH COMPANY NO. 1, Coalville, Utah; Irrigated Acreage: 51 (1990); Incorporation Date: NA

ISLAND DITCH COMPANY NO. 2, Coalville, Utah; Irrigated Acreage: 18.0 (1990); Incorporation Date: NA

ISLAND DITCH COMPANY NO. 3, Coalville, Utah; Irrigated Acreage: 44.30 (199); Incorporation Date: NA

RIVERS WATER COMPANY, Park City and Salt Lake City, Utah; Irrigated Acreage: 39.25 (1990); Incorporation Date: NA

JEREMY WATER COMPANY, Salt Lake City, Utah; Irrigated Acreage: NA; Incorporation Date: NA

L. TOONE DIVERSION DITCH, Henefer, Utah; Irrigated Acreage: 84.8 (1990); Incorporation Date: NA

MAIN CANYON CREEK IRRIGATION COMPANY, Henefer, Utah; Irrigated Acreage: 83.9 (1990); Incorporation Date: NA

MANOR LANDS WATER DISTRICT #1 COMPANY (CULINARY), Ogden and Salt Lake City, Utah; Irrigated Acreage: NA; Incorporation Date: NA

MARCHANT-MILE EXTENSION IRRIGATION COMPANY, Peoa, Utah; Irrigated Acreage: 562 (1990); Incorporation Date: NA

MARION LOWER DITCH, Kamas, Utah; Irrigated Acreage: 508 (1990); Incorporation Date: NA

MARION UPPER DITCH COMPANY, Kamas, Utah ; Irrigated Acreage: 551.6 (1990); Incorporation Date: NA

MARION WATER WORK COMPANY, Kamas and Marion, Utah; Irrigated Acreage: NA; Incorporation Date: NA

MIDDLE CHALK CREEK, Coalville, Utah; Irrigated Acreage: 166.2 acres; Incorporation Date: NA

MILL CANAL WATER COMPANY, Oakley, Utah; Irrigated Acreage: NA; Incorporation Date: NA

NEW FIELD & NORTH BENCH IRRIGATION COMPANY, Oakley and Peoa, Utah; Irrigated Acreage: 1396.5 (1990); Incorporation Date: 4/23/1904
NORTH CHALK CREEK IRRIGATION COMPANY, Oakley, Utah; Irrigated Acreage: 257.8 (1990); Incorporation Date: 4/25/1904
NORTH NARROWS IRRIGATION COMPANY, Coalville, Utah; Irrigated Acreage: 197.6 (1990); Incorporation Date: 7/6/1946
PARK WEST WATER ASSOCIATION, Salt Lake City, Utah, and Minneapolis, Minnesota; Irrigated Acreage: NA; Incorporation Date: NA
PARLEY'S PARK MUTUAL WATER COMPANY AKA GORGOZA MUTUAL WATER,Park City, Holladay and Syracuse, Utah; North Hollywood, California; Irrigated Acreage: NA; Incorporation Date: NA
PEOA PIPELINE COMPANY, Peoa, Utah; Irrigated Acreage: NA; Incorporation Date: NA
PEOA SOUTH BENCH CANAL AND IRRIGATION COMPANY, Peoa and Oakley, Utah; Irrigated Acreage: 826 (1990); Incorporation Date: 10/22/1926
PINE MEADOW RANCH WATER COMPANY, Salt Lake City, Utah; Irrigated Acreage: NA; Incorportion Date: NA
PINE MOUNTAINS MUTUAL WATER COMPANY, Salt Lake City and West Valley City, Utah; Irrigated Acreage: NA; Incoropration Date: NA
PINE PLATEAU WATER SYSTEM (CULINARY), Salt Lake City and Ogden, Utah; Irrigated Acreage: NA; Incorporation Date: NA
PINE VIEW IRRIGATION COMPANY NO. 1, Coalville and W. Valley, Utah; Irrigated Acreage: 361.2 (1990); Incorporation Date: NA
RICHARDS & COMPANY, Oakley, Utah; Irrigated Acreage: 205 (1990); Incorporation Date: NA
ROBINSON BROTHERS DITCH COMPANY, Coalville, Utah; Irrigated Acreage: 168.7 (1990); Incorporation Date: 7/6/1926
ROCKPOINT WATER COMPANY, Salt Lake City, Utah; Irrigated Acreage: NA; Incorporation Date: NA
RODEBACK IRRIGATION COMPANY, Coalville, Utah; Irrigated Acreage: 172 (1990); Incorporation Date: NA
RUST DITCH, Henefer, Utah; Irrigated Acreage: NA; Incorporation Date: NA
SAGE BOTTON DITCH COMPANY, Peoa, Utah; Irrigated Acreage: 400 (1990); Icorporation Date: NA
SAMAK COUNTRY ESTATES WATER ASSOCIATION INC. (CULINARY) Salt Lake City and Kamas, Utah; Irrigated Acreage: NA; Incorporation Date: NA
SHILL AND DAVIS IRRIGATION COMPANY, Henefer, Utah; Irrigated Acreage: 103.6 (1900); Incorporation Date: NA
SILVER CREEK IRRIGATION COMPANY, Salt Lake City, Utah; Irrigated Acreage: 835 (1990); Incorporation Date: NA
SLAUGHTER HOUSE DITCH COMPANY, Coalville, Utah; Irrigated Acreage: 32.7 (1990); Incorporation Date: NA
SMITH-MOOREHOUSE RESERVOIR COMPANY, Oakley, Utah; Irrigated Acreage: NA; Incorporation Date: 5/16/1926
SOUTH BENCH CANAL COMPANY, Peoa, Utah; Irrigated Acreage: 855 (1990); Incorporation Date: NA
SOUTH BOULDERVILLDE DITCH COMPANY, Kamas, Utah; Irrigated Acreage: 213 (1990); Incorporation Date: NA
SOUTH KAMAS IRRIGATION COMPANY, Kamas and Coalville, Utah; Irrigated Acreage: 270 (1990); Incorporation Date: NA

SOUTH UPPER CHALK CREEK IRRIGATION COMPANY, Coalville, Utah; Irrigated Acreage: 56 (1990); Incorpration Date: NA

STEPHENS RANCH, Henfer, Utah; Irrigated Acreage: 70 (1990); Incorporation Date: NA

SUMMIT COUNTY WATER USERS ASSOCIATION, Coalville, Utah; Irrigated Acreage: NA; Incorporation Date: NA

SUMMIT PARK MUTUAL WATER CORPORATION, Salt Lake City and Park City, Utah; Irrigated Acreage: NA; Incorporation Date: NA

SUMMIT WATER DISTRIBUTION COMPANY, Salt Lake City, Utah; Irrigated Acreage: 111 (1990); Incorporation Date: NA

TAYLORS DITCH, Henefer, Utah; Irrigated Acreage: NA; Incorporation Date: NA

THACKERAY-CROUCH DITCH, Henefer, Utah; Irrigated Acreage: 71 (1990); Incorporation Date: NA

UINTA DRIVE WATER USERS COMPANY (CULINARY), Ogden, Morgan and Salt Lake City, Utah; Irrigated Acreage: NA; Incorporation Date: NA

UPPER CHALK CREEK IRRIGATION COMPANY, Coalville, Utah; Irrigated Acreage: 318.4 (1990); Incorporation Date: NA

UPPER HENEFER IRRIGATION COMPANY, Henefer, Utah; Irrigated Acreage: 273.77 (1990); Incorporation Date: 4/12/1902

UPTON WATERWORKS COMPANY, Coalville, Utah; Irrigated Acreage: NA; Incorporation Date: NA

WANSHIP COTTAGE WATER COMPANY (CULINARY), Coalville and Salt Lake City, Utah; Irrigated Acreage: NA; Incorporation Date: NA

WANSHIP IRRIGATION COMPANY #2, Wanship, Utah; Irrigated Acreage: 212.6 (1990); Incorporation Date: NA

WANSHIP MUTUAL WATER COMPANY (CULINARY), Coalville, Utah; Irrigated Acreage: NA; Incorporation Date: NA

WASHINGTON IRRIGATION COMPANY, Kamas, Utah; Irrigated Acreage: 1200 (1990); Incorporation Date: 2/18/1891

WEST HOYTSVILLE IRRIGATION COMPANY, Coalville, Utah; Irrigated Acreage: 310 (1990); Incorporation date: 7/19/1943

WEST WANSHIP IRRIGATION COMPANY, Coalville, Utah; Irrigated Acreage: NA; Incorporation Date: 2/9/1948

WILDE A. DITCH COMPANY, Coalville, Utah; Irrigated Acreage: 49.5 (1990); Incorporation Date: NA

WOODENSHOE PIPELINE COMPANY, Peoa, Utah; Irrigated Acreage: NA; Incorporation Date: NA

WOODLAND HILLS MUTUAL WATER COMPANY, Salt Lake City and Kamas, Utah; Irrigated Acreage: NA; Incorporation Date: NA

WOODLAND MUTUAL WATER COMPANY, Kamas, Utah; Irrigated Acreage: NA; Incorporation Date: NA

WOODLAND WATERWORKS ASSOCIATION, Woodland, Utah; Irrigated Acreage: NA; Incorporation Date: NA

YOUNGS DIVERSION; Irrigated Acreage: NA; Incorporation Date: NA

WEBER COUNTY

ALDER CREEK IRRIGATION COMPANY, Ogden, Utah; Irrigated Acreage: 93 (1990); Incorporation Date: 2/26/1909

ANDERSON DITCH; Irrigated Acreage: NA; Incorporation Date: NA

ANDERSON-FELT DITCH COMPANY, Huntsville, Utah; Irrigated Acreage: 35 (1990); Incorporation Date: NA

ANDERSON-WINTER DITCH, Huntsville, Utah; Irrigated Acreage: 130 (1990); Incorporation Date: NA

BERTINOTTI IRRIGATION COMPANY, Ogden, Utah; Irrigated Acreage: 81 (1990); Incorporation Date: 12/5/1904

BEUS CREEK WATER COMPANY, INC., Ogden, Utah; Irrigated Acreage: NA; Incorporation Date: 12/4/1936

BONA-VISTA WATER IMPROVEMENT DISTRICT, Ogden, Utah; Irrigated Acreage: NA; Incorporation Date: NA

CHAMBERS DITCH, P.B.; Irrigated Acreage: 73 (1990); Incorporation Date: NA

CHARLES STORY DITCH, Liberty, Utah; Irrigated Acreage: 20 (1990); Incorporation Date: NA

CHERRY HILL WATER ASSOCIATION, Perry, Utah; Irrigated Acreage: NA; Incorporation Date: NA

COLD WATER IRRIGATION COMPANY NO. 2, Ogden and Brigham City, Utah; Irrigated Acreage: 724 (1990); Incorporation Date: NA

COOP FARM IRRIGATION COMPANY, INC., Huntsville, Utah; Irrigated Acreage: 344.5 (1990); Incorporation Date: 4/2/1913

CROOKED CREEK IRRIGATION COMPANY, Huntsville, Utah; Irrigated Acreage: NA; Incorporation Date: NA

DAVE MOORE DITCH COMPANY, Ogden, Utah; Irrigated Acreage: NA; Incorporation Date: NA

DEXTER FARR, ; Irrigated Acreage: NA; Incorporation Date: NA

DINSDALE WATER COMPANY, Ogden, Utah; Irrigated Acreage: 273.7 (1990); Incorporation Date: 5/19/1911

DOWNS DITCH WATER COMPANY, Huntsville, Utah; Irrigated Acreage: 80 (1990); Incorportion Date: 2/27/1914

DUNN CANAL COMPANY, Ogden, Utah; Irrigated Acreage: 106.5 (1990); Incorportion Date: 8/21/1906

EAST WILLOW SPRINGS IRRIGATION COMPANY, Ogden and No. Ogden, Utah; Irrigated Acreage: NA; Incorporation Date: NA

EDEN HILLS WATER COMPANY, Ogden, Utah; Irrigated Acreage: NA; Incorporation Date: NA

EDEN IRRIGATION COMPANY, Ogden and Eden, Utah; Irrigated Acreage: 2200 (1990); Incorporation Date: 10/9/1919

EDEN WATERWORKS COMPANY, Eden, Utah; Irrigated Acreage: NA; Incorporation Date: NA

EMIL ROBERTS DITCH COMPANY, Liberty, Utah; Irrigated Acreage: NA; Incorporation Date: NA

EMMERTSEN IRRIGATION COMPANY, Huntsville, Utah; Irrigated Acreage: 101 acres; Incorporation Date: 5/6/1912

FARR-ENOCH DITCH, Ogden, Utah; Irrigated Acreage: NA; Incorportion Date: NA

FELT, PETERSEN, SLATER DITCH COMPANY, Huntsville, Utah; Irrigated Acreage: 210 (1990); Incorporation Date: 1/20/1906

GLASMANN FARMS, Ogden, Utah; Irrigated Acreage: NA; Incorporation Date: NA

GLENWOOD DITCH COMPANY, INCORPORATED, Ogden, Utah; Irrigated Acreage: NA; Incorporation Date: 6/10/1941

GREAT BASIN WATER COMPANY, Ogden, Utah; Irrigated Acreage: NA; Incorporation Date: NA

HABERTSON DITCH COMPANY, Ogden, Utah; Irrigated Acreage: 27.6 (1990); Incorporation Date: NA

HOLMES-FERRIN IRRIGATION ASSOCIATION, Libery, Utah; Irrigated Acreqge: 251.7 (1990); Incorporation date: NA

HOOPER IRRIGATION COMPANY, Hooper and Ogden, Utah; Irrigated Acreage: 8442 (1990); Incorporation Date: 1/5/1903

HUNTSVILLE IRRIGATION COMPANY, Huntsville, Utah; Irrigated Acreage: 3000 (1990); Incorporation Date: 4/26/1939

HUNTSVILLE SOUTH BENCH CANAL COMPANY, Ogden, and Huntsville, Utah; Irrigated Acreage: 300 (1990); Incorporation Date: 6/4/1929

HUNTSVILLE WATERWORKS CORP., Huntsville, Utah; Irrigated Acreage: NA; Incorporation Date: NA

IDEAL ROCK PRODUCTS, Ogden, Utah; Irrigated Acreage: NA; Incorporation Date: NA

JONES DITCH COMPANY, Ogden, Utah; Irrigated Acreage: 73.6 (1990); Incorporation Date: NA

KNIGHT IRRIGATION COMPANY, Plain City and Ogden, Utah; Irrigated Acreage: NA; Incorporation Date: NA

LAKEVIEW WATER CORPORATION, Ogden, Utah, and Palm Beach, Florida; Irrigated Acreage: NA; Incorporation Date: NA

LDS SLATERVILLE WARD; Irrigated Acreage: NA; Incorportion Date: NA

LEWIS SHAW DITCH COMPANY, Ogden, Utah; Irrigated Acreage: NA; Incorporation Date: NA

LIBERTY IRRIGATION ASSOCIATION, Libery, Utah; Irrigated Acreage: 1000 (1990); Incorporation Date: NA

LITTLE MISSOURI IRRIGATION COMPANY, Ogden, Utah; Irrigated Acreage: 110 (1990); Incorporation Date: 6/25/1910

LYNNE IRRIGATION COMPANY, Ogden, Utah; Irrigated Acreage: 967 (1990); Incorporation Date: 5/17/1930

MARRIOTT IRRIGATION COMPANY, Ogden, Utah; Irrigated Acreage: 600 (1990); Incorporation Date: 7/5/1895

MIDDLE FORK IRRIGATION COMPANY, Eden and Harrisville, Utah; Irrigated Acreage: 300 (1990); Incorportion Date: 2/10/1919

MONTGOMERY SLOUGH DITCH #1, Liberty, Utah; Irrigated Acreage: NA; Incorporation Date: NA

MOUND FORT IRRIGATION COMPANY #1, Ogden, Utah; Irrigated Acreage: 270 (1990); Incorporation Date: 6/25/1935

MOUND FORT IRRIGATION COMPANY #2, Ogden, Utah; Irrigated Acreage: NA; Incorporation Date: NA

MOUND FORT IRRIGATION COMPANY #3, Ogden, Utah; Irrigated Acreage: NA; Incorporation Date: NA

MOUND FORT IRRIGATION COMPANY #4, Ogden, Utah; Irrigated Acreage: NA; Incorporation Date: NA

MOUND FORT IRRIGATION COMPANY #5, Ogden, Utah; Irrigated Acreage: NA; Incorporation Date: NA

MOUND FORT IRRIGATION COMPANY #6, Ogden, Utah; Irrigated Acreage: NA; Incorporation Date: NA

MOUNTAIN CANAL IRRIGATION ASSOCIATION, Ogden and Huntsville, Utah; Irrigated Acreage: 1200 (1990); Incorporation Date: 10/19/1939

MOUNTAIN CANAL & IRRIGATION COMPANY, Huntsville, Utah; Irrigated Acreage: 1200 (1990); Incorporation Date:10/19/1939

MOUNTAIN WATER IRRIGATION COMPANY, Ogden and Brigham City, Utah; Irrigated Acreage: NA; Incorporation Date: NA

NORDIC VALLEY WATER COMPANY, Ogden and Provo, Utah; Irrigated Acreage: NA; Incorporation Date: NA

NORTH OGDEN IRRIGATION COMPANY, Ogden and North Ogden, Utah; Irrigated Acreage: 3000 (1990); Incorportion Date: 3/26/1909

NORTH SLATERVILLE IRRIGATION COMPANY, Ogden and Slaterville, Utah; Irrigated Acreage: 540 (1990); Incorporation Date: 2/21/1905

OGDEN BENCH CANAL & WATER COMPANY, Ogden, Utah; Irrigated Acreage: NA; Incorporation Date: 7/17/1891

OGDEN BRIGHAM CANAL (NORTH), Ogden, Utah; Irrigated Acreage: NA; Incorporation Date: NA

OGDEN RIVER DISTRIBUTION SYSTEM, Ogden, Utah; Irrigated Acreage: NA; Incorporation Date: NA

OGDEN RIVER RESERVOIR COMPANY, Ogden, Utah; Irrigated Acreage: NA; Incorporation Date: 4/27/1912

OGDEN RIVER WATER USERS ASSOCIATION, Ogden, Utah; Irrigated Acreage: 500 (1990); Incorporation Date: 11/3/1933

OLD WILSON CANAL ; Irrigated Acreage: NA; Incorporation Date: NA

PATIO SPRINGS WATER COMPANY, Provo and Eden, Utah; Irrigated Acreage: NA; Incorporation Date: NA

PERRY IRRIGATION COMPANY, Ogden, Utah; Irrigated Acreage: 478 (1990); Incorporation Date: NA

PINE CANYON DITCH COMPANY, INC., Liberty and Free Mont, Utah; Irrigated Acreage: NA; Incorporation Date: NA

PINE VIEW HOME OWNERS WATER DEVELOPMENT, INC., No. Ogden, Sandy and Centerville, Utah; Irrigated Acreage: NA; Incorporation Date: NA

PINEVIEW WATER SYSTEMS, Ogden, Utah; Irrigated Acreage: 11,000 Incorporation Date: 1933

PINEVIEW WEST WATER COMPANY, Ogden and Salt Lake City, Utah; Irrigated Acreage: NA; Incorporation Date: NA

PIONEER IRRIGATION CANAL COMPANY, Ogden, Utah; Irrigated Acreage: 100 (1990); Incorporation Date: 1/9/1895

PIONEER LAND & IRRIGATION COMPANY, Ogden, Utah; Irrigated Acreage: NA; Incorporation Date: NA

PLAIN CITY IRRIGATION COMPANY, Plain City, Utah; Irrigated Acreage: 2324 (1990); Incorporation Date: 9/19/1958

PLEASANT VIEW CULINARY WATER ASSOCIATION, Ogden, Utah; Irrigated Acreage: NA; Incorporation Date: NA

POLE PATCH IRRIGATION COMPANY, Ogden and Pleasant View, Utah; Irrigated Acreage: NA; Incorporation Date: NA

RICH ACRES IRRIGATION COMPANY, Ogden, Utah; Irrigated Acreage: NA; Incorporation Date: NA

RICHARDSON & COWAN DITCH COMPANY, Ogden and Plain City, Utah; Irrigated Acreage: NA; Incorporation Date: NA

RIVER CREEK IRRIGATION COMPANY, North Ogden, Utah; Irrigated Acreage: NA; Incorporation Date: NA

RIVERDALE BENCH CANAL COMPANY, Ogden, Utah; Irrigated Acreage: 330.3 (1990); Incorporation Date: NA

ROLLO-JOHNSON-DOWNS DITCH, Huntsville, Utah; Irrigated Acreage: NA; Incorporation Date: NA

ROY WATER CONSERVANCY SUBDISTRICT, Roy, Utah; Irrigated Acreage: NA; Incorporation Date: NA

SCHRIBNER DITCH COMPANY, Roy, Utah; Irrigated Acreage: NA; Incorporation Date: NA

SHUPE MIDDLETON DITCH, Ogden, Utah; Irrigated Acreage: NA; Incorporation Date: NA

SMOUT & HOLLEY IRRIGATION COMPANY, Layton and Ogden, Utah; Irrigated Acreage: NA; Incorporation Date: NA

SOUTH OGDEN CONSERVANCY DISTRICT, Ogden, Utah; Irrigated Acreage: 2000 (1990); Incorporation Date: NA

SOUTH SLATERVILLE IRRIGATION COMPANY, Slaterville and Ogden, Utah; Irrigated Acreage: 959.3 (1990); Incorporation Date: 1/4/1904

SOUTH WEBER DIVERSION CANAL COMPANY, South Weber, Utah; Irrigated Acreage: NA; Incorporation Date: NA

SOUTH WEBER IMPROVEMENT DISTRICT, South Weber, Utah; Irrigated Acreage: NA; Incorporation Date: NA

SOUTH WEBER IRRIGATION COMPANY, South Weber, Utah; Irrigated Acreage: 378 (1990); Incorporation Date: 12/19/1921

SOUTH WEBER RESERVOIR CANAL COMPANY, Ogden, Utah; Irrigated Acreage: NA; Incorporation Date: NA

SPRING MOUNTAIN WATER COMPANY, Ogden, Utah; Irrigated Acreage: NA; Incorporation Date: NA

TRIANGLE B RANCH; Irrigated Acreage: NA; Incorporation Date: NA

UINTAH CENTRAL CANAL COMPANY, Ogden, Utah; Irrigated Acreage: NA; Incorporation Date: 5/11/1895

UINTAH MOUNTAIN STREAM IRRIGATION COMPANY, Ogden, Utah; Irrigated Acreage: NA; Incorporation date: 6/9/1955

UPPER CLUB PLAIN CITY, Plain City, Utah; Irrigated Acreage: NA; Incorporation Date: NA

WARREN IRRIGATION COMPANY, Ogden, Utah; Irrigated Acreage: 3052.7 (1990); Incorporation Date: 4/5/1907

WEBER BASIN WATER CONSERVANCY DISTRICT, Layton, Utah; Irrigated Acreage: See Appendix C for water usage; Incorporation Date: 6/26/1950

WEBER CANAL WATER COMPANY, Ogden, Utah; Irrigated Acreage: NA; Incorporation Date: NA

WEBER RIVER DISTRIBUTION SYSTEM, Ogden, Utah; Irrigated Acreage: NA; Incorporation Date: NA

WEBER RIVER WATER USERS ASSOCIATION, Ogden, South Weber, Layton, and Morgan, Utah; Irrigated Acreage: NA; Incorporation Date: 1/9/1926

WEBER-BOX ELDER CONSERVANCY DISTRICT, Ogden, Utah; Irrigated Acreage: NA; Incorporation Date: NA

WESTERN IRRIGATION COMPANY, Ogden, Utah; Irrigated Acreage: 2700 (1990); Incorporation Date: 3/31/1903

WILSON IRRIGATION COMPANY, Ogden and Hooper, Utah; Irrigated Acreage: 3000 (1990); Incorporation Date: 3/26/1903

WOLF CREEK IRRIGATION COMPANY, Eden, Utah; Irrigated Acreage: NA; Incorporation Date: NA

WOODLAND BENCH WATER COMPANY, INC., Ogden, Utah; Irrigated Acreage: NA; Incorporation Date: NA

Appendix C

 Allocation of Weber Basin Water, 1992 (Acre-Feet)[1]

TABLE 1
Municipal and Industrial Water Use

Contracting Entity	Contract Amount	Amount Used	Percentage Used
Untreated Water			
Chevron, USA	1200.00	777.74	64.81
Great Salt Lake Minerals	4100.00	5078.00	123.85
Ogden City	1500.00	822.00	54.80
North Salt Lake City	30.00	14.35	47.83
Weber Basin Job Corps	60.00	35.89	59.82
Treated Water			
Black Hawk Mineral	5.00	1.01	20.20
Bona Vista Water Imp. Dist.	1657.00	1658.38	100.08
Bountiful City	1000.00	1033.86	103.39
Centerville City	500.00	503.89	100.78
Chevron, USA	2000.00	1352.14	67.61
Clearfield City	4258.00	4171.7	97.97
Clinton City	1411.00	1376.99	97.59
Davis County Solid Waste	300.00	295.92	98.64
Farmington City	501.00	462.20	92.26
Fruit Heights City	300.99	295.92	98.64
Geneva Rock	42.00	30.46	72.52
Great Salt Lake Minerals	330.00	476.61	144.43
Highlands Water Company	5.00	.23	5.60
Hill Air Force Base	1018.79	1012.74	99.41
Hooper Water Imp. Dist. (WID)	5.00	0.00	0.00
Kaysville City	2500.00	2165.62	86.62
Layton City	3789.00	3607.80	95.22
Mutton Hollow WID	113.00	150.25	133.00
North Salt Lake City	130.00	385.85	296.81
Ogden City	6500.00	5709.97	87.85
Parsons	22.00	22.00	100.00
Riverdale City	819.00	821.37	100.29

Roy City	3468.00	3333.16	96.11
South Davis County WID	360.00	356.38	98.99
South Weber City [2]	260.00	233.31	89.73
South Ogden City[2]	785.00	785.00	100.00
Sunset City	1400.00	1276.30	91.16
Syracuse City	725.00	540.31	74.53
Taylor-West Weber WID	150.00	151.48	100.99
Uintah Highlands WID	237.00	224.54	94.74
Uintah Town	216.00	208.22	96.40
Union Pacific Railroad	5.00	5.00	100.00
Washington Terrace City	1000.00	992.38	99.24
Webbs Canyon Water Co.	9.00	9.54	106.00
West Bountiful City	525.00	501.99	95.62
West Point City	550.00	542.79	98.69
West Warren-Warren WID	100.00	89.55	89.55
Western Zirconium	560.00	340.43	60.79
Woods Cross City	100.00	99.91	99.91
Totals	**44,545.79**	**41,989.70**	**94.29%**

TABLE 2
Irrigation Company Water Use

Contracting Entity	Contracted Amount	10% Loss	Net Useable	Used	% Used
Bountiful Water Subdist.	17,500	1,600	15,900	11,57	672.8
Centerville Deuel Creek	2,891	264	2,627	1,750	66.6
Chalk Creek Irrigation	643	64	579	501	86.5
Co-op Farms Irrigation	300	30	270	270	100.0
Croyden Irrigation	450	45	405	405	100.0
Downs Creek Irrigation	100	10	90	90	100.0
East Porterville Irrigation	200	20	180	180	100.0
East Wanship/Gibbons & Pace	100	10	90	80	88.9
Eden Irrigation	1,200	120	1,080	1,146	106.1
Emmertsen Irrigation	100	10	90	90	100.0
Farmington Area Pressure Irr.	4,330	433	3,897	1,684	43.2
Felt, Peterson, Slater Irr.	100	10	90	90	100.0
Haights Creek Irrigation	6,970	697	6,273	5,240	83.5
Hill A.F.B. Golf Course	779	64	715	499	69.8
Hill Field at 193	139	0	139	115	82.7
Hooper Irrigation	5,500	0	5,500	5,500	100.0
Huntsville Irrigation	600	60	540	540	100.0
Huntsville So. Bench Irr.	600	60	540	540	100.0
Kays Creek Irrigation	2,000	200	1,800	1,322	73.4
Kaysville Irrigation	1,775	178	1,597	1,196	74.9

Lagoon Amusement Park	225	23	202	93	46.0
Layton Canal & Irr. Co.	5,403	540	4,863	4,224	86.9
Littleton-Milton Irrigation	300	30	270	270	100.0
Middle Fork Irrigation	820	82	738	738	100.0
Mountain Valley Canal Irr.	1,303	130	1,173	1,064	90.7
North Morgan Irrigation	160	16	144	144	100.0
North Round Valley	160	16	144	135	93.8
Oakridge Country Club	500	50	450	325	72.2
Ogden River Wtr. Users Assoc.	3,420	0	3,420	3,420	100.0
Peterson Irrigation	974	97	877	559	63.7
Salmaho Irrigation	167	17	150	42	28.0
South Davis County WID	3,210	321	2,889	1,798	62.2
South Morgan Water Co.	400	40	360	360	100.0
South Ogden Cons. Dist.	2,345	234	2,111	1,826	86.5
South Weber WID	2,506	0	2,506	2,212	88.3
Syracuse City	1,000	100	900	274	30.4
Uintah Mountain Streams	200	20	180	190	105.6
Valley View Golf Course	267	27	240	223	92.9
Warren Irrigation	500	0	500	500	100.0
Weber Basin Job Corps	300	30	270	51	18.9
Weber-Box Elder Cons. Dist.	2,194	219	1,975	1,579	79.9
Weber Canal Company	200	20	180	180	100.0
Welch Field Ditch	240	24	216	209	96.8
West Hoytsville Irrigation	300	30	270	270	100.0
West Wanship Irrigation	150	15	135	61	45.2
Wilson Irrigation	1,500	0	1,500	1,500	100.0
Individual Users	2,126	0	2,126	2,126	100.0
Totals	77,147	5,956	71,191	57097	80.2%

1. From *Weber Basin Water Conservancy District—1992 Summary of Operations.*
2. South Ogden City has 873.91 acre-feet of Birch Creek water treated.

Index

Compiled by Lucile H. Anderson

A. Guthrie Company (construction), 127
Abbott, Fred M., 160, 199, 200, 202
Adams, D. C., 66
Adams, Elias, 63, 64
Adams, Elias, Jr., 63
Adam's Canyon, 63, 234n4
Agee, A. W., 118
Alder Creek, 34
Alexander, Henry, 91
Alexander, Thomas, 5, 109
Allen, Thomas L., 98–99
Anderson, John, 78
Anderson, Olof B., 82
Andrews, Joseph B., 162
Andrus, Joseph B., 97
Anglo California Bank, 123–24
Appleby, Charles S., 88
Armstrong, W. W., 126
Army Corps of Engineers, 10, 165, 201, 207–8
Army supply depot, 139
Arrington, Leonard J., 16, 229n4
Artesian basins, 142
Artesian Park, 132–33
Artesian wells: in Kaysville, 62; in North Og-
    den, 35; in Ogden Valley, 33, 44, 132–34,
    178–79; in Plain City, 40; in South Bounti-
    ful, 55-56; in Syracuse, 65-66; in Warren
    and West Warren, 41; in West Bountiful,
    52-54; under Pineview Reservoir, 132–34,
    178–79. *See also* Underground water;
    Wells
Arthur, Jake, 81
Arthur V. Watkins Dam and Reservoir. *See*
    Willard Bay Dam and Reservoir
Ashley, William H., fur company, 27, 43, 87
Associated Civic Clubs of Northern Utah,
    140, 150
Atkinson, William N., 55–56
Aurora, 161
Ayers, James M., 193
Bachelor, William, 88
Bachelor's Canyon, 88–89
Bachelor's Creek, 89
Baer Canyon, 63

Baer Creek, 234n4
Bailey, Reed W., 136, 137
Bair, John, 63
Baker, George, 90
Baker, William, 43
Bald Rock Canyon, 88
Ballantyne, Richard, 44
Bamberger Electric Railroad, 198
Bambrough Canal, 65
Bannigan, Joseph, 121–22
Bannister, Charles K., 121
Barlow, John, 124
Barlow, Woodrow, 52
Barnard Creek, 60, 234n4
Barnes, Emily, 61
Barnes, H. J., 224
Barnes, John R., 118, 126
Barton, Joseph, 61
Barton, William G., 136
Barton Creek, 50, 51, 54, 234n4
Bear Lake, 108
Bear Lake and River Water Company, 32
Bear Lake and River Water Works and Irriga-
    tion Company, 76, 107–8
Bear River, 125, 129, 154
Bear River Canal, 6, 76, 107–9
Bear River Canyon, 76, 108
Bear River City, 108
Bear River Irrigation and Ogden Water
    Works, 76, 108-9
Bear River Irrigation Company, 6
Bear River Project, 12, 17, 228
Bear River Water Company, 76
Beard, George, 93
Beaver Creek, 80, 86, 90, 95, 234n4
Beckstead, Gordon, 83
Beckwith, E. G., 87
Beesley, William, 63
Belnap, Gilbert R., 41
Belnap, Hyrum, 41–42
Bennett, Wallace F., 179, 210
Bennett Springs, 85
Benson, Ezra T., 19, 61
Berrett, Robert, 34

Beus, J. R., 129
Beus Canyon, 234n4
Big Cottonwood irrigation canal, 16–17
Big Field (Brigham City), 71
Big Field (Salt Lake Valley), 16–17
Bigelow, A. P., 115, 126, 131
Bingham, J. R., 128
Bingham, Thomas, 46
Bingham Fort, 33
Bishops: land distributed by, 57; water dis-
    putes settled by, 36, 44-45, 52; water man-
    aged by, 18, 58–59, 100, 103
Bishop's Canyon, 88
Black Hawk Indian War, 88
Blamires, Lamber, 63
Blanch, George T., 201
Blood, Henry H., 61, 132, 136
Bohman, Victor R., 79
Bond, Nephi, 89
Bone, Thomas, 62
Bonneville Canal, 50, 54, 56, 142
Bonneville Canal Company, 117, 121, 122–25
Bonneville Irrigation District, 122–24
Bosone, Reva Beck, 160, 164, 166, 203, 211
Bostaph, William M., 120, 129
Bostaph Electric Plant, 120
Bothwell, John R., 107
Bountiful, 149, 153, 194, 204, 205; assess-
    ment of, 157; canal to, 117, 123, 141; fil-
    tration plant at, 219; settlement of, 1, 50–
    52; water rates in, 161
Bountiful Irrigation Company, 55, 160
Bountiful Mill Creek Irrigation Company, 55,
    223
Bouwhuis, Chris, 117
Bowns, Norton, 129
Box Elder Canyon, 71–72, 75
Box Elder County, 69–76, 107, 129, 134
Box Elder Creek, 69–71, 75
Boyce, Doren B., 151, 157
Boyden, Maurice, 198, 199, 216
Boyle, Peter, 30
Bradberry, Thomas, 94
Brewer, Charles, 89
Bridge Canyon, 88
Bridger, Jim, 87
Brigham City, 69–72, 75–76, 108, 131, 161
Brigham City Electric Company, 72
Broad Hollow, 234n4
Bromley, James, 90
Brooks, William D., 69
Broom, John, 33
Brough, Charles, 19, 22, 229n4
Brough, Samuel, 80–81
Brown, Alexander, 27, 29
Brown, James, 27, 29, 30
Brown, Jesse, 27, 29

Brown, M. P., 129
Brown, Samuel, 50
Brown, William, 55–56
Brownsville, 27
Buckland, Aldonus D. L., 51
Bud King Construction Company, 187
Bull, Daniel, 84
Bundy, Ora, 133
Burch, Daniel, 29, 37
Burch, James, 37
Burch, Robert, 37
Burch Creek, 37, 234n4
Bureau of Reclamation, 7, 9–12, 14, 109, 137,
    139; creation of, 104, 110; East Canyon
    dam built by, 121; Echo Dam built by,
    126–28, 139; involvement of, in Weber
    Basin Project, 142–67, 169–70, 172, 179,
    181, 190, 192–95, 197, 201–2, 209, 212–
    13, 219; Pineview Dam built by, 130, 139,
    178
Burningham, John H., 140
Burningham, Thomas, 124
Burns, James E., 124
Burrows, J. A., 217, 219
Bybee, David, 63
Bybee, John T., 126

C. M. C. Construction Company, 187
Caldwell, Richards and Sorenson (engineering
    firm), 155–56
Call, Anson, 52, 117
Callister, E. R., 115
Camp Kiesel Road, 187–88
Campbell, Arlie, 134
Canfield, Cyrus, 29
Canfield Creek, 29
Cannon, George Q., 121
Cannon, Sylvester Q., 136
Carey Act of 1894, 6, 24, 125
Carpenter, Joe, 81
Carrigan, Irvin, 79
Carrigan, James W., 79
Carter, Jimmy, 11
Carter, Samuel, 82
Carter, William, 15
Cartio, George, 99
Carver, Elmer, 152, 157, 198
Castle Gate, 161
Causey dam and reservoir, 130, 182, 187–88,
    248n68
Cedar City, 161
Centerville, 54, 59–60, 78, 80, 122–23, 149,
    153, 204; flooding in, 136; water rates in,
    161; settlement of, 50, 59
Centerville Canyon, 234n4
Centerville Deuel Creek Irrigation Company,
    60

Central Canal Company, 117
Central Pacific Railroad, 31, 107
Central Utah Project, 11, 12, 17
Central Valley Project of California, 10
Chalk Creek, 86, 87, 93, 95, 234n4; dam site, 170
Chalk Creek-Hoytsville Water Users' Corporation, 94
Chapman, Oscar, 203
Chard, William H., 45
Chase, Ezra, 29, 33
Chipman, Stephen L., 98
Chittenden, Hiram M., 8
Chrisman, Charles, 124
Christensen, Orson, 150
Cisterns, 53
City Creek, 15, 19
Civilian Conservation Corps (CCC), 132, 136
Clark, Horald G., 144, 155, 162, 197, 198, 206
Clark, Tom, 209
Clayton, William, 15–16
Clearfield, 64, 121, 149; settlement of, 50, 66; water rates in, 161; water sold to, 221
Clearfield Naval Supply Depot, 7
Cleveland, Henry A., 136
Clinton, 50, 65
Clyde, George D., 142, 173
Clyde, W. W., Construction Company, 172, 175, 184
Coalville, 86, 91, 92–94, 197
Coalville-Hoytsville Canal, 94
Coffin, Nathan, 43
Cold Spring, 70
Cold Water Canyon, 234n4
Cold Water Creek, 34
Collinston, 76, 108
Colorado River, 9, 11, 12, 141–43
Columbia River Basin Project, 10
Como Springs, 83–85
Compton, George, 85
Condie, Thomas, 84
Conley, Solomon, 81
Connor, Patrick, 101
Conrad, Joshua, 62
Coolidge, Calvin, 126
Corbett Creek, 234n4
Corinne, 108
Corinne Mill Canal and Stock Company, 107–8
Cottonwood Creek, 83, 234n4
Cowley, Charles G., 205
Cox, Richard, 193
Cragun, Simeon, 34
Crandall's Creek, 86, 96
Croft, A. R., 136, 137
Croft, Russell, 148
Croseclose, William R., 193
Croydon, 83, 106

Croydon Irrigation Company, 84
Cutler Creek, 135

Dalton, Matthew W., 70, 75
Dalton Creek, 78–79
Dank's Canyon, 88
Danton, George, 217
Daughters of Utah Pioneers, 218
Davis, Daniel C., 49
Davis, William, 69
Davis and Butler Construction Company, 173
Davis and Weber Canal, 43, 49, 62, 64, 65–66, 116, 147; construction of, 104, 117–21; extension of, 141; planned in 1856, 59, 117
Davis and Weber Canal Company, 117–20, 127, 142–43, 182, 191–92, 205, 225
Davis Aqueduct, 172, 173–76, 182, 194
Davis Canal Company, 59
Davis Canyon, 57
Davis County: bond election in, 215–18; produce from, 122; special election in, 220–22; value of irrigated land in, 200; water claims in, 106, 107; water development in, 49–67; water survey assessment for, 157
Davis County Commissioners, 164
Davis County Experimental Watershed Project, 136
Davis County Flood Control Committee, 136
Davis County Water Users Association, 140–42, 145, 148, 151, 153, 164, 210
Davis Creek, 135, 234n4
Davis-Weber Counties Municipal Water Development Association, 151–52, 155–59, 160, 162–64, 197–98, 203, 204, 212, 213, 219, 225
Dawson, Alexander, 63
Dawson, William, 140, 147–49, 152–54, 159, 160
de la Baume, Charles, 106
Deep Creek, 77, 78, 106
Deer Creek Reservoir, 98
Denver and Rio Grande Railway, 31
Denver and Rio Grande Western Railroad, 198
Department of Agriculture, 166–67, 169, 202, 212
Dern, George H., 136
Deseret, state of, 16, 19
Deseret Land and Livestock Company, 224
Desert Land Act of 1877, 6, 23, 106
Deuel, Osmyn M., 59
Deuel, William, 59
Deuel Creek, 59–60, 234n4
Deuel Creek Irrigation Company, 60
Devil's Gate, 28, 76, 118
Devil's Slide, 76, 83, 188, 192
Deweyville, 76, 108

Dexheimer, Wilbur A., 173
Dickson, Bill, 81
Dinsdale, Jeffrey, 33
Dominy, Floyd E., 224
Droughts, 128, 226
Dry Canyon, 234n4
Dry Creek, 51, 151
Dry farming, 65
Dugway Proving Grounds, 7
Dunbar, Robert G., 2, 22, 112, 229n4
Dunn, Simeon A., 70
Dunn, Thomas, 36
Dunn Canal, 65
Dunn's Canyon, 70
Dye, Samuel G., 131

E-W Construction Company, 191
East, Thomas, 151–52, 158, 162
East Bountiful Pumping Plant, 175
East Canyon, 80, 81, 87
East Canyon Creek, 76, 77, 80–82, 84–85,
    118–19, 151, 234n4
East Canyon Dam, 49, 62, 64–66, 104, 117–
    21, 171, 248n68; enlargement of, 191–94
East Canyon Reservoir, 43, 171; enlargement
    of, 169, 182; recreation at, 210, 226
East Layton, 64–65, 204
East Porterville Canal Company, 80
Eccles, David, 46
Echo, 86, 90
Echo Canyon, 86–88, 90–92, 234n4
Echo Creek, 90, 224
Echo Dam, 10, 40, 62, 90, 98, 104, 117, 121,
    171, 203;   construction of, 125–28
Echo Reservoir, 78, 83, 115, 146; construction
    of, 139; use of, for flood control, 171,
    207; water from, 41, 194, 205, 223–24
Eden, 43, 44
Eden Irrigation Company, 44, 114
Ekins, Ernest R., 158–59
Eldridge, Clarence, 122
Electric Power and Light Company, 41
Electric railways, 121
Elkhorn Creek, 95
Elkhorn Ditch Company, 95
Ellis, Alma, 151–52, 157
Ellis, Steven H., Sr., 53
Ellison, E. P., 118, 126
Ellison, Harold, 151, 157–59, 161, 163, 166,
    173, 197–200, 203, 206, 210, 212, 219
Emigration Canyon, 87
Enterprise, 77, 83
Environmentalists, 11
Ephrata Pre-Mix Company, 191
Evans, James, 99

F. S. Smithers and Co. (bonding company),

219
Farmington, 149, 153, 175–76, 194; settle-
    ment of, 49, 50, 57–59; water rates in, 161
Farmington Bay Refuge, 166
Farmington Bay Wasteway, 176
Farmington Canyon, 57, 135, 137, 234n4
Farr, Enoch, 33
Farr, John, 30
Farr, Lorin, 29, 33, 35, 46
Farr West, 35
Farr's Fort, 29
Ferrel, William, 80–81
Ferrin, Josiah, 46
Ferrin, Samuel, 46
Fife Construction Company, 186
Filtration plants, 221, 223
Fire Canyon, 88
Fire protection, 31–32
Fish and wildlife, 159, 166, 214, 226–27
Fish and Wildlife Service, 154
Fish Lake Reservoir Company, 97
Fisher Springs, 45
Fjeldsted, Ezra J., 1, 121, 144, 150–51, 155,
    157–60, 162; secretary of Davis-Weber
    Counties Municipal Water Development
    Association, 152, 162–67, 197, 199, 200,
    202, 204, 211, 212–13; secretary-manager
    of Weber Basin Conservancy District,
    205, 219–20, 224
Flaming Gorge, 11
Flint, Ivan, 224
Flint, John, 118
Flood control: at Echo Dam, 127, 207; at
    Pineview Dam, 132; in the Weber Basin,
    134–37, 147, 158–60, 166, 205, 208, 213
Flooding: in Box Elder County, 73–75; in
    Davis County, 54; in Summit County, 89–
    90; in Weber County, 37, 38, 40–41, 46; in
    1952 and 1953, 171–72
Florence, Henry, 80
Florence, George, 197–98
Flour mills, 37, 75
Ford, Eugene, 143, 153
Ford Canyon, 135, 234n4
Forsling, Clarence L., 136, 137
Fort Bridger, 87
Fort Buenaventura, 24, 27
Fort Creek, 90–91
Fortier, Samuel, 108, 120, 129–30, 191
Four Mile Creek, 35, 39, 41
Fowles, Thomas, 129
Fox, Feramorz, 229n4
Fox, Jesse W., survey by, 34, 51, 83–84, 86,
    89, 96
Francis, Esther Charlotte Emily Weisbrodt, 85
Francis, Frank, 131
Francis, 95

Franklin, Thomas Job, 88
Franklin Canyon, 88
Franklin Creek, 89
Fruit Heights, 50, 157
Fry, Richard, 84
Fuhriman, Walter U., 201
Fuller, Edward, 38
Fuller, Enoch, 46
Fuller, Henry, 46

Garland, William, 108
Garner Canyon, 234n4
Gateway Canal, 172–74, 181, 194, 227
Gateway Power Plant, 182, 226
Gateway Tunnel, 170–72, 174–76, 209, 212, 226–28
Geary, Edward, 82
Geddes, William, 114
Gemmell, Robert C., 109
General Land Office, 23
General Mining Law of 1866, 4–5, 23
Geneva Steel Corporation, 7
George M. Brewster and Son (construction company), 183
Gibbons, Albrey, 96, 97
Gibbons, Glen, 97
Gibbons and Reed Construction Company, 185, 190, 207
Glen Canyon, 11
Godfrey, J. Morris, 140
Goodale, Isaac, 1, 30, 33
Goodyear, Miles, 24, 27, 228
Gordon Creek, 83
Gorlinski, Joseph S., 158
Goshen, 161
Grand Coulee Dam, 10
Granger, Walter K., 152, 162, 164–66, 203, 211
Grant, George, 77
Grant, Jedediah Morgan, 77–78
Grant, Joseph H., 98
Grant, L. M., 54
Grass Creek, 94
Great Salt Lake, 41, 49, 54–55, 64, 70, 87, 141, 176, 205; survey of, 50, 145; Willard Bay adjacent to, 69, 144, 154, 183
Green, John H., 61
Green, William, 126, 129–139, 144
Greenhalgh, Rex, 194
Green River, 203
Gristmills, 19, 20, 29, 33, 60, 62–63, 75, 81, 91, 94, 104
Ground water, 169, 180–89, 209. *See also* Underground water
Grover, Joseph, 44
Grover, Thomas, 59
Gunnison, John W., 50, 87

H and M Construction Company, 173
Haight, Hector C., 50, 57, 60
Haight Bench Irrigation Company, 176
Haight's Bench, 141
Haight's Creek, 60, 62, 121
Haight's Creek Irrigation Company, 62, 121
Hales, Ray A., 143
Hamblin, Richard, 66
Hammond, Wendell B., 124
Hancock, C. B., 46
Hansen, James, 78
Harding, Robert, 141–42
Harding, Thomas, 126
Hardscrabble Canyon, 77, 80
Hardscrabble Creek, 76, 80, 81
Harris, D. D., 115, 128, 140, 142, 144, 162, 207–8
Harris, M. H., 203–4
Harris, Martin, 35
Harrisville, 1, 34, 35, 100
Harrisville Canal, 33
Harrisville Irrigation Company, 114
Harvey, William "Coin," 230n10
Hasler and Smith Construction Company, 183
Hastings, Kester L., 139
Hastings, Lansford W., 87
Hatch, Mark T., 223
Hatch, Stearns, 53
Hayden, Carl, 211
Heiner, Daniel, 98
Heiner, Martin, 85
Heiner Creek, 224
Hemingway, Jonathan, 81
Henderson, David E., 81
Hendricks, John A., 205
Henefer, James, 88
Henefer, William, 88
Henefer, 86–89
Henefer Irrigation Company, 89
Henefer Pipe Line Company, 89
Henry, Andrew, fur company, 27, 43, 87
Hercules Powder Corporation, 11
Hermitage Hotel, 46–47
Heslop, Charles L., 128
Hess, John W., 59
Hess, Joseph W., 98
Heywood, A. R., 130
Hiawatha, 161
Higley, George, 83
Hill, James, 64
Hill, Joseph (Cap), 64
Hill, William, 35
Hill Air Force Base, 7, 139
Hinman, Henry, 81
Hinman, Morgan, 81
Hobb's Canyon, 234n4

Hobbs Reservoir, 64
Hodson, Heber, 118
Hogge, Robert, 85
Holbrook, Joseph, 52
Holbrook, Moses L., 124
Holbrook, Ward, 145, 148, 205, 206, 211
Holbrook Canyon, 51, 234n4
Holmes, Samuel Oliver, 60
Holmes Creek, 60–61, 63, 64, 234n4
Holmes Creek Irrigation Company, 62
Holmgreen; Homer, 140
Homestead Act of 1862, 4, 23
Hooper, John, 42
Hooper, William H., 41, 117
Hooper, 28, 39, 41–43, 118, 237n41
Hooper Canal, 39, 42, 66, 106, 180
Hooper Canyon, 234n4
Hooper Irrigation Company, 38, 43, 115, 127
Hoover Dam, 9
Horner Construction Company, 173–74
Hot Springs (Davis County), 34, 35, 70
Hot Springs (Weber County), 17, 39, 59
Howard, Amasa R., 160
Howard, Harold R., 158
Howard, James, 124
Howard, Samuel, 123
Howell, J. A., 115, 150, 161–62, 164, 197,
    199, 204, 206
Hoyt, Joseph, 96, 97
Hoyt, Samuel P., 94, 95
Hoyt's Canyon, 96
Hoytsville, 94, 95
Hudson's Bay Company, 43
Hughes, Owen, 163
Hunsaker, Israel, 76
Hunt, Hollis, 163
Hunt, Hyrum, 43
Hunt, Jefferson, 43–44
Hunt, Joseph, 43
Hunt, Marshall, 43
Hunter, Edward, 82
Huntsville, 43–44; cemetery, 179; water rates
    in, 161
Hurst, Hollis, 198
Hydrants, 32
Hydroelectric power, 121–22, 173, 182, 226

Iakisch, J. R., 132
Ickes, Harold L., 132, 133
Ideal Cement Company, 198
Indians, 15, 30, 42, 55, 90, 95; troubles with,
    61, 72, 76, 83, 96
Internal Revenue Service, 11
Irrigation conservancy districts, 162
Irrigation districts, 13, 20, 104, 113
Israelson, Orson W., 129, 142

J. S. Lee and Sons (construction company),
    169
Jacob Creek, 79
Jacobs, L. Henry, 62
Jarvis-Conklin Mortgage and Trust Company,
    107–8
Jenkins, Ab, 139
Jenkins, David, 44
Jenkins, Lawrence, 163
Jennings, William, 63, 117
Jensen, Ben, 159
Jensen, H. P., 72
Jensen, Keith, 193, 224
Jeremy dam site, 151, 163, 248n68
Jerman, Reid, 143, 151, 156
John H. Burningham, 140
Johnson, A. V., 115
Johnson, David H., 116–17
Johnson, Joseph, 1, 121, 140–57, 159–60,
    162, 164–66, 197, 200, 202–5, 210
Johnson, O. D., 117
Johnson, R. C., 154
Johnson, Wallace M., 136
Jones, Robert A., 90
Jones, T. R., 126, 129
Jordan River, 17, 54, 122–24, 142, 154
Joseph, Harry S., 94
Jump Off Canyon, 234n4

Kamas, 86, 87, 95
Kamas Valley, 223
Kanesville, 39, 116
Kap, John, Sr., 116–17
Kay, William, 61
Kay's Creek, 61, 63, 64, 66, 234n4
Kay's Creek Irrigation Company, 62
Kaysville, 149, 153; canals to, 59, 118, 121;
    settlement of, 50, 60–63; water rates in,
    161
Kaysville Irrigation Company, 121
Kaysville Irrigation Reservoir, 64
Kearns Air Base, 7
Kelsey, Frank C., 125
Kennecott Corporation, 7
Kiesel, Fred J., 66, 230n10
Kimball, Heber C., 19, 59
Kirkendall, P. F., 126
Kirland, R. O., 140
Knight, George, 83
Korean War, 207
Koziol, F. C., 142
Kraling Construction Company, 179
Kreek, Morgan, 163

L. D. Shilling Company (construction), 191
Lagoon Resort, 135
Lake Bonneville, 49

Lambert, Dan, 99
Lambert, Jade, 99
Lamoreaux, Warwick C., 141
Larrabee Creek, 86
Larrabee dam site, 125, 169, 170, 202, 216, 249n20
Larson, E. O.: federal engineer, 132, 139, 143–44; regional director of Bureau of Reclamation, 125, 128, 143–44, 146–47, 149, 153, 155–57, 159, 165, 198, 199, 206, 209, 213
Laws. *See* Utah water laws
Layton, Christopher, 61, 63
Layton, 66, 149, 153, 204; settlement of, 50, 60–64; water rates in, 161
Layton Canal, 180, 185–86, 228
Layton pump intake channel, 180
Layton Pumping Plant, 186
Ledingham, John, 124
Lee, J. Bracken, 210
Lee Moulding Construction Company, 179
LeGrand Johnson Construction Company, 190
Lehi, 39
Leonard, Glen, 58–59
Leonard's Canyon, 88
Lewis, Morgan, Jr., 216
Lewis's Grove, 46
Liberty, 43, 45
Liberty Irrigation Company, 45
Lienhard, Frederick, 50
Line Creek, 77, 78, 106
Line Creek Irrigation Company, 78
Linke, Harold, 200, 203
Little, Feramorz, 117
Little, Jesse C., 24, 78
Littleton, 77, 78
Litton Corporation, 11
Logan, 161
Lone Tree Canyon, 88
Lost Creek, 76, 83–84, 106, 234n4
Lost Creek Dam and Reservoir, 151, 159, 163, 169, 182, 188–90, 226
Loveland, Chester, 51
Lyman, Francis Marion, 95
Lyman, Richard R., 126
Lynn Irrigation Company, 114
Lynne, 33

Mabey, Charles R., 52–53, 56–57, 122
Mabey, Joseph T., 54
Macbeth, James, 32
Magpie Creek, 187
Magpie dam site, 130, 151, 163, 169, 187, 202, 209, 248–68
Main Canyon, 88–89
Malad River, 17
Manti, 19

Manwaring, Lorus, 124
Marchant, Jack, 99
Margets, S. G., 129
Marion, 95–96
Marriotsville Dam, 40
Marriott, J. Willard, 36-37
Marriott, John, 33, 36
Marriott, 33, 37, 40, 41
Marsh Spring, 70
Marshall, H. L., 148
Mathias Creek, 70
Maughan, J. Howard, 142
Maw, John, 126
Maxwell, George H., 8
Maxwell, John, 99
May, Dean L., 229n4
Maycock, M. W., 151
McBride, Brice, 129
McFarland, Archibald, 38
McKay, Donald D., 116, 129, 130, 134, 206, 219
McKee, Mary Chard, 45
McLaughlin, W. W., 129
McMasters, J. S., 199
Mead, Elwood, 107, 108, 229n4
Merrill, David, 229n4
Merrill, Steven, 128
Middleton, Charles A., 33
Miles, Ad, 99
Miles, Calvin, 63
Miles, George J., 60
Military bases, 11
Military Springs, 139
Mill Creek, 33, 37, 51–55, 234–4, 235–19
Millard, James, 58–59
Miller, William H., 65
Mills, John M., 131
Milton, 77, 78, 82, 85
Miners, 96, 204
Mining, 7, 21, 23, 198
Missouri River Basin, 10
Miya Brothers Construction Company, 191
Montgomery Springs, 34
Moore Ditch, 33
Morgan, 81, 84, 106, 120, 208; water rates in, 161
Morgan Canyon, 88
Morgan County, 3, 69, 115, 127, 143, 150, 153–54, 157, 197–98, 204; bond election in, 215–18; special election in, 220–22; value of irrigated land in, 200; water claims in, 106; water development in, 76–86
Morgan County Water Users Association, 197
Morgan Valley, 154
Mormon church,, 121, 123; water managed by, 2–4, 16–17, 21, 49, 103–4, 137. *See*

also Pioneers
Morris, Isaac, 81
Morrison Knudsen Construction Company, 132, 173–74
Moss, Lawrence, 125
Mound Fort, 27, 33
Mound Fort Water Association, 114
Mountain Fuel Supply Company, 198
Mountain Green, 77, 82–83, 151, 170
Mountain men, 27, 43
Mountain States Construction Company, 180
Mower, Henry, 34
Mower, John, 34
Mueller Park Canyon, 54
Muir, E. O., 124
Muir, Frank, 62
Muir, William, 54
Murdock, Joseph R., 98, 126
Murdock, Neil, 192
Muskrat Springs, 41, 43
Myrick, Joseph, 96, 97

Nalder, W. Alvin, 153
National Irrigation Congress, 8–9, 110, 112
National Park Service, 209
Naylor, Henry, 56
Neeley, P. H., 115
Nelson, G. Lowery, 229n4
Nelson, Geneva, 74
Nelson, Jens, 53
Nelson Brothers Construction Company, 175
Newell, Frederick Haynes, 8
Newey, LaVerna, 44–46
Newlands, Francis G., 9
Newlands Reclamation Act (1902), 9, 110–11
Nichols, DeLore, 1, 121, 124, 136, 140–47, 149–50, 153, 157, 159, 162, 197, 199, 202, 204–5, 213
Nixon, Stephen, 91
Nixon, Thomas, 91
North Bench Canal, 114
North Canyon, 50, 51, 54, 234n4
North Canyon Creek, 51
North Cottonwood Creek, 57–58
North Morgan, 77
North Ogden, 34, 131; water rates in, 161
North Ogden Canal, 33; list of those who dug, 236–22
North Ogden Canyon, 234n4
North Ogden Irrigation Company, 35, 114–15, 127
North Salt Lake, 50, 51, 54, 158
North Willow Creek, 69, 72
Norton, Alanson, 92
Norwood, Richard, 84
Nye, Ephraim, 46
Nye, J. C., 118

Nye, Stephen, 46

Oakley, 86, 91, 96–99, 126
Odell, Fred, 122
Ogden, Peter Skene, 27, 43
Ogden Arsenal, 211
Ogden Bay Refuge, 166
Ogden Bench Canal, 1, 30-31, 235n11
Ogden Brigham Canal, 72, 131, 133
Ogden Canyon, 46, 130
Ogden Canyon Hot Springs Resort, 46
Ogden City, 76, 131, 139, 153, 157, 158; artesian wells in Ogden Valley for, 44, 132, 161; canal water to, 108, 117–18; electricity to, 121–22; filtration plant for, 221; flood control in, 208; growth of, 140; settlement of, 19, 27–29; water rates in, 203
Ogden City Board of Commissioners, 160
Ogden City Chamber of Commerce, 164
Ogden City Council, 35, 37
Ogden City Water Works, 109
Ogden Irrigating Company, 30
Ogden River, 27, 31, 33, 40, 43, 44, 108, 114–15, 125, 128–34; main tributaries of, 27–28, 234–4
Ogden River Valley, 28
Ogden River Water Users Association, 11, 144, 200, 206, 208, 223, 225; founding of, 117, 130–31; Pineview Dam contracted by, 131–33, 178–79, 218
Ogden Valley, 33, 43–46, 126, 161, 216, 227
Ogden Valley Canal, 186
Ogden Valley Commercial Club, 129
Ogden Valley Diversion Dam and Canal, 186
Ogden Water Company, 31
Ogden Water Works Company, 6
Ogden Western Irrigating Company, 35
Olsen, Alf N., 131
Olsen, C. J., 197
Olsen, Norman T., 131, 139, 144, 154
Olsen, Reinhardt, 79
Olsen Construction Company, 190
One Horse Canyon, 234–4
Oregon Shortline Railroad, 135
Owens, William J., 88
Owens Canyon, 88

Page, William, 55
Palmer, James, 197
Park City, 86, 92, 96, 99, 198, 199, 204, 207
Parker, A. F., 120, 129, 191
Parker, Dean, 151
Parker, W. J., 118
Parkin, Burt, 124
Parkin, H. B., 140
Parley's Canyon, 87
Parley's Park, 87, 91, 92

Parowan, 19
Parrish, Samuel, Sr., 60
Parrish Canyon, 135–36
Parrish Creek, 60, 136-37, 234n4
Parson Construction Company, 183
Pearson, Drew, 166
Pearson, Levi, 115, 126
Penrod, O. A., 45
Peoa, 90–92, 95, 97
Perdue Canyon, 99
Perdue dam site, 151, 163, 170, 206, 209, 216, 219, 248n68, 249n20
Perham Brothers and Parker construction company, 119
Perkins, Jesse N., 55
Perry, John, 50, 55
Perry, Lorenzo, 73
Perry, O. A., 75
Perry, 69–75
Perry Canyon, 135
Perry Power and Light Company, 72
Peterson, Charles S., 16, 229n4
Peterson, Charles Shreeve, 77, 79
Peterson, Leroy, 126
Peterson, William, 129, 229n4
Peterson, 77, 79, 106
Peterson Creek, 79, 106
Petterson, Hans, 38
Phelps, W. W., 90
Phillips, Caroline, 90
Phillips, Edward, 61
Pierce, Thomas, 69
Pineview Dam, 10, 31, 46, 72, 76, 104, 117, 125, 221; construction of, 128–34; enlargement of, 163, 178–79, 209, 218
Pineview Reservoir, 11, 31, 139, 194, 223; expansion of, 169, 216; recreation at, 210
Pioneer Canal, 37–38
Pioneer Electric Company Plant and dam, 117, 121–22, 130, 235n19
Pioneers, 12, 14, 15–24, 27, 43, 50, 72, 87, 99–101, 103, 171, 223, 228; canals built by, 18–19; diversion dams built by, 19
Pisani, Donald J., 21
Plain City, 39–40, 163
Plain City Canal, 40, 237n37
Plain City Irrigation Company, 40, 114–15, 127
Pleasant View, 34
Polygamy, 18, 104, 230n10
Porter, Sanford, Jr., 80
Porter, Sanford, Sr., 59, 80
Porter, Warriner, 80
Porter Spring, 70, 73
Porterville, 77, 80, 81, 106
Powell, John Wesley, 3, 6–7, 13, 230-31n16
Pratt, Orson, 16

Pratt, Parley, 87
Prior appropriation, 43, 51, 58; established in Western United States, 2, 21–22, 103; supported by law, 2, 5, 23–24, 104, 111, 113
Provo, 19, 161
Provo Reservation Water Users Company, 127
Provo River, 17, 86, 95, 97–98, 125, 126, 223–24
Provo River Project, 127
Provo Valley, 126
Provost, Etienne, 87
Public Works Administration (PWA), 132
Pumping plants, 178

R. A. Heintz Construction Company, 187
R. W. Coleman Company (construction), 194
Railroad: transcontinental, 3, 5, 28, 87, 101, 104; water for operations, 31
Randall, James E., 131
Randall Springs, 34
Reclamation Act of 1902, 10, 13, 104, 153
Reclamation policies, 13–14
Reclamation Service, 9, 110, 125. *See also* Bureau of Reclamation
Recreation, 11, 46–47, 60, 85, 159, 182–83, 209, 213–14, 226–27
Red Cross, 136
Reeder, W. H., Jr., 115
Reese, John, 157
Remington Arms, 7
Rhees, R. T., 129
Rhoads, Thomas, 95
Rice Creek, 34, 234n4
Rich, John H., 81
Rich, Thomas, 81
Richards, A. Z., 96
Richards, I. F., 139
Richards, Ralph A., 97–98, 128
Richards, Willard, 70, 72–73
Richins, Charles, 88
Richins, Leonard, 88
Richville, 77, 81, 84
Richville Irrigation Canal Company, 81
Ricks Creek, 60, 194, 234n4
Riddle, Isaac, 34
Riddle, John, 34, 44
Ririe, James, 44, 132
Ritter, Willis R., 179
Riverdale, 29, 34, 37
Robb, James, 38
Robb, John, 38
Roberts, Richard C., 56
Robison, Daniel, 85
Robson, J. L., 129
Rockport, 86, 96
Rockport Reservoir, 96, 125, 170, 172
Rockwell, Porter, 73

Roosevelt, Theodore, 9
Roper, Henry, 91
Rose, Alton P., 151–52, 162
Rose, John H., 82
Roskelly, C. O., 129
Roswell Canyon, 83
Round Valley, 77, 82
Rowe, W. H., 108
Roy, 43, 66, 161
Rudd Canyon, 57
Rudd Creek, 234n4
Rundquist, C. A., 118
Russell, Osborn, 50

Salt Lake Aqueduct, 141
Salt Lake Chamber of Commerce, 159
Salt Lake City, 12, 17, 19, 24, 50, 121–22, 218
Salt Lake County, 128
Salt Lake Growers Market, 122
Salt Lake Valley, 15–19, 228
Salter, George, 124
Sanborn, John, 159
Sant, Paul, 165
Sargent, Amos, 95
Sargent, Nephi, 95
Sargent Lakes, 95
Saunders, Commissioner, 160
Savage, Levi, 83
Sawmills, 19, 20, 37, 46, 63, 75, 78–81, 83, 99
Scott, David, 144, 198, 199, 205
Scott, George M., 230n10
Sessions, Leland, 124
Sessions, Orson, 124
Sessions, Perregrine, 1, 50–51, 57
Sewage systems, 32
Shattuck Construction Company, 123
Shaw, Ambrose, 33, 114
Shaw, John, 45
Shaw, Oscar, 46
Sheep Canyon, 80
Sheffield, H. J., 63
Shepard Canyon, 57
Shepard Creek, 234n4
Sherwood, Henry G., 16
Shill, Charles, 84
Shill, George, 83, 84
Shingle Mill Creek, 80
Shoemaker, Jezreel, 51
Shupe, John W., 33
Shurtleff, Lewis W., 98
Shurtliff, Francis M., 46
Shurtliff, Luman, 1, 35–36, 100
Silver Creek, 86, 87, 91–92, 234n4
Simmons, George, 84
Simon, Ben, 82–83
Skeen, E. J., 198
Slaterville, 34–37

Slaterville Canal, 180
Slaterville Diversion Dam, 180, 183, 185
Slaterville Irrigation Company, 114
Smeath, George, 150
Smith and Morehouse Creek, 86, 96, 99, 234n4
Smith and Morehouse dam and reservoir, 96, 97, 125, 206–7, 224
Smith Creek, 78, 79
Smith, Joseph F., 98
Smith, LeRoy B., 206
Smith, Roy, 140
Smith, William H., 92
Smith, William R., 117
Smythe, William, 8, 13
Snow Creek, 63
Snow, Lorenzo, 71
Snyder, George S., 91
Snyder's Mill, 92
Snyderville, 99
Sorensen, Horace A., 173
Sorensen, Marion, 95
Sorenson, Edward, 144, 155, 198, 200, 206, 216
Sorenson, Phil, 144
South Bountiful, 50, 51, 54, 55, 123
South Davis Pumping Plant, 175
South Hooper, 65
South Morgan, 77
South Ogden, 161
South Ogden Conservation District, 131
South Ogden Highline Canal, 133
South Ogden Water Conservancy District, 11
South Slaterville Ditch, 37
South Weber, 50, 65, 161, 204
Southern Pacific Railroad, 41, 185, 198
Southwick, Joseph, 46
Spaulding, Lewis, 115, 206
Sperry Rand Corporation, 11
Split Canyon, 203
Sprague, Artemus, 29
Spring Canyon, 95
Spring Creek, 37–38, 45, 234n4
Springville, 161
Staker, John, 39
Stalling, J. H., 99
Stallings, H. B., 129
Standing, Arnold, 137
Stannard, Jay D., 107
Stansbury, Howard, 16, 50, 87
State engineer, 13, 24, 109–13, 115, 122
Steed, Thomas J., 118
Steed Canyon, 57, 135
Steed Creek, 234n4
Steelman, John R., 204
Steenberg Construction Company, 188–90
Steiner, Heber J., 63

Stephens, William P., 131
Stevens, Roswell, 77, 79, 83
Stevenson, George V., 64
Stevenson, Richard, 64
Stevenson, William, 90
Stewart, Clyde, 201
Stinson, Willis, 99
Stoddard, Judson L., 82
Stoddard, 82, 83
Stoddard Diversion Dam, 170, 173–74, 227
Stoddard Spring, 82
Stone, Amos, 33, 114
Stone Creek, 51, 52, 54, 194, 234n4
Storey, Boyd, 213
Strauss, Michael, 148, 149, 153, 209
Strawberry Creek, 83
Strawberry Valley Project, 10, 110
Strong Construction Company, 185
Strong's Canyon, 32, 234n4
Sugar Pine Creek, 46
Summit County, 3, 69, 115, 128, 153, 170,
    197–99, 227; bond election in, 215–18;
    land acquisition in, for Wanship Dam,
    172; special election in, 220–22; value of
    irrigated land in, 200; water claims in,
    106; water development in, 86–99
Sundwall and Company (construction), 191
Sunset, 50, 66, 139, 204
Swan, George, 62
Syblon-Reid Construction Company, 185–86
Syracuse, 42, 149, 153, 161, 180; settlement
    of, 50, 65–66

Taggart, George W., 81
Tax revolt, 134
Taxes, levied for water projects, 104, 113,
    137, 167, 198–99, 222–23
Taylor, John, 99
Taylor, Joseph, 81
Taylor, Joseph A., 114
Taylor, Moses W., 98
Taylor, 39
Taylor Canyon, 234n4
Teele, Roy, 22
Templeton, Win, 150, 155, 158–59, 161-63,
    197, 207
Thiokol Chemical Corporation, 11
Thomas, Arthur L., 8
Thomas, Elbert, 145, 147–48, 152, 163–64,
    166–67, 203
Thomas, George, 5, 106, 107, 109, 229n4
Thomas, Harold E., 145
Thomas, W. P., 129
Thompson, Charles, 117
Thompson, John, 237n41
Thorn, Asael, 75
Thorn, Isaac, 74–75

Thorn, Rebecca, 74
Thornley, John, 63
Three Mile Canyon, 73, 91, 96
Three Mile Creek, 70, 73–75
Thurston, Cordelia, 77–78
Thurston, Thomas J., 77–78, 84
Timber Culture Act of 1873, 6
Tippetts, William Plummer, 73
Titan Steel Company, 191
Tolman, Eugene, 143, 153
Tooele Ordinance Depot, 7
Toone, W. H., 84
Tourism, 11
Townsend, H. S., 90
Tracy, Moses, 33
Trappers, 27, 43, 50, 82, 87
Tri-State Drilling Company, 191
Truckee-Carson Ditch, 9
Truman, Harry: objections of, to Weber Basin
    Project, 201, 202, 204; Weber Basin Pro-
    ject Bill signed by, 166–68; Weber Basin
    Project funded by, 169, 204, 207, 211, 213
Tullidge, Edward, 32, 92
Turner, Bert, 81
Turner, Charles, 84
Turner, George, 81
Turpin, R. L., 166
Tuttle, Newton, 52

Uinta Mountains, 28, 86, 91, 99, 144, 155
Uintah, 34, 37, 106, 176, 208
Uintah Central Canal, 38
Underground water, 40, 62, 129, 145, 147. *See
    also* Artesian wells; Ground water; Wells
Union Light and Power Company, 122
Union Pacific Railroad, 80, 85–86, 88, 127,
    154, 176, 198
United Concrete Pipe Corporation, 175
United States Army Arsenal, 139
United States Congress, 162–66, 169, 203,
    210–13
United States Geological Survey (USGS), 7–
    10, 110–11, 125, 142, 145, 147, 209
United States House of Representatives, 147,
    153
United States Senate, 147, 210
Ursenbach, Octave, 119
Utah Central Railroad, 31
Utah Conservancy Law, 223
Utah Construction Company, 120, 132, 170,
    172, 179
Utah Copper Company, 7
Utah County, 128
Utah Fish and Game Commission, 166
Utah General Depot, 7
Utah Irrigation District Act of 1865, 4
Utah Oil Transportation Company, 198

Utah Power and Light Company, 46, 73, 94, 122, 124, 175, 176, 198
Utah Power and Water Board, 170
Utah State Agricultural College, 201, 202, 207, 212
Utah State Board of Health, 194
Utah State Fish and Game Department, 154
Utah State Legislature, 23–24
Utah State Parks and Recreation Division, 226
Utah State Water and Power Board, 149
Utah Taxpayers Association, 203
Utah Territorial Legislature, 4, 20, 23, 63
Utah War, 59, 101
Utah Water and Power Board, 203
Utah water laws, 3–13, 17–24, 103–17; of 1852, 20, 103–4; of 1865, 20, 51–52, 103–4, 110; of 1880, 5–6, 10, 23, 32–33, 104–9; of 1897, 109–10; of 1901, 24; of 1903, 10, 104, 111–12; of 1919, 104, 109, 112–13; of 1941, 11, 113, 126
Utah Water Storage Commission, 125, 130
Utah Water Users Association, 142, 145, 146, 197, 203, 225
Utah-Idaho Sugar Company, 108, 150

Val Verda, 50, 54
Vetter, Carl, 131

Waddoups, Vee, 151, 152
Wade, Lester A., 116
Wadsworth, T. S., 85
Waite, James, 54
Waldron, Gillispie Walter, 81
Walker, James, 83
Walker, William, 73
Walker War, 72
Wallace, William R., 126, 133, 142, 146, 170
Wanship, 86, 91–92
Wanship Dam, 90, 96, 181, 209, 216, 248n68; construction of, 172–73
Wanship Power Plant, 176, 182, 226
Wanship reservoir, 169–70, 172, 206, 212; recreation at, 210
War industries, 7
Ward, O. H., 131
Ward Canyon, 51, 234n4
Wardell, George, 99
Wardleigh, Raymond E., 235n11
Warnick, Francis, 144, 154–55, 162–63, 165, 194, 198, 204, 206, 218
Warren, Francis E., 9
Warren, 40–41
Warren Act (1911), 9
Warren Canal, 37, 185
Warren Irrigation Company, 41, 127
Water conservancy districts, 113, 117, 126, 137

Water districts, 20
Water laws. *See* Utah water laws
Water lines, in Ogden, 31–32
Water measurement, 105–7, 112, 156
Water rates, 161, 203–4, 212
Water users associations, 137, 146
Waterfall Canyon, 32, 135, 234n4
Watermasters, 18, 51, 58–59
Watkins, Arthur V., 145–46, 150, 152, 158, 173, 179, 210; Weber Basin Project legislation sponsored by, 162–66, 203–5
Watkins, Arthur V., Dam and Reservoir. *See* Willard Bay Dam and Reservoir
Watson, Ed H., 141, 142, 145, 154, 159
Wayment, Samuel, 40
Weaver, Pauline, 87
Webb, James, 61
Webb Canyon, 61, 234n4
Weber, John H., 27, 87
Weber Aqueduct, 172–74, 176, 182
Weber Basin Canal, 79
Weber Basin Job Corps Conservation Center, 194
Weber Basin Project, 11, 17, 49, 54, 56, 60, 62, 69, 70, 72, 90, 104, 113–14, 121, 137; bond election for, 214–18; bonds issued for, 219; construction of, 169–95; contract for,   169, 213–14; first sale of water from, 221–23; funding for, 162, 164, 197–214; land purchases and right-of-ways for, 218–19; special 1961 election for, 220–22; water   delivered by, during 1957, 177
Weber Basin Water Conservancy District, 125, 151, 167–68, 169, 172, 175, 193–95; directors of, 206, 224; formation of, 197–206; secretary-managers of, 224
Weber Canal, 29–30, 37, 84–85
Weber Canyon, 28, 59, 99, 139, 169; flooding at mouth of, 172, 208
Weber City, 77
Weber County, 3, 107, 115, 127, 143, 146, 153; bond election in, 215–18; special election in, 220–22; value of irrigated land in, 200; water survey assessment for, 157
Weber County Commission, 157, 158
Weber County Farm Bureau, 129
Weber County Water Development Executive Committee, 151
Weber County Water Users Association, 164
Weber River, 24, 27, 37, 40–42, 59, 63, 76, 83–86, 90, 94, 99, 112, 115, 125; main tributaries of, 27–28, 234n4
Weber River Irrigation Project, 125
Weber River Survey, 147–62
Weber River Valley, 28
Weber River Water Users Association, 43, 117, 207–8, 126–28, 144, 146, 218–19,

223, 224
Weber State College, 213
Weber Water Users' Association, 11
Weber-Box Elder Conservation District, 11,
    131, 134
Weber-Provo Canal, 95, 98, 128, 223–24
Weinel, John, 62–63
Welch, Harold, 162
Welch, T. R. G., 84
Wells, 31, 43, 55, 62, 65, 145, 227–28. *See
    also* Artesian wells; Underground water
Wells, Heber M., 111
Welsh, James, 99
Wendover Air Base, 7
West, C. H., 95
West, Chauncey W., 35, 46
West Bountiful, 50–54, 123
West Bountiful Mill Creek Irrigation Com-
    pany, 55
West Hoytsville Irrigation Company, 95
West Layton, 64
West Point, 50, 65
West Richville Irrigation and Canal Company,
    82
West Warren, 39–41
West Weber, 38
West Weber Canal, 39
West Weber Irrigation District, 38
Western Irrigation Canal, 35
Western Irrigation Company, 114, 127
Wheeler, Levi, 46
Wheeler, R. A., 158
Wheeler Canyon, 46, 122, 234n4
Wheeler Creek, 46
Wheelwright Construction Company, 176, 186
Wheelwright Plumbing Company, 82
Whipple, Edison, 18
White, Barnard, 73–75
White, W. Rulon, 203, 206, 213, 216
White's Spring, 70, 73
Whitear, Albert, 79
Whitear, Arthur, 79
Whitesides, Emil M., 149
Widdison, Howard, 158, 198
Wilbert, Harry E., 139, 147, 154
Wilde, George, 99
Wilde, Keith D., 249n23
Willard, 69–73, 75, 131, 144
Willard Bay Dam and Reservoir, 69, 144, 154,
    163, 169, 180, 182–85, 228; recreation at,
    210, 226–27

Willard Canal, 180, 183, 185
Willard Canyon, flooding in, 135
Willard Municipal Power Plant, 135
Willard Pumping Plant, 185
Willey, Israel E., 122
Williams, Andrew, 92
Williams, Daniel, 85
Williams, Henry B., 65
Williams, Joseph, 99
Williams, Joshua, 79
Willow Creek, 70, 72–73, 75
Willow Springs, 35
Wilson, Billie, 46–47
Wilson, 39
Wilson Canal, 39, 106, 186
Wilson Irrigation Company, 39, 106, 127
Wilson, M. T., 142
Winegar, Wayne, 193, 224, 226–27
Winsor, L. M., 129
Wittfogel, Karl, 1, 14, 117, 129
Wolfe Creek, 44, 46
Wood, Daniel, 51
Wood, James G., 118
Wood, John, 81
Wood, Joseph, 43
Wood, Oscar, 124
Woodruff, Wilford, 15, 24
Woods, Clinton D., 194, 218, 219
Woods Cross, 123, 157, 194; settlement of,
    50, 51, 53, 54
Wooley, Ralf, 142
Work, Hubert, 126
Works Progress Administration (WPA), 65
World War I, 122, 125
World War II, 2, 3; government facilities
    from, 7, 121, 139
Wright, Angus T., 129
Wright, Joseph E., 131
Wright, Walter, 135
Wright's Spring, 70
Wrighton, William, 75

Young, Brigham, 61, 70, 72, 75; colonization
    directed by, 15–18, 29, 30, 51, 78, 80, 81,
    88, 90, 94, 95; death of, 5, 23; water pol-
    icy guided by, 2, 4, 15–18, 59, 100–1, 103–
    5, 117
Young, Willard, 125

Zeimer, Charles, 33

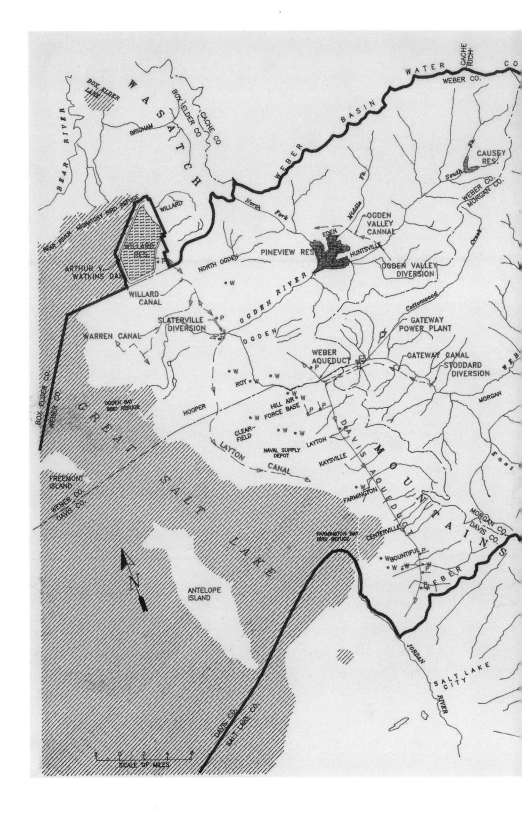